Introduction to
Membrane Noise

Introduction to Membrane Noise

Louis J. DeFelice

Emory University School of Medicine
Atlanta, Georgia

PLENUM PRESS · NEW YORK AND LONDON

Library of Congress Cataloging in Publication Data

DeFelice, Louis J
Introduction to membrane noise.

Includes index.
1. Membranes (Biology)—Electric properties. 2. Electrophysiology. I. Title. II. Title: Membrane noise. [DNLM: 1. Membranes. 2. Electrophysiology. QH 601 D313i]

QH601.D4	574.19'127	80-16163

ISBN-13: 978-1-4613-3137-7   e-ISBN-13: 978-1-4613-3135-3
DOI: 10.1007/978-1-4613-3135-3

© 1981 Plenum Press, New York
Softcover reprint of the hardcover 1st edition 1981
A Division of Plenum Publishing Corporation
227 West 17th Street, New York, N.Y. 10011

Preface

I started working on membrane noise in 1967 with David Firth in the Department of Physiology at McGill University. I began writing this book in the summer of 1975 at Emory University under a grant from the National Library of Medicine. Part of the writing was also done at the Marine Biological Laboratory Library in Woods Hole and in the Library of the Stazione Zoologica in Naples.

I wrote this book because in the intervening years membrane noise became a definable subdivision of membrane biophysics and seemed to deserve a uniform treatment in one volume. Not surprisingly, this turned out to be much more difficult than I had imagined and some areas of the subject that ought to be included have been left out, either for reasons of space or because of my own inability to keep up with all aspects of the field.

This book is written for biologists interested in noise and for physicists and electrical engineers interested in biology. The first three chapters attempt to bring both groups to a common point of understanding of electronics and electrophysiology necessary to the study of noise and impedance in membranes. These chapters arose out of a course given over a period of six years to electrical engineers from the Georgia Institute of Technology and biologists from Emory University School of Medicine.

K. S. Cole (1979) has pointed out that to some extent "the computer succeeds calculus as a way of intellectual life," and I have to agree with him. Nevertheless I have stressed the analytic solution whenever possible. In doing so, I have tried to make the book mathematically self-contained. The principles behind integral and differential calculus are assumed, but otherwise the mathematics are developed as needed. For those readers who require it, I recommend the excellent self-study review of elementary calculus found in the first few chapters of Franklin's *Advanced Calculus*.

The centerpiece of the mathematics used in this book is the Fourier transformation. In developing this subject I have followed the outstanding book *The Fourier Transform and Its Applications* by Bracewell. Fourier transformation is a special class of integral transformation; other types of transforms arise throughout this text and as an aid to the reader I have collected the most important equations from four major works in an appendix.* For those unacquainted with integral equations, some time will have to be spent on the purely mathematical aspects of this book to become familiar with the use of tables like those in the Appendix. The introductory chapter in Campbell and Foster's *Fourier Integrals for Practical Applications* is highly recommended as is Bracewell's practical dictionary of transforms at the end of his book. For a more advanced treatment, I recommend the elegant development of *Integral Equations* by Hochstadt. I have included many problems and exercises in this book that require mathematical solutions. The problems can be worked out using only the material presented in this volume and in some cases parallel examples given in the text. The exercises will require outside reading and may assume a mathematical background beyond that presented in this book.

The first use of noise analysis to study biological membranes is certainly not obvious. The idea appears to have been around for some time and one often finds oblique reference to the notion in unexpected places. Consider the figure below taken from Bacq (1976). The figure represents Bacq's view of chemical transmission taken from an earlier work of his published in 1934. Although I am probably reading more into Bacq's figure than he intended, I can almost hear the noise from the postganglionic cell in his figure, and I think any of the earlier workers on noise in electronic tubes would probably agree with me. One finds more explicit references to membrane noise analysis from sources not normally associated with the field but obviously very much in sympathy with the idea. From Frankenhaeuser (1960) we have the following comparison of the membrane of the node of Ranvier and the squid giant axon: "The toad fiber is clearly

---

* For convenience, text references to the equations from the four major works are given in forms such as CF # I-201, BMP # 1.1-2; the initials stand for the work from which the equation is taken [CF for Campbell and Foster (1948), BMP for Bateman Manuscript Project (1954), GR for Gradshteyn and Ryzhik (1965), and BH for Bierens de Haan (1939)], and the numbers are those used in the reference work (and in the Appendix for equations listed there). For example, BH # 151-9 is equation (9) of Table 151 in Bierens de Haan (1939).

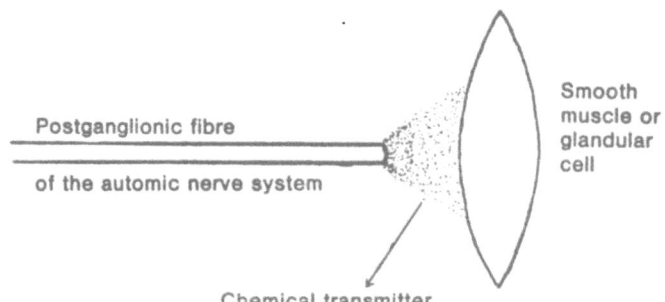

Schematic presentation of chemical transmission proposed by Bacq in 1934. (From Bacq, 1976.)

more permeable than the squid fiber. There is at present no way of deciding whether the difference depends on differences in the number of sites or in the efficiency of single sites." I'm not sure that Frankenhaeuser would agree, but I think we now know that the difference is basically in the number of more or less similar sites. The strongest evidence for this comes from noise analysis.

Some subjects only touched upon in this book obviously deserve more attention, for example, the use of inherent membrane noise to measure intercellular communication. The measurements are difficult and the theory is not completely worked out, so it may be some years before this technique is used regularly.

Another topic that I have not developed as far as I would have liked is the relationship between membrane noise and membrane impedance. The two are derived from the kinetics of channel conductance, and the formula that relates membrane noise to impedance, $S_V = |Z|^2 S_I$, implies linearity. Yet even for small perturbations, excitable membranes may be significantly nonlinear, as shown in Figure 109.2. The above equation is a reasonable approximation in some cases, but it is unclear whether it holds for all voltage and frequency ranges or for all types of excitable membranes, some of which may be more nonlinear than others.

The most significant recent advance in membrane noise analysis is the measurement of currents through single channels in membranes. Bean *et al.* (1969) were the first to do this in model membranes. The technique was immediately exploited by Ehrenstein *et al.* (1970), who described the mechanism of the voltage-dependent conductance in this model membrane system at the level of single channels. Not long afterward, single-channel currents were measured in biological membranes (Neher and Sakmann,

1976a). The influence of this work on membrane noise analysis and membrane biophysics in general has been enormous.

The study of single-channel currents is just beginning. Preliminary reports of currents from single channels in excitable membranes have already been given at several recent conferences and the publications will be out before this book appears. David Clapham and I have been working on the measurement of several channels working together in small patches of cardiac membrane. The theoretical and experimental question of an action potential or an action current constructed from only a few channels is an unexplored area of noise analysis relevant to small cells or portions of cells in which the density of channels is low.

The channel models we develop from noise measurements or single-channel data must also describe macroscopic phenomena. The question is always the same: Is the macroscopic property being considered also a property of the single channel, or does the property result from an ensemble of channels, not to be found in any one channel taken individually? An outstanding example of this type of problem is the explanation of the time variant conductance. I have taken a particular viewpoint in Figure 74.1, but the question is by no means resolved. One deep-seated prejudice that such models share is that all channels are alike and that the fluctuations we observe are due to statistical variations of identical subunits. This is implicit in the statistics we use and is essentially the same assumption one makes about electrons when analyzing Johnson noise.

Such problems will have to wait. In the meantime, I hope that the present work fulfills the expectations of those who supported it. I would like to thank certain people who, though they are not responsible for the shortcomings of this book, have influenced me and to whom I would like to express my appreciation: Hans Plendl; Cyril Challice; David Firth; Bert Verveen; Hans Michalides; Chuck Stevens; Enzo Wanke; Franco Conti; my mentor and friend, Alex Mauro; Bob DeHaan for his fruitful collaboration and for the Tender Heart Club in which many of the ideas presented in this book were first tested; Bill Adelman and Alberto Monroy for their laboratories and for the libraries of their institutions, the Marine Biological Laboratory in Woods Hole and the Stazione Zoologica in Naples, in which much of this book was written; Susan Clapham for typing, editing, and proofreading the manuscript and for helping not only with the organization of the book but also with the organization of the Erice Conference on Noise in Biological Membranes, held in Sicily in 1977; and David Clapham, Barry Sokol, Dick Ypey and John Clay, who helped me with many aspects of this work. I would also like to thank Vanni Taglietti for organizing a course on Noise Analysis in Biological Membranes in Pavia in 1978 that in some ways served as a model for the final form of this book, and Jerry Sutin for providing the environment in which this work could be done. This work was supported in part by NIH Grant LM02505 from the National Library of Medicine. Lastly, I wish to thank my family, Louie and Evelyn, Rachel, Emile and Anna Catherine, and Jean and Louis, for years of support and understanding.

It seemed appropriate to end the book more or less where it began, with the use of equivalent circuits to describe bioelectric phenomena. The papers in the last chapter were selected to give the reader a sense of the development of the subject as well as a summary of data. The list is incomplete but I hope adequate to the task.

Louis J. DeFelice

Contents

Introduction to Membrane Noise

1

Animal Electricity

Although we have come, in many instances, to an understanding of natural phenomena or effects, what we actually mean by "understanding" is unclear. I believe we often mean nothing more than familiarity: if a new phenomenon can be compared to something we are already used to, we seem to understand it and we feel free to go on to the next step. The set of ideas and the phenomena or the effects we are used to depend on the age in which we live and the field in which we work, and no doubt vary from person to person. But such differences, it seems to me, are of secondary importance; the science of other periods and fields and individuals really differs from our own only in where this reduction to the familiar happens to stop.

In this sense, human understanding operates largely by analogy, and the feeling of comfort that accompanies understanding is achieved largely by repetition. Understanding music is something very similar to this process; we come to appreciate music when we recognize patterns in it and begin to expect certain things to happen. But this requires a great deal of listening and a certain amount of simple repetition. Much of scientific understanding comes about in the same way; rewarded expectation is the criterion of understanding, and the unknown becomes known by reducing the problem to one we have seen before.

This has certainly been true in the development of electrophysiology as a science. For example, we use electronic circuits at all levels of complexity to represent biological systems—from a single resistor for cytoplasm to a computer for the brain. I have no doubt that some analogies, such as the last, have led us astray, but rightly or wrongly this is what we do.

1

As nearly as I can tell, this all began long before Galvani's frog experiments, but it is in his works that the analogies first become explicit and thus they serve as a prime example of what I mean to say. Basically Galvani explained the flow of electricity between the frog's spinal nerves and leg muscles, through the bimetallic arc he placed there, by analogy with the discharge of a Leyden jar. In effect, Galvani was using a capacitor as an equivalent circuit to help explain his results. The analogy was crude, but his basic program for understanding electrophysiology is the same one we use today.

In this chapter I will present some early examples of correlations made between various effects in animal electricity and the physical devices available to investigators in those times. The logic behind these comparisons parallels the more specific connections we will make in the following chapters and underlies the modern view of the way in which membranes work.

1. Newton's Opticks and Ganot's Physics

I. Bernard Cohen has said that the 18th century began in light and ended in electricity. His remark summarizes a century that saw both the publication of Newtons' *Opticks* in 1704 and the remarkable set of events that transformed electricity into a science and culminated in the invention of the Voltaic pile in 1800.

The study of electricity began with an electric spark. It took a great deal of effort and several curious accidents to convert that spark into the controlled flow of electricity from a battery. In a sense, we have made the same sort of conversion for nuclear energy in our own century. Any such development depends to a great extent on preliminary work.* The development that concerns us most, and the one we will pursue for the rest of this chapter, is the link that was made between certain phenomena in electrophysiology and the physical models that were used to represent these phenomena.

Consider an early model of neuromuscular transmission made by Isaac Newton. Newton's *Opticks* contains a number of Queries that describe unfinished work set down "in order to a farther search to be made by others." Querie 23 in the *Opticks* deals with vision and with hearing.

---

* Park Benjamin discusses many of the ideas that led up to the invention of the battery in his detailed *History of Electricity* (Benjamin, 1898).

Querie 24, reproduced below, proposes a model for animal motion (Newton, 1730, pp. 353–354):

> Is not Animal Motion perform'd by the Vibrations of this Medium (Aether), excited in the Brain by the power of Will, and propagated from thence through the solid, pellucid and uniform Capillamenta of the Nerves into the Muscles, for contracting and dilating them? I suppose that the Capillamenta of the Nerves are each of them solid and uniform, that the vibrating Motion of the Aethereal Medium may be propagated along them from one end to the other uniformly, and without interruption: For Obstruction in the Nerves create Palsies. And that they may be sufficiently uniform, I suppose them to be pellucid when view'd singly, tho' the Reflexions in their cylindrical Surfaces may make the whole Nerve (composed of many Capillamenta) appear opake and white. For opacity arises from reflecting Surfaces, such as may disturb and interrupt the Motions of this Medium.

Although Newton's model is not very useful in modern terms, it seems obvious that what he is trying to do is explain something he does not understand—the propagation of impulses along nerves—in terms of something he does understand—light.

When electrophysiology came to be set down in textbooks, this was often done by simply including a chapter on animal electricity, along with chapters on gravitation, heat, sound, etc., in physics books; electrophysiology was thought of as a subdivision of physics. One popular physics text of the 17th century was Ganot's *Eléments de Physique.**

Ganot divided electricity into two types, frictional and dynamic. Book IX deals with static charge induced in bodies by friction (Ganot, 1890, p. 688). Not all substances, the text explains, are capable of holding the same level of electrification. Poor conductors like glass or sealing wax hold the charge well; metals spread the electricity over their entire surface. Thus a glass rod, held on one end and rubbed with silk on the other, will hold its charge, whereas a metal rod will not. Bodies were classified according to their ability to conduct electricity (Ganot, 1890, p. 690). In order of decreasing conductance, the conductors listed were metals, well-burnt charcoal, graphite, acids, aqueous solutions, water, snow, vegetables, animals, soluble salts, linen, and cotton. The semiconductors were alcohol and ether,

---

* The book went through many editions in French and was eventually translated into German, Spanish, and, in 1863, into English by E. Atkinson. The English translation, titled *Elementary Treatise on Physics*, was popularly known as Ganot's *Physics* and served as a textbook for schools and colleges for many years. The following material is from the 13th edition, published in 1890 by William Wood and Co., New York and will be referred to as Ganot (1890).

powdered glass, flour of sulfur, dry wood, paper, and ice at 0°C. The nonconductors were dry oxides, ice at —25°C, lime, caoutchouc, air and dry gases, dry paper, silk, diamonds and precious stones, wax, sulfur, resins, amber, and shellac.

Early investigators interspersed animals, plants, and minerals in their lists of good and poor conductors. But they went much further than this. They often included themselves, to the point of acting as the sensor in electrical experiments, and scientific conclusions on the nature of electricity were often reached through the subjective experience of an electric shock.

Book X of Ganot's *Physics* deals with dynamic electricity, that is, batteries, currents, electromagnetic fields, electrical instruments, anything we would include in circuit theory today. Surprisingly, Ganot credits the discovery of dynamic electricity to Galvani (Ganot, 1890, p. 770):

> The fundamental experiment which led to the discovery of dynamical electricity is due to Galvani, Professor of Anatomy in Bologna ... who observed that when the lumbar nerves of a dead frog were connected to the crural muscles by a metallic conductor, the latter became briskly contracted.

It is difficult to imagine a modern textbook of physics or electrical engineering opening a chapter on circuit theory in this way. Although biologists use physics all the time, only rarely do physicists take their examples from biology. And certainly the time has passed when no distinction at all was made between electrophysiology and dynamic electricity.

2. Pre-Galvani Experiments and the Leyden Jar

Galvani was not the first investigator to connect bimetallic strips to living tissue.* For example, Jan Swammerdam in the *Byble der Nature* in 1658 reported that brass and silver caused contraction of frog leg muscle when the metals were placed in sliding contact with the muscle and the crural nerve (Dibner, 1952, pp. 7–9). It is unclear whether contraction depended on mechanical stimulation due to the sliding contacts, or to purely electrical stimulation due to the bimetallic strips. Swammerdam also suggested that cardiac muscle should respond to similar stimulation. About 100 years later, J. G. Sulzer performed a similar experiment using taste as a sensor in place of muscle contraction (Dibner, 1952, pp. 7–9). He joined a strip of lead to one of silver and applied the open ends to opposite sides

---

* Some earlier experiments are described by B. Dibner (1952) and by H. E. Hoff (1936).

of the tongue. This produced a strong unpleasant taste. Applying disconnected strips of metal produced no such sensation. This may be verified with a number of different metals. In both experiments, the metallic circuit was completed through living tissue and that same tissue acted as the sensor for the flow of current. But what was the source of this current? Was the source inherent in the animal or was it somehow induced by the metallic circuit?

Electricity was applied to animal tissue from undisputed sources of electricity, such as the Leyden jar, for purposes of stimulation. For example, Antoine Louis in 1747 showed that paralyzed muscle could be stimulated to contract but atrophied muscle could not (Dibner, 1952). But the crucial question of whether or not living tissue was itself an inherent source of electricity remained unanswered. Before going on to the experiments that helped decide this issue, let us examine some early electrical devices of the period that were to play an important role in the interpretation of Galvani's and Volta's experiments.

Perhaps the most popular electrical device of this period was the Leyden jar—a glass bottle filled with metal shot and wrapped with metal foil. Figure 2.1 is a drawing of one common form of this device.

PROBLEM. Assume the Leyden jar is a simple capacitor. With the help of equivalent circuits, describe the discharge of the Leyden jar illustrated in Figure 2.1a. First, contact the *outer coating* before touching the knob, then do the reverse. Also describe the slow discharge illustrated in Figure 2.1b.

The metal shot inside the jar is connected to a metal knob or hook by a chain passing through the stopper. The Leyden jar was invented by Dean von Kleist in 1745 in Pomerania near Leipzig, Germany, in the form of a nail in a medicine vial (Benjamin, 1898, pp. 513–515). The famous experimentalist van Musschenbroeck of Leyden invented it independently a year later, and gave the instrument both its name and its renown. The capacitance of the Leyden jar is charged by holding the outer metal foil in one hand and touching the knob to a piece of glass that had acquired a charge through rubbing. The jar is discharged by touching the other hand to the knob. If the jar was well charged, the shock that resulted could be painful. van Musschenbroeck reports (Benjamin, 1898, p. 519):

Suddenly I received in my right hand a shock of such violence that my whole body was shaken as by a lightening stroke. The vessel, although of glass, was not broken, nor was the hand displaced by the commotion: but the arm and body were affected in a manner more terrible than I can express. In other words, I believed I was done for ...

The thinner the glass, the stronger its effects; and the charge may be stored for three to eight days. Since the accumulated charge could exceed that of the machines used to produce it, its power was tested in many ways. Birds and beetles were killed to prove its potency. To enhance its effects even further, Gralath combined many jars in parallel (Benjamin, 1898, p. 523).

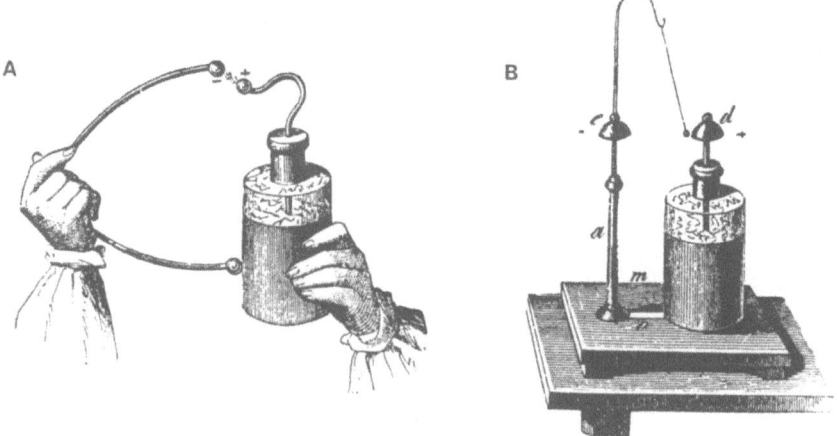

Figure 2.1. The Leyden jar. From Ganot (1890), Figures 707–708. The jar is of any convenient size and shape. The so-called "French shape" is shown here. The interior is coated with tinfoil, or the jar is filled with thin leaves of copper or gold leaf. The outside is wrapped in tinfoil. Through the cork in the neck, a brass (or other metal) rod communicates with the metal on the inside. The jar is charged by connecting one of the coatings to ground and the other to a source of electricity. (One common source is shown in Figure 3.1.) When the jar is held by a hand and the knob is touched to a source of static electricity, the charge acts inductively across the glass dielectric to either repel or collect electrons from the ground. The jar may be discharged by connecting the inner and outer plates. The outer (held by the hand) must be contacted before the upper part of the discharged rod touches the knob, as illustrated in Figure 2.1a. This causes a rapid discharge. The reverse order of contacting the jar will shock the experimenter. Figure 2.1b illustrates a method of slowly discharging the Leyden jar. The inner coating terminates on a small bell (d). The outer coating is connected to a similar bell (e) through (o) and the upright support (a). The pendulum is attracted to (d) and is immediately repelled, striking the second bell (e), and partially discharging the jar. Being again neutral, the pendulum is attracted to (d), and so on, until the jar is completely discharged.

3. Benjamin Franklin and the Magic Square

Benjamin Franklin's *Experiments and Observations on Electricity*, originally published in 1751, sought a unified theory of many diverse experiments done with a Leyden jar (Franklin, 1774). According to Franklin, every body contains a natural or normal amount of electrical fluid mixed with the "common matter" in bodies. Rubbing or friction may alter the normal amount of electrical fluid; an excess was defined as "plus" and a deficit as "minus." Franklin divided material bodies into "electrics" (nonconductors) and "nonelectrics" (conductors); classification was contingent on the material's tendency to retain charge induced by rubbing. He also developed the idea of the conservation of charge, which helped explain how the Leyden jar worked. Franklin realized that a charged Leyden jar had not acquired a net charge but only a separation of electricity. The metal shot and the external foil have an equal and opposite charge so there is no more electricity in a charged Leyden jar than in an uncharged one. Touching the knob to rubbed glass (Figure 3.1) induces a separation of charge only when the outer foil is returned to ground, e.g., through the experimenter's body. The current that flows is a displacement of current; excess charge on the knob is transmitted to the metal shot only when an equal but opposite charge is displaced through ground through the outer circuit. No current flows through the glass. Removing the knob from the rubbed surface opens the circuit, and charge is retained since there is no return pathway. If the experimenter provides one by touching the knob with his other hand a current flows and nullifies the charge displacement in the jar.

Franklin invented an alternate form of the Leyden jar called a "magic square" or plate. It is illustrated in Figure 3.2.

Franklin also made other discoveries that influenced the early theories of animal electricity. Both Galvani and Volta were familiar with his *Experiments and Observations on Electricity*, which was translated into Italian in 1774. Franklin showed that sharply pointed conductors would lose their electrification more rapidly than smooth conductors. He also verified that lightning is an electrical discharge, and investigated and tested the lightning rod. Galvani used electrification of the atmosphere to stimulate his preparations and used Franklin's description of the discharge of a Leyden jar as an analogy of muscle contraction (Section 6).

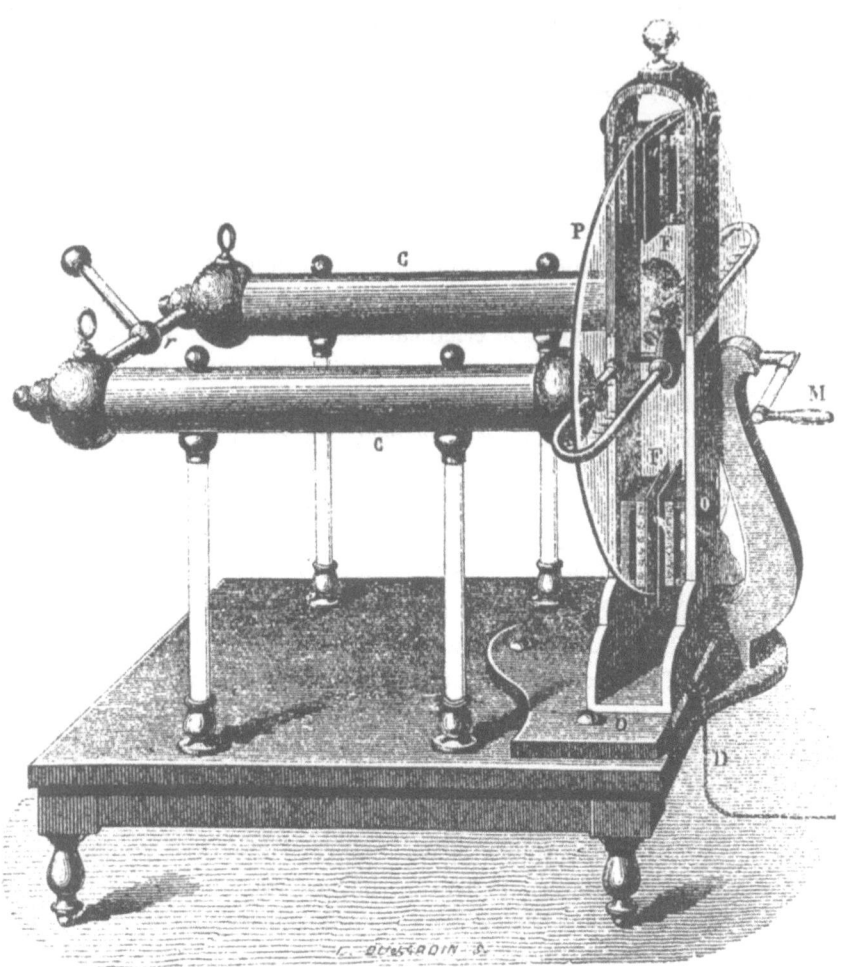

Figure 3.1. Plate electric machine. From Ganot (1890), Figure 683. The original electric machines were made with glass or other nonconducting materials formed into cylinders. The machine shown here is a modification due to Ramsden, which was a glass plate (P) supported by wood and turned on its axis by a handle (M). The plate moves against cushions (F) of leather or silk on either side of the plate, and above and below the handle. The cushions are connected to ground by their support (O) and the chain (D). The pressure of the cushions against the plate may be adjusted. The plate also passes through two U-shaped brass rods. The rods have a row of needles on one side close to the glass, and are connected to two large metallic cylinders (C) on the other side. These large metallic cylinders are connected by a brass rod (r), and are insulated from ground by the glass feet on which they stand. When the glass turns, charge is removed from the plate and returned to ground via the cushions and the chain. The plate induces a charge on the cylinders (C), which then discharges to the plate via the needles to neutralize the positive charge on the plate. This process repeats itself until a large negative charge is removed from the cylinders, leaving them with a net positive charge. This represents a true imbalance of charge, i.e., a net charge deficit, on the cylinders, that may be used to charge the capacitance of a Leyden jar.

Figure 3.2. Franklin's plate. From Ganot (1890), Figure 705. Franklin's plate is also known as the "magic square" of Franklin. It is actually a simple form of the condenser and is capable of giving relatively strong shocks. Franklin's plate is a pane of glass in a wooden frame by which it may be held. One sheet of tinfoil is attached to one side of the glass, leaving a margin between it and the wooden frame. Another sheet of tinfoil is attached to the other side of the glass pane. This second sheet of tinfoil is also attached to a chain that may be in contact with the earth. To charge Franklin's plate, the insulated sheet of tinfoil is brought close to an electrostatic generator (see Figure 3.-1) when the other is connected to ground through the chain. The "magic square" may be discharged in the usual manner by connecting the two charged plates. When the connection was made through the hands, a violent shock could be felt through the entire body.

4. Volta's Electrophorus

The electrophorus was invented in 1775. It is, in effect, a Leyden jar with one removable plate. A flat cake of some "electric" such as wax or resin lies on a "nonelectric" (conducting plate) connected to earth. The resin may be charged by rubbing. A second conductor is placed on top of the cake. If the two plates are connected by an external conductor (e.g., through the fingers of the hand or through ground), the upper plate acquires a charge. The upper plate may then be removed (with an insulated handle) and the process repeated over and over without ever rubbing the plate again. The electrophorus is illustrated in Figure 4.1.

PROBLEM.   Explain how the cover of the electrophorus is charged and how the cover is used to charge a Leyden jar. Use equivalent circuits. Explain why the cake holds its charge so well once it has been rubbed (see Figure 4.1).

The electrophorus led eventually to Volta's condensing electroscope

Figure 4.1. The electrophorus. From Ganot (1890), Figures 681 and 682. This instrument is described as one of the simplest and least expensive means of collecting large supplies of static electricity. The resin cake (B) is about 12 in. in diameter and 1 in. thick. The cake rests on a metal surface. The smaller metal plate (A), called the cover, may be moved about with an insulating glass handle. The cake is briskly rubbed with silk or catskin and acquires a negative charge. When the cover is placed on the charged cake, and is connected to the cake either by a mutual ground or directly with two fingers of the hand, a charge is induced on the plate of the cake. The cover need never actually touch the cake; irregularities in the surface prevent most of the area from being in direct contact. By raising the cover with the insulating handle, the induced positive charge on the cover may be physically transported. If the charged cover is brought near a conductor connected to ground, a sharp spark will pass between them. Once the electrophorus is charged, it remains in that state for some time. Sparks may be drawn from it over and over again. Similarly, a Leyden jar may be given repeated small charges from the electrophorus, causing the jar to acquire a sufficient charge for further experimentation.

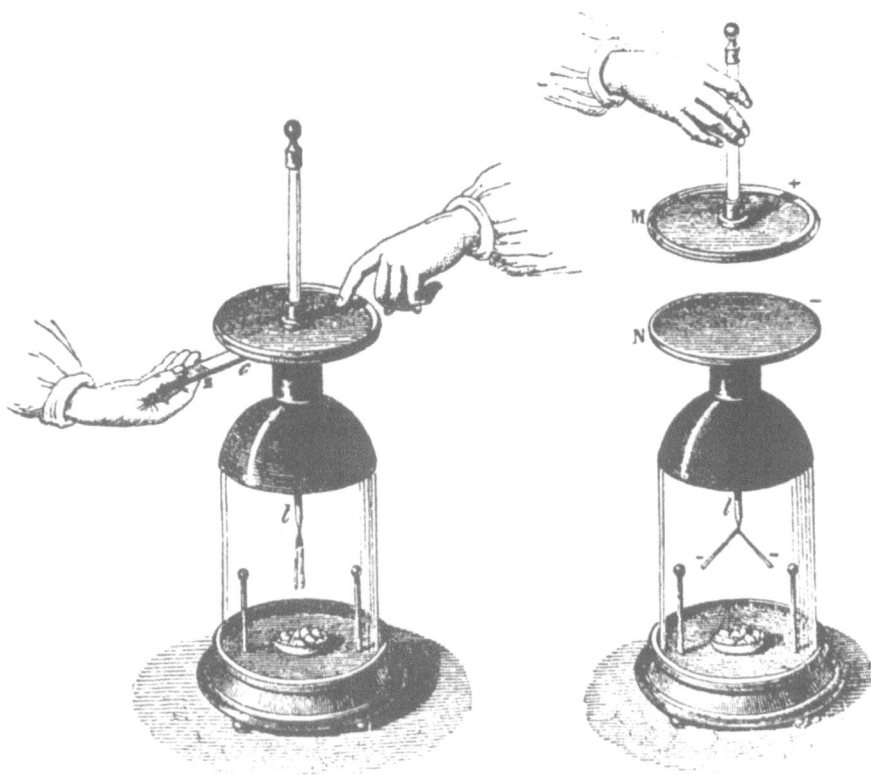

Figure 4.2. Volta's condensing electroscope. From Ganot (1890), Figures 716 and 717. This instrument was invented by Volta around 1781. It is, in effect, a combination of the ordinary gold-leaf electroscope and Volta's own variable plate capacitor. The gold leaves are connected electrically to a flat metal plate (N). A metal plate of the same size (M) rests on N and may be removed with a glass insulating handle. Both metal plates are covered with a layer of varnish or shellac, which insulates them from one another. The condensing electroscope can detect minute quantities of electricity. In the left diagram, a bimetallic strip of zinc and copper is held on one hand by the zinc end, while the upper end touches the bottom, or *collecting plate*. The other hand touches the upper *condensing plate*. The hand is removed from the condensing plate, and then the bimetallic strip is removed. The leaves do not diverge. However, if the upper plate is removed, as in the right diagram, the gold leaves move far apart (see text for explanation).

(Figure 4.2). The condensing electroscope is a variation of the ordinary gold leaf electroscope but far more sensitive. The gold leaves terminate on a disk. Another disk of the same size (and with an insulating handle) rests on the lower one. Both disks are covered with shellac at their interface. Let the lower plate be touched by a body suspected of carrying a small net charge; at the same time let the upper plate be connected to ground. There is no sensible change in the position of the leaves. Remove the ground and the leaves still do not diverge. Upon separating the plates with the insulated handle, the electroscope shows a perceptible response. The two plates form a condenser and when the plates are separated the capacitance is decreased and the voltage across them is increased. The instrument works by applying a given charge when the capacitance is large and measuring it when it is small. Volta called this new instrument a "microelectrometer"; he used the electrometer in his critical analysis of Galvani's experiments in animal electricity.

PROBLEM. Explain the action of the condensing electroscope with equivalent circuits. Use the equations relating charge, potential, and capacitance in the explanation.

5. Galvani's First Experiment

It is uncertain why Galvani began to experiment with the electrical properties of biological tissue. He describes the events that led to his initial work as follows (Galvani, 1791, pp. 45–57):

> Having in my mind other things, I placed the (dissected) frog on the same table as an electric machine ... so that the animal was completely separated from ... the machine's conductor. When one of my assistants lightly applied the point of scalpel to the inner crural nerves ... of the frog, suddenly all the muscles of the limbs were seen so to contract that they appeared to have fallen into violent tonic convulsions. Another assistant who was present when we were performing electrical experiments thought he observed this phenomenon when a spark was discharged from the conductor of the electrical machine ...

Galvani explains that when this phenomenon was brought to his attention, he set aside other things and devoted himself to the experiment. Sparks were drawn from an electric machine similar to that in Figure 3.1. The scalpel was applied first to one nerve and then to another. This process left no doubt that the tetanus and the electrical discharge were correlated. Many control experiments were performed to verify this correlation and it was discovered that if one touched the bone handle of the scalpel, instead of its metal parts, the effect was not present. Galvani concluded that the phenomenon resulted from allowing "...the electric fluid to flow (by whatever means) into the frog..." (Galvani, 1791, p. 48).

One explanation of the effect observed by Galvani and his assistants is as follows. The frog appears to have been insulated from ground during these experiments. For example, Galvani states that the greatest effects were observed when the frog was on the same table as the electrical machine, the table being covered with an oily layer (Galvani, 1791, p. 51). As the charge accumulated on the electrical machine, a charge separation is induced in the frog. When the source of this charge imbalance is removed (i.e., when a spark occurs in the electric machine), a local current is induced in the nerve. This current fires a volley of action potentials toward the muscle and causes it to contract.

Galvani also used other forms of electricity for his experiments. For example, he substituted the magic square of Franklin (Figure 3.2) for the frictional plate machine (Galvani, 1791, p. 52). However, a spark from a charged magic square produced no effect in the muscle. The cause of this failure is unclear but a similar experiment with the Leyden jar and the electrophorus proved successful and thereby established the generality of the phenomenon. Any kind of electricity that produced a spark seemed to work, though in some cases the contractions were very slight. The source of sparks was not limited to so-called "artificial electricity." Lightning was also used. Galvani was puzzled by the results obtained with lightning because there was no direct electrical contact between the frog and the spark. He created a vacuum around the experimental animal but the effect still occurred and the muscle contracted whenever the spark discharged.

PROBLEM. Draw a circuit equivalent to explain current flow in the frog when a distant electrostatic machine is discharged. Galvani's first experiment is similar to the phenomenon known as "return shock." Return shock is a violent, often fatal electric shock that may occur in an animal or person at great distances from the actual place where the lightning strikes. Return shock is described in Ganot (1890), p. 1026.

6. Galvani's Second Experiment

These earlier experiments led Galvani to what he regarded as his greatest scientific discovery, an inherent animal electricity. The second series of experiments began by observing contractions of frog muscle that had no apparent relationship to changes in the electrical state of the atmosphere. Galvani writes (Galvani, 1791, p. 59):

> ... when I brought the animal into a closed room, placed it on an iron plate, and began to press the hook which was fastened to the spinal cord against the plate, behold!, the same contractions and movements occurred as before.

Nothing happened when the connection between muscle and nerve was made by a nonconductor such as glass or dry wood. According to Galvani the results "surprised us greatly and led us to suspect that the electricity was inherent in the animal itself"; further experiments led him to the conclusion that there was an "electrical outflowing, as it were, of nerve fluid" through an external conductor. However, no ordinary conductor would do (Galvani, 1791, pp. 60–61). (See Figure 6.1.)

To understand the phenomenon Galvani drew upon Franklin's analysis of the Leyden jar and likened the discharge of the jar to the flow of animal

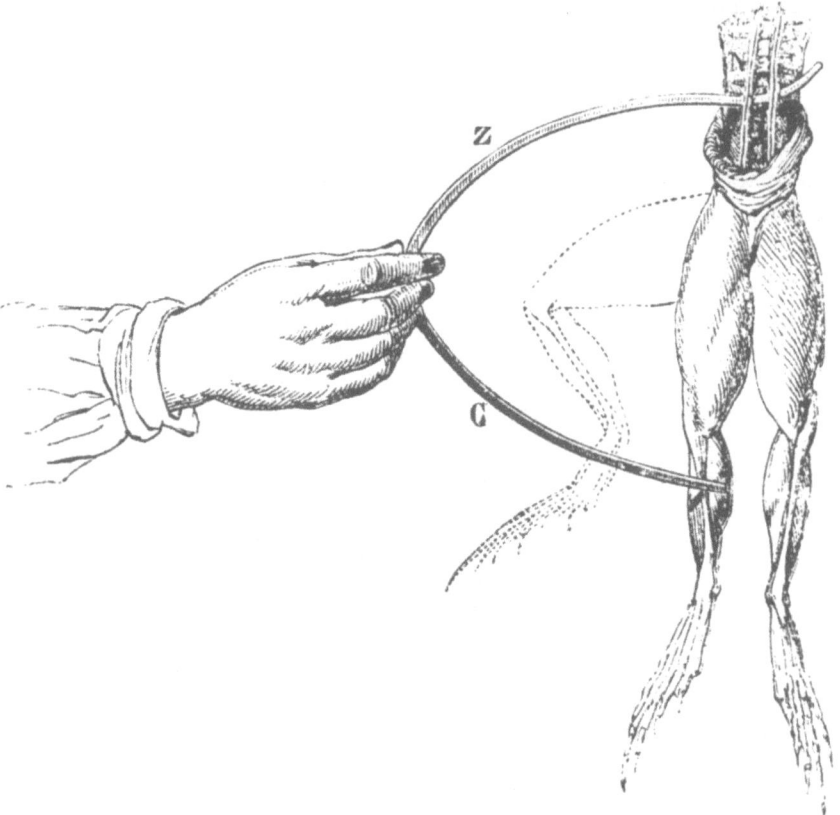

Figure 6.1. Galvani's second experiment. From Ganot (1890), Figure 740. The fundamental observation is this: when the lumbar nerves of a frog are connected by a bimetallic circuit to the crural muscle, the leg kicks outward. In this example, the metal arc is composed of zinc and copper. Each time contact is made, the leg moves, as indicated in the diagram. Galvani's first series of experiments showed analogous motions could be produced in frog's legs under the influence of nearby electrostatic machines. His second series of experiments (illustrated here) proved, to him, that an inherent electricity existed in the animal. Galvani believed that this electricity, which he called a *vital fluid*, passed from the nerves to the muscles through the metallic arc, and that this caused muscles in the leg to contract and the leg to kick outward. This leaves the use of a bimetallic arc unexplained.

electricity through the arc (Galvani, 1791, p. 62):

> From the discovery of a circuit of nerve fluid, (an electric fire, as it were), it naturally seemed to follow that a two-fold and a dissimilar, or rather an opposite, electricity produces this phenomenon in the same way that the electricity in the Leyden jar or the magic square is two-fold, whereby it releases in these bodies its electric fluid in a circuit. For, as the natural philosophers have shown, a flow of electricity in a circuit can take place only in a restoration of equilibrium and occurs chiefly between two opposite charges. That these charges were embodied in one and the same metal seemed contrary to nature ...

Galvani is proposing an equivalent circuit to describe his electrophysiological observations. That he had only partial understanding of the discharge of a capacitor seems less important than the attempt to make rational, in terms of a model, his observations. On the basis of this argument he insisted that electricity must reside in the animal. The conclusion, if not the argument, is correct. Galvani used other analogies. For example, he enhanced the effects illustrated in Figure 6.1 by "arming" the exposed nerve and muscles of his dissected frog by wrapping them in tinfoil, "even as the natural philosophers were accustomed to do with their magic squares and the Leyden jar..." (Galvani, 1791, p. 63).

Galvani extended his observations to other parts of the frog and eventually to warm-blooded animals. He knew he was dealing with a general phenomenon of immense practical significance. But he always viewed muscles and nerves as his prime examples of animal electricity and he used them to arrive at some remarkable conclusions. For Galvani, a "twofold" electricity was present in animals, each completely separated from the other by nature. He of course knew nothing of cell membranes, for which this condition actually exists and the analogy with the condenser is valid.

Galvani thought that the nerve and the muscle were like the two conductors of the Leyden jar, each with its own sign of electricity, and that both types of electricity are found in the muscle. In some way, the metallic arc between the nerve and the muscle discharged the muscle and produced a contraction. Galvani also suggested that animal electricity is similar to ordinary electricity: a muscle fiber is "something like a small Leyden jar..." and a nerve is "in some measure, the conductor of the jar "(Galvani, 1791, p. 74). In this way, Galvani likens the entire muscle to a large array of Leyden jars. Because he knew that the Leyden jar is most easily discharged by a metallic arc, Galvani understood that the metallic arc is also the most effective means of exciting muscle contractions.

A flaw in this logic, one I have been unable to understand by reading Galvani's works, is his explanation of the need for two different metals in the arc. He offers no good reason why one metal will not do, as would seem to follow from his equivalent circuit. Volta, on the other hand, focused all his attention on the bimetallic strip, and it was Volta's experiments on the properties of metallic junctions that led to his invention of the battery.

7. Animal Electricity Described in Ganot's Physics

Before turning to Volta and the controversy that followed Galvani's work, let us consider the 1889 description of animal electricity found in Ganot's *Physics*. After a brief historical account, and some words of caution about the galvanometer and the electrodes to be used,* a description of the currents flowing in muscle at rest follows (Ganot, 1890, pp. 989–990):

In describing these experiments the surface of the muscle is called the *natural longitudinal section*; the tendon the *natural transverse section*; and the sections obtained by cutting the muscle longitudinally or transversely are respectively the *artificial longitudinal* and *artificial transverse sections*.

If a living irritable muscle be removed from a recently killed frog, and the clay of one electrode be placed in contact with its surface, and of the other with its tendon, the galvanometer will indicate a current from the former to the latter; showing, therefore, that the surface of the muscle is positive with respect to the tendon. By varying the position of the electrodes, and making various artificial sections, it is found—

1. That any longitudinal section is positive to any transverse section.
2. That any point of a longitudinal section nearer the middle of the muscle is positive to any other point of the same section farther from the centre.
3. In any artificial transverse section any point nearer the periphery is positive to one nearer the centre.
4. The current obtained between two points in a longitudinal or in a transverse section is always much more feeble than that obtained between two different sections.
5. No current is obtained if two points of the same section equidistant from its centre be taken.
6. To obtain these currents it is not necessary to employ a whole muscle, or a considerable part of one, but the smallest fragment that can be experimented with is sufficient.
7. If a muscle be cut straight across, the most powerful current is that from the centre of the natural longitudinal section to the centre of the artificial transverse; but if the muscle be cut obliquely, as in Fig. [7.1], the most positive point is moved from *c* towards *b*, and the most negative from *d* towards *a* ("*currents of inclination*").

---

* A zinc wire in a $ZnSO_4$ solution is recommended. The sulfate solution is separated from the muscle by a china-clay plug moistened with a common salt solution.

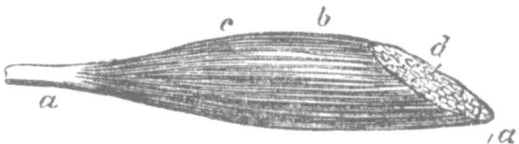

Figure 7.1. Frog muscle. From Ganot (1890), Figure 961. A dissected frog muscle cut obliquely to demonstrate the flow of electrical current under various conditions.

To explain the existence and relations of these muscular currents, it may be supposed that each muscle is made up of regularly disposed electromotor elements, which may be regarded as cylinders whose axes are parallel to that of the muscle, and whose sides are charged with positive and their ends with negative electricity; and, further, that all are suspended and enveloped in a conducting medium. In such a case [Figure 7.1] it is clear that throughout most of the muscle the positive electricities of the opposed surfaces would neutralise one another, as would also the negative charges of the ends of the cylinders; so that, so long as the muscle was intact, only the charges at its sides and ends would be left to manifest themselves by the production of electromotive phenomena; the whole muscle being enveloped in a conducting stratum, a current would constantly be passing from the longitudinal to the transverse section, and, a part of this being led off by the wire circuit, would manifest itself in the galvanometer.

This theory also explains the currents between two different points on the same section; the positive charges at *b*, for instance [Figure 7.2], would have more resistance to overcome in getting to the transverse section than that at *d*,

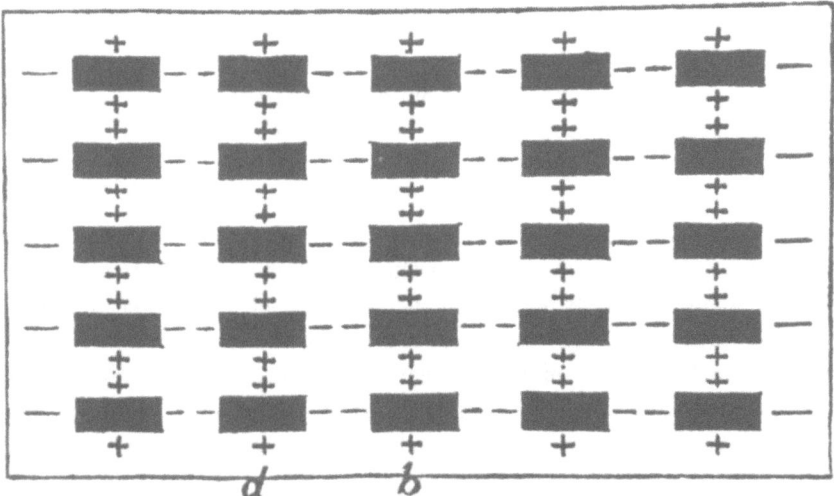

Figure 7.2. Equivalent circuit. From Ganot (1890), Figure 962. The equivalent circuit of the existence and the relationship of the observed muscular currents.

therefore it has a higher tension; and if *b* and *d* are connected by the electrodes, *b* will be found positive to *d*, and a current will pass from the former to the latter. What are called *currents of inclination* are also explicable on the above hypothesis, for the oblique section can be represented as a number of elements arranged ... so that both the longitudinal surfaces and the ends of the cylinders are laid bare, and it can thus be regarded as a sort of oblique pile whose current adds itself algebraically to the ordinary current and displaces its poles as above mentioned.

A perfectly fresh muscle, very carefully removed, with the least possible contact with foreign matters, sometimes gives almost no current between its different natural sections, and the current always becomes more marked after the muscle has been exposed a short time; nevertheless, the phenomena are vital, for the currents disappear completely with the life of the muscle, sometimes becoming first irregular or even reversed in direction.

Electric currents in nerve are described as being similar to muscle although weaker. The above passage shows that a popular physics textbook of the late 1800's still used the equivalent circuits introduced by Galvani nearly 100 years earlier.

PROBLEM.   Give a modern explanation of the effects described in Figure 7.1.

8. *The Voltaic Pile and the Electric Fish*

Galvani and Volta focused the attention of Western scientists on the emerging electrical science: Galvani, the anatomist and defender of electrophysiology, and Volta, the physicist and discoverer of a new source of electricity. What was Volta's contribution to this story and how did the Galvani–Volta controversy affect the development of animal electricity as a subject of study?*

Until Volta's time, electricity was studied by amateurs or by scientists who considered the subject incidental to their main research. Volta is the first scholar to spend his entire career studying electrical phenomena and to make fundamental contributions to the subject over several decades.

The electrophorus (Figure 4.1), reported in a letter to Joseph Priestly in 1775, was Volta's first major discovery. The device could be used in conjunction with some means of storing electrical charge, such as the Leyden jar, to amass large quantities of electricity. After he moved to Pavia in 1781,

---

* Some of the material in this section is taken from the excellent film *Volta and Electricity* (1974), History of Physics Laboratory, Barnard College, New York, edited by Samuel Devons.

Volta made his second major discovery. This new invention measured very weak sources of electricity and was called a "condensing electroscope" (Figure 4.2). Volta used the condensing electroscope to argue that Galvani's second experiment results from simulations by bimetallic arc and not from an inherent animal electricity. By the time Volta learned of Galvani's work, he had been studying electricity for nearly twenty years and had already received international recognition, including a fellowship in the Royal Society and a corresponding membership in the French Academy. His works had been translated into English and appear to have been widely read. The inventions established Volta's reputation throughout Europe and supported his initial persuasive arguments against the findings of Galvani.

In 1791 Volta received an account of animal electricity from Galvani. He substantially altered Galvani's hypothesis of an animal electricity discharged via external conductors. Rather than use physical analogs, as Galvani had done, Volta conjectured that the entire source of the phenomenon lay in the physical objects associated with the experiment, and he guessed that there must be some source of electricity in the bimetallic arc.

First, Volta proved to himself that known sources of electricity may indeed cause nerve–muscle or other preparations to respond. The physicist-turned-electrophysiologist performed experiments on every form of animal: vertebrates and invertebrates, hot-blooded and cold-blooded species, and finally, himself. Thus Volta arrived at his famous experiment on the sensation of light and showed that light can be experienced the moment one end of a bimetallic strip touches the corner of the eye while the other touches the hand or mouth.

Volta theorized that each metal has its own electromotive force. In the region of contact of two dissimilar metals, the equilibrium of these two forces is disturbed. A closed circuit of two different metals gives no current because the forces at the two junctions cancel each other. But suppose the circuit contains two different metals in contact with a "wet conductor" as in Galvani's second experiment. According to Volta, the cancellation which occurred in the all-metal loop no longer applies, and a continuous circulation of current is possible. The frog in Galvani's second experiment was simply a passive wet conductor and only by chance a sensitive detector of electrical current induced by the metallic junction.

Volta then began to study the new source of electricity he had postulated. He discarded the frog and studied the physical phenomenon itself. Because the source of electricity lacked strength, and because he had removed from the experiment the sensitive frog leg that had detected the electricity, he turned to his own invention, the condensing electroscope. When the strips are removed, the plate separates and the leaves diverge. The experiment is illustrated in Figure 4.2, where the hand acts as the wet conductor.

Although Galvani's concept of animal electricity was discarded, Volta's explanation of the weak effects of the wet conductor was not acceptable to everyone. To further his theory, Volta returned to the electrophorus (Figure 4.1). The copper plate, held by an insulating handle, was first brought into sharp contact with the zinc and then removed and touched to the upper plate of the condensing electroscope. After repeating this process many times, Volta opened the electroscope to magnify the charge that had accumulated. The success of this experiment secured Volta's claim to the discovery of a contact electricity in metals.

After proving that this weak source of electricity existed, Volta combined many bimetallic elements, each separated by a wet conductor, to produce a source of electricity sensible to anyone. The array of copper, wet pasteboard, and zinc is repeated dozens of times to produce the voltaic pile. Hundreds of variations are possible, each with the same effect. The pile produced electricity that could be detected and used at will.

The device illustrated in Figure 8.1 is similar to Volta's original design. Although Volta's theory of electricity came under attack, the pile worked and was rapidly improved upon and used throughout the Western world. It made electricity available in steady and usable quantities. Soon after its invention, the pile was used to decompose water into hydrogen and oxygen, to isolate metallic sodium and potassium, and to produce light.

Although Volta insisted that Galvani had not demonstrated an inherent source of animal electricity, he tacitly admitted the existence of animal electricity by comparing his own invention to the organ of the electric fish torpedo, even going so far as to refer to the pile as an "artificial electric organ" (Mauro, 1969). Volta was using a physical device as a model of an electrophysiological phenomenon, just as Galvani had done earlier.

Apparently the electric fish was an undisputed source of electricity, used long before the Volta–Galvani controversy ever began. If everyone knew that animals could produce electricity, what then was the controversy all about? We can only speculate that except for the bimetallic strip, Volta might never have challenged the idea that the nerves and muscles of Galvani's second experiment might also be inherent sources of current, and that Volta's views of this specific effect might never have been taken, however unwittingly, as a general refutation of animal electricity.

But criticism had its effect, and Galvani set out to prove his point by doing just the reverse of what Volta had done—he removed the metallic

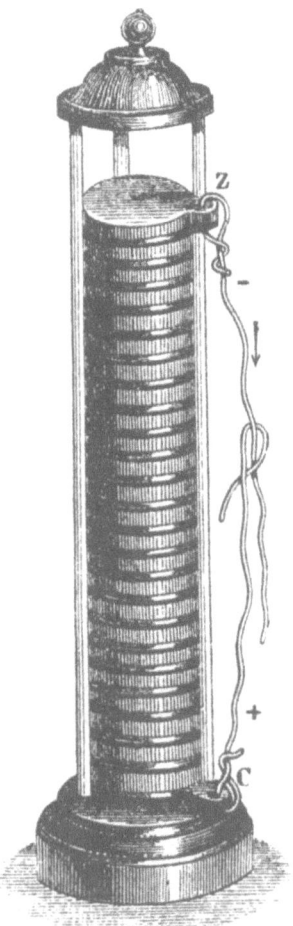

Figure 8.1. The Voltaic pile. From Ganot (1890), Figure 744. This early arrangement follows a design by Volta himself. At the bottom of the wooden frame is a disk of copper; this is followed by cloth moistened with brine, then a disk of zinc and the pattern is repeated over and over again, ending in a zinc disk at the top. The vertical rods which hold the disks are of glass. The original form of the Voltaic pile, or battery, is inconvenient, and a rapid decrease of the current occurs with use. The first "constant" batteries were invented in 1836 by Daniell.

arc. Hoff refers to this as Galvani's third experiment. The preparation was held by one leg in such a way that the vertebral column and sciatic nerve touched muscles of the other leg. The muscles contracted. No metal was present, so the tissue itself was capable of producing the effect and was, by inference, an inherent source of electricity. These experiments were repeated by Humbolt in 1797 with apparently little effect on Volta's position (Mauro, 1969, pp. 148–149).

9. Examples

The following exercises have been adapted from Ganot (1890).

PROBLEM. 100 in. of Cu wire weighing 100 gr has a resistance of 0.1515 Ω. What is the resistance of 50 in. weighing 200 gr?

ANSWER. 0.01894 Ω.

PROBLEM. The terminals of a battery are connected to a conductor of 1 Ω producing a current of 1.32 A. The same battery connected to a conductor of 5 Ω produces a current of 0.33 A. What is the emf and internal resistance of the battery?

ANSWER. $R = 1/3$ Ω, $E = 1.76$ V.

PROBLEM. What is the best arrangement of six cells, each of 2/3 Ω internal resistance, against an external resistance of 2 Ω? (Note: There are *two* possible answers.) Best implies maximum current.

PROBLEM. Let battery have an emf E and internal resistance 3 Ω. Let each arrangement in Figure 9.1 be connected to an external resistance of 12 Ω. Show that the current that flows in the external resistance is $(6/30)E$, $(6/33)E$, $(6/42)E$, and $(6/75)E$ in the four cases illustrated. Show that the external current into two ohms is now $(6/20)E$, $(6/13)E$, $(6/12)E$, $(6/15)E$. Notice that in the last case, the best arrangement (the largest external current) is number 3.

PROBLEM. Let N be the total number of cells available for a given combination, and let n be the number of cells arranged in series. Thus N/n elements will be arranged in parallel. If E is the emf of one cell, and R is its internal resistance, show that the maximum current through an external resistor r is obtained when

$$n = \left(\frac{Nr}{R} \right)^{1/2}$$

In the first case of the preceding problem, $N = 6$, $R = 3$, and $r = 12$. Thus $n = 24^{1/2} = 4.9$, which is closest to the series number 6. In the second case of the preceding problem, $N = 6$, $R = 3$, and $r = 2$. Thus $n = 4^{1/2} = 2$, which is the arrangement 3.

These examples parallel similar problems in ionic conductors that appear in Chapter 2. Note that the Siemens unit, which has become standard in noise theory as the pS unit of single-channel conductance (10^{-12} Ω^{-1}), was originally defined as the resistance (Ω) of a column of mercury 1 m long and 1 mm in cross section (Ganot, 1890, p. 966). This is 0.9534 Ω at room temperature.

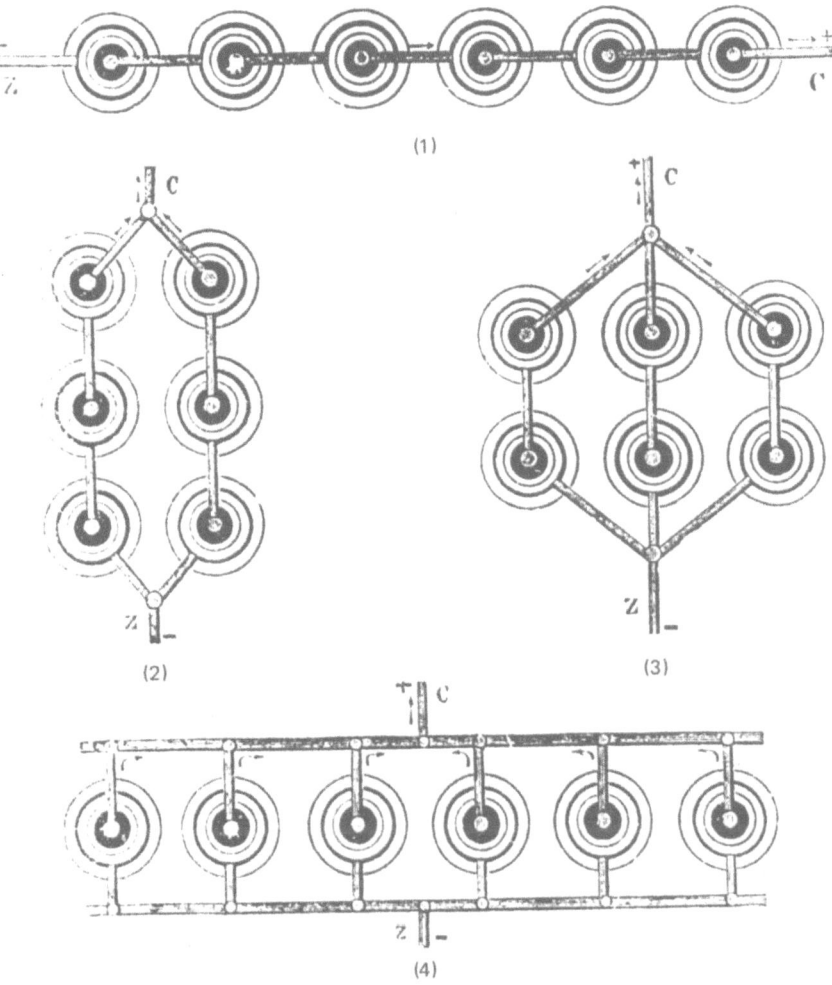

Figure 9.1. Six batteries in four arrangements. From Ganot (1890), Figures 667–770. (1) In the single series arrangement, the zinc of each element is connected to the copper of the second, and so on. (2) A system of three double elements. (3) A system of two triple elements. (4) One large element composed of all six batteries.

10. *Models and Analogies Used in Electrophysiology*

In the preceding sections we saw that Galvani and Volta both used equivalent circuits to help understand their experiments. Galvani used the Leyden jar to explain his frog experiments and Volta used the electric fish to explain the pile. These equivalents are not circuits in the modern sense of the word but are models or analogies that in some way stand for the phenomena.

In this section I wish to distinguish between models and analogies. A model represents an existing object and in science often implies an idealization of that object. Analogy is a resemblance of attributes only and does not imply essential similarity. Explanation by analogy depends only on similarity of relationships whereas explanation by a model is an argument from example.

I do not know whether Galvani and Volta meant their equivalents to be models or to be analogies. I would, however, like to be clear about our own use of equivalent circuits in the remainder of this book. In all cases but one, the equivalent circuits we use are models of the phenomena. Cytoplasm is a resistor and the membrane lipid substrate is a capacitor in the sense that they have the material properties and obey the physical laws that the symbols R and C ideally represent. The equivalent circuit that represents the subthreshold behavior of excitable membranes contains an rL branch which is not a model but is an analogy of a resistor and an inductor in series. The resistor r represents more than a simple resistivity in the membrane (membrane kinetics are also involved) and there is no coil or magnetic field in the membrane that L stands for. The membrane simply acts in a certain limited sense like the Ohm's law resistor and the Faraday's law inductor that the symbols r and L represent.

Confusion may arise because there actually are local current loops and magnetic fields in the membrane. The separation of currents and the propagation of the action potential in Hodgkin and Huxley's 1952 description of excitability require this. These, however, are not the inductors we are talking about in this book. Ours are merely analogies to the time variant conductance originally described by Hodgkin and Huxley.

It should be recognized that some authors have used real inductors arising from local ionic currents to describe propagation in axons. Lieberstein (1973), for example, has reformulated the differential equations that underlie excitability by associating a line inductance to toroidal current sheets that sweep along the axon during propagation. Our own treatment of the $RrLC$ transmission line in Section 88 should not be confused with Lieberstein's approach.

Once one arrives at an equivalent circuit, either by example or by analogy, it is a useful tool to describe the behavior of the thing it represents. Membrane noise is an example. Nyquist not only related Johnson noise to thermodynamics, he also showed us how to describe noise sources with equivalent circuits. A real resistor can be replaced by an ideal noiseless resistor and a Johnson noise generator. Membrane impedance can be calculated from the membrane's equivalent circuit and the noise can be determined by using Nyquist's theorem.

Do the analogical components of the membrane's equivalent circuit have the same noise representation as the model components? Does the equivalent circuit of an excitable membrane derive only from a population of channels or is it legitimate to speak of the impedance of a single channel? These are some of the questions we shall attempt to answer in the later chapters of this book.

2

Basic Electrophysiology

11. Salt Water Conducts Electricity

As a solid, common salt forms a well-ordered lattice structure. For example, NaCl has the repeating structure shown in Figure 11.1. When this highly ordered crystal goes into aqueous solution, the order is almost completely broken down and the individual ions form a randomly moving, gaslike substance immersed in the water. One expresses this reaction (shown in Figure 11.2) by writing

$$NaCl \rightleftharpoons Na^+ + Cl^-$$

Within the imaginary box drawn in Figure 11.2, there is an equal number of positive and negative ions on the average. At any instant, however, there may be an imbalance of charge. Suppose that we could count the ions in a certain size box and keep track of their total number, N, in time. Figure 11.3 might be the result. Let $\langle N(t) \rangle$ represent the average number. One way to describe the deviations from the average is to form the quantity

$$\langle [N(t) - \langle N(t) \rangle]^2 \rangle^{1/2}$$

The meaning of this expression is shown in Figure 11.4.

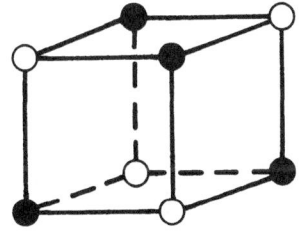

Figure 11.1. Diagram of the internal structure of a NaCl crystal. The solid circle represents Na and the open circle represents Cl.

27

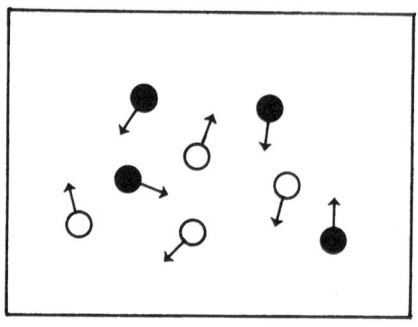

Figure 11.2. Diagram of the random movement of Na and Cl ions dissolved in water.

The standard deviation, or root mean square, of the random variable $N(t) - \langle N \rangle$ describes the average fluctuation present in $N(t)$. If the individual ions were perfectly independent of one another (ours are not since they are charged), then it would be found that

$$\langle (N - \langle N \rangle)^2 \rangle = \langle N \rangle$$

In words, the variance is proportional to the mean. (Notice that this equation drops the explicit time dependence $N(t)$, although time dependence is still implied.) If this expression holds, it says that the standard deviation varies as the square root of the mean, i.e., as the size of the box increases, the relative noise decreases. Thus

$$\frac{\langle (N - \langle N \rangle)^2 \rangle^{1/2}}{\langle N \rangle} = \frac{1}{\langle N \rangle^{1/2}}$$

Often the symbol N by itself stands for the average value of N. This may lead to confusion; in long discussions that only deal with average properties this shorthand form will be used.

There are several ways of describing how many ions are present, on the average, per unit volume of water. One mole (M) of a substance implies the molecular weight of that substance in grams. For example, KCl has a molecular weight of 74.56; one mole of KCl weighs 74.56 g. One mole of

Figure 11.3. The total number of ions in a specified volume; the dashed line is the average number, and the solid line is the actual number.

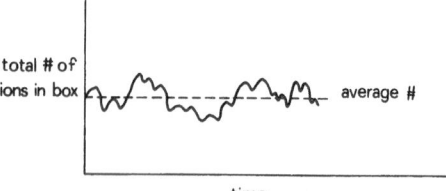

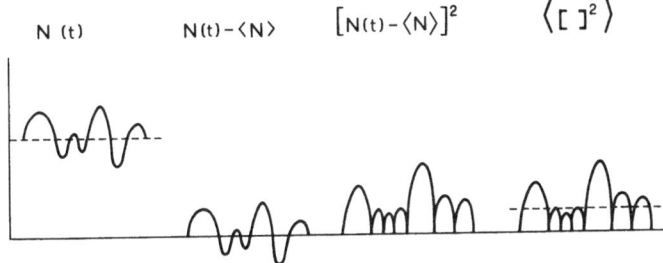

Figure 11.4. A diagram to illustrate the operations of removing the average (to focus on the fluctuation), squaring the fluctuation, and taking the average of the square fluctuation (dashed line).

any substance contains one Avogadro's number of molecules of the substance; 74.56 g of KCl contain 6.02×10^{23} molecules of KCl. There are this number of K ions and of Cl ions when one mole of KCl is dissolved in water. When one mole of a substance is dissolved so that the final volume of water is 1 liter, we have a concentration of 1 molar (M). Because of the biological range of concentrations, we usually use millimolar (mM) as the unit of ionic concentration. Let n stand for the number of moles per liter of water of a certain substance (usually given as a subscript). Then

$$n_{KCl} = \frac{\text{No. of moles KCl}}{1 \text{ liter } H_2O} \ominus \frac{\text{No.}}{\text{volume}}$$

The symbol $\ominus$ stands for the phrase "has the units of."

To a first approximation, the resistivity of an ionic solution is inversely proportional to the concentration of ions. Table 11.1 gives the resistivity of KCl at 25°C for three concentrations.

Thus, the inverse relation between ϱ and n is not strictly valid; as n increases by a factor of 10, ϱ does not decrease by the same factor. It seems intuitive that as more and more ions are available the resistance should

TABLE 11.1. *Resistivity of KCl at 25°C*

ϱ ($\Omega \cdot$cm)	nKCl
708	0.01
77.6	0.10
8.95	1.00

TABLE 11.2. *Equivalent Conductances at 25°C* [a]

Λ_{HCl}	Λ_{KCl}	Λ_{NaCl}	n
426.16	149.86	126.45	0.000
421.36	146.95	123.74	0.001
415.80	143.55	120.65	0.005
412.00	141.27	118.51	0.01
407.24	138.34	115.76	0.02
399.09	133.37	111.06	0.05
391.32	128.96	106.74	0.1

[a] Taken from MacInnes (1961), p. 339.

decrease. However, as more ions occupy the same volume, their effective number is reduced by their interaction.

Electrochemists have introduced the concept of an equivalent conductance, Λ, which is related to ϱ through the formula

$$\Lambda = \frac{1000}{\varrho n}$$

The ideal behavior $\varrho \propto 1/n$ is expressed by $\Lambda = \text{const}$. The 1000 is introduced simply to give a convenient range of Λ. Consider Table 11.2. The values of Λ given in this table are approximately described by the theoretical expression

$$\Lambda = \Lambda_0 - [0.2273\Lambda_0 + 59.78]n^{1/2}$$

where Λ_0 is the equivalent conductance when $n = 0$ (see MacInnes, 1961, p. 328). This expression is based on the theoretical work on Onsager for very dilute solutions of strong electrolytes. The expression may be rearranged to give

$$\frac{\Lambda + 59.78n^{1/2}}{1 - 0.2273n^{1/2}} = \Lambda_0$$

Letting $\Lambda_0 \to \Lambda_0 + Bn$ gives a better approximation at higher concentrations. B is an empirical constant which, for the three electrolytes considered, has the value

$$B_{HCl} = 169, \qquad B_{KCl} = 141.9, \qquad B_{NaCl} = 95.79$$

Higher-order correction terms exist.

Notice that ionic species and electrical conduction are not simply related. In the example given, the H ion < Na ion < K ion (in size), yet this is not the order of conduction. Also, HCl is a much better conductor than either KCl or NaCl.

PROBLEM. Give a brief intuitive interpretation that KCl is inherently a better conductor than NaCl, and the meaning of Λ_0. Why does HCl seem to be a completely different type of conductor in aqueous solutions?

In spite of the evident complexity of the relationship between ionic species, ionic concentration, and electrical conduction in water, it is often convenient to use the formula

$$\varrho = \left(\frac{1000}{\Lambda}\right)\frac{1}{n}$$

as if Λ had the constant value Λ_0.

There is a limit to the extent a substance may be dissolved in a fixed volume of water at given temperature and pressure. Under normal conditions, the saturation point of KCl is about 4.5 M. Putting in more KCl than this results in the re-formation of crystals. The physiological range of dissolved salts is much below their saturation levels.

12. Resistance of Salt Water

The average resistance of a material object is given by

$$R = \varrho\,\frac{l}{A}$$

where l is the length of the object and A is its cross-sectional area. The quantity $\varrho \ominus \Omega$ cm is called the resistivity and is an intensive parameter of the material. The shape of the object is contained entirely in the term (l/A).

To describe the resistance of aqueous salt solutions it is necessary to know how ϱ depends on the type of salt and on its concentration. To a crude approximation

$$\varrho \propto \frac{1}{n}$$

where the proportionality is constant for a particular electrolyte. However, actual values of $\varrho(n)$ may vary from this simple approximation, especially at higher concentrations. For example, consider the resistivity of KCl in aqueous solution as a function of ionic concentration (calculated from tables given in MacInnes, 1961, p. 339).

PROBLEM.   Estimate the resistance of an agar bridge (use Λ_0) prepared with 0.1 M KCl. The dimensions of the cylindrical bridge are 10 cm in length and 1 mm in diameter. Compare this result with the more accurate estimate using $\Lambda(n)$.

PROBLEM.   Estimate the size of the fluctuation of the resistance due to the thermal motion of the ions, assuming the bridge is in contact with an aqueous reservoir.

13. How Ions Move: The Flux Equation

In order to conduct electricity, the ions must be mobile. Their thermal motion has already been introduced in Section 11. Here we shall be concerned only with their average motion due to some force.

To help describe the motion of ions in solution, it is useful to define the flux of particles. Let

$$\phi = \frac{\text{No. of moles}}{\text{area} \cdot \text{time}}$$

be the flux: this definition holds for both charged and uncharged particles and describes the average number of particles that move across an area in a certain time. We may imagine that some water molecules move along with the particles. The bulk of the water, however, is still. The symbol ϕ stands for the average movement of the particles relative to the bulk of the water.

If the particles are ions, their average movement is an electric current. However, the flux of ions and the electric current must be distinguished. For example, one may encounter zero net current with a nonzero flux (see Figure 13.1). Also, according to our definition, one may have zero net flux, but nonzero current (see Figure 13.2).

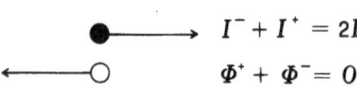

$$I^- + I^+ = 2I$$
$$\Phi^+ + \Phi^- = 0$$

Figure 13.1. The average movement of ions in an electric field. Positive ions (filled circle) move opposite to negative ions (open circle) in response to the same electric field. I is the current due to one ion.

Figure 13.2. The average movement of ions in a concentration gradient. Positive- and negative-charged ions move in the same direction in response to the same gradient. ϕ is the flux due to one ion.

$$I^+ + I^- = 0$$

$$\Phi^+ + \Phi^- = 2\Phi$$

There are two basic principles behind the average motion of particles. The first is diffusion which is general; the second applies only to charged particles such as ions in solution.

Simple diffusion is described by Fick's law. Consider two groups of particles of unequal concentration in communication with each other. As a consequence of the thermal motion of the particles there will be a tendency for the two groups to mix. A net vector of motion will exist from the dense to the sparse group. Figure 13.3 has been drawn as if the two groups contained, on the average, equal numbers of positively and negatively charged ions. However, the phenomenon of diffusion does not rely on this. Two groups of uncharged particles might have been considered with the same results.

Fick's law states that

$$\phi_{\text{Fick}} \propto -\frac{dn}{dx}$$

where $n = n(x)$ is the distribution of concentration in the direction x. The minus sign simply gives the correct (observed) direction of the motion. Thus, in Figure 13.3

$$\frac{dn}{dx} < 0 \quad \text{and} \quad \phi_{\text{Fick}} > 0$$

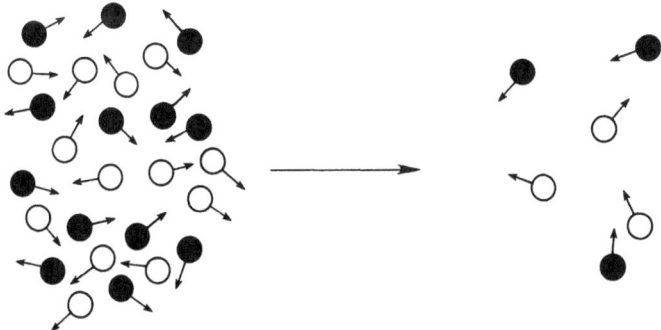

Figure 13.3. Diagram of the net movement of ions from a region of high concentration to a region of low concentration. This movement is a consequence of the thermal motion of ions in solution; in the region of mixing there is a net vector of motion to the right.

The net flow is from left to right. The proportionality constant in Fick's law is the diffusion constant D. Using this constant we write

$$\phi_{\text{Fick}} = -D \frac{dn}{dx}$$

to describe the average one-dimensional movement of particles due to diffusion; $n(x)$ is the value of the concentration n at x.

It is known that the average flux is proportional to temperature. This is intuitive if one accepts diffusion as a consequence of the thermal motion of the particles. The dependence on temperature is described by

$$D \propto kT$$

The constant k is Boltzmann's constant and T is the absolute temperature. The constant of proportionality is called the mobility u. The resulting form of Fick's law is then

$$\phi_{\text{Fick}} = -ukT \frac{dn}{dx}$$

[An interesting historical account of the derivation of the relationship $D = ukT$ can be found in Moelwyn-Hughes (1960), Chap. 1.]

PROBLEM. Show that the mobility u as defined above has the units

$$u \ominus \frac{\text{velocity}}{\text{force}}$$

Another common definition of mobility, used exclusively for charged particles, is

$$u' = \frac{\text{velocity}}{\text{electric field}}$$

The two definitions are related by $u' = eu$, where e is the electronic charge.

Ohm's law describes the net motion of charged particles in an electric field. The flux of particles is proportional to the field, thus

$$\phi_{\text{Ohm}} \propto \pm \frac{dV}{dx}$$

depending on the sign of the charge. $V = V(x)$ is the value of the potential V at x. Let the sign of the charge be included through the valence z (i.e., $z = \pm 1, \pm 2$, etc.). The flux is also proportional to the charge density en

(where e is the electronic charge) and to the mobility u. Finally

$$\phi_{\text{Ohm}} = -zenu\,\frac{dV}{dx}$$

This is Ohm's law for the movement of ions in solution.

The total flux of ions in solution due to diffusion and electronic forces is then

$$-\phi = ukT\,\frac{dn}{dx} + zenu\,\frac{dV}{dx}$$

When more than one type of ion is involved, subscripts are written to distinguish them. Thus, the flux of the ith ion is written

$$-\phi_i = u_i kT\,\frac{dn_i}{dx} + z_i en_i u_i\,\frac{dV}{dx}$$

Notice that V (like the temperature, T) is not given a subscript since there may be only one voltage at any given point.

The movement of ions, due either to density or to voltage gradients, implies an electric current. The relationship between the flux and the current is

$$I = zeA\phi \ominus \text{amperes}$$

where A is the cross-sectional area to which ϕ (and I) is perpendicular.

A density gradient and a voltage gradient across ions in solution affect their movement differently, as illustrated in Figure 13.4. In Figure 13.4, the density gradient of KCl, for example, tends to move the KCl to the right. The voltage gradient tries to separate the charge; this separation in opposed by the charge on the particles themselves. The interdependence of these forces is the basis of elementary theories of bioelectricity.

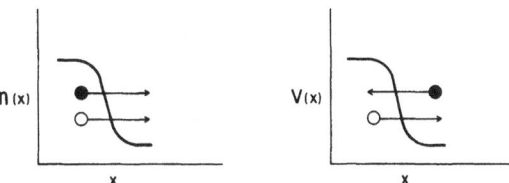

Figure 13.4. $n(x)$ is the concentration of ions as function of distance. $V(x)$ is the voltage as a function of distance. Positive and negative ions move down the concentration gradient but, depending on their sign, move up or down a voltage gradient. The solid circle represents the positive ion.

PROBLEM.   A uniform resistor of length l and cross-sectional area A has a linear constant field across it. Starting from the formula $\phi_{Ohm} = -zenu(dV/dx)$ show that the form of the resistivity of the resistor must be

$$\varrho = \frac{1}{z^2 e^2 nu} \ominus \Omega \text{ cm}$$

PROBLEM.   Let $l = 1$ cm and $A = 0.01$ cm² in the above problem. Suppose the resistor is composed of 70% (by volume) 0.015 M NaCl and 30% 0.03 M KCl. What is the resistance?

14. First Application of the Flux Equation: The Nernst Relation

Consider two solutions of a salt at different concentrations separated by a membrane. Let the membrane pass only the cation and be completely impermeable to the anion. This will be referred to as a perfect membrane.

For example, consider two KCl solutions separated by a perfect K membrane. In Figure 14.1 the concentration of KCl on either side of the membrane is different, but on each side the same number of cations and anions exist. Furthermore, in arbitrarily small volumes, the average charge is zero. We wish to consider the case that there is no net current. Thus

$$\sum z_i \phi_i = 0$$

This formula states that in some arbitrary volume the charge that enters must equal the charge that leaves. Notice that the valence of the ith ion is

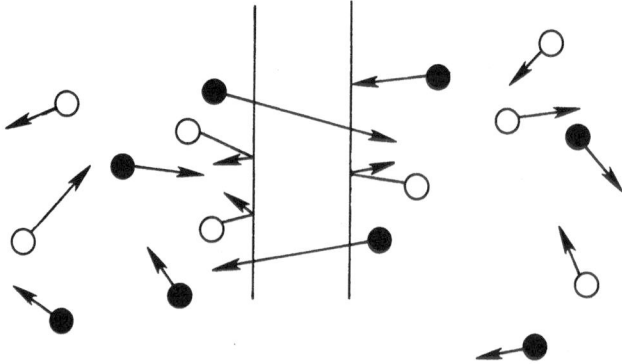

Figure 14.1. Diagram of a K membrane. K ions (filled circles) pass through the membrane but Cl ions (open circles) do not.

taken into the sum; for example, one positive ion and one negative ion moving into the same volume obey the rule. The sum only constrains the average movement of charge and makes no other restriction. (Obviously in the limit of atomic dimensions other constraints exist.)

The sum $\sum z_i \phi_i = 0$ is equivalent to Kirchhoff's rule for electronic circuits and simply describes the steady-state flow of current without the buildup of charge. Applied to the perfect K membrane, Kirchhoff's rule is

$$\sum z_i \phi_i = z_1 \phi_1 + z_2 \phi_2 = 0$$

where the subscript 1 stands for Cl and 2 stands for K. $z_1 = -1$ and $z_2 = +1$. Within the perfect K membrane, $\phi_1 = 0$. Our conditions reduce to the expression $\phi_2 = 0$, or

$$u_1 kT \frac{dn_1}{dx} + e n_1 u_1 \frac{dV}{dx} = 0$$

Recall that $n_{KCl} = n_K = n_{Cl}$ in any small volume. We may drop the subscript and consider n to be the concentration of KCl. There are two values for the concentration of KCl: n'' on one side of the membrane and n' on the other.

Rearranging the above expression gives the Nernst equation

$$\frac{dV}{dx} = - \frac{kT}{en} \frac{dn}{dx}, \qquad dV = - \frac{kT}{e} \frac{dn}{n}$$

By integrating between the two limits of the membrane thickness, we obtain

$$V'' - V' = - \frac{kT}{e} \ln \frac{n''}{n'}$$

The ratio kT/e will appear often. It is useful to realize that

$$\frac{kT}{e} = \frac{AkT}{Ae} = \frac{RT}{F} \simeq 25 \text{ mV at room temperature}$$

where A is Avogadro's number, R is the gas constant, and F is Faraday's constant.

Had we retained the identity and valence of the ion passing the perfect K membrane, the equation would be

$$V'' - V' = - \frac{kT}{e z_K} \ln \frac{n_K''}{n_K'}$$

This is called the Nernst potential for the K ion. If the membrane passed only Cl, a similar expression would hold with Cl replacing K.

PROBLEM. Calculate the Nernst potential across a perfect Cl membrane separating 50 mM CaCl$_2$ from 1.0 mM CaCl$_2$.

15. *Second Application of the Flux Equation: The Diffusion Potential*

Consider a membrane permeable to more than one ion. For example, let the membrane in Section 14 admit Cl as well as K. From the flux equation and the condition $\sum z_i \phi_i = 0$, considering only that the thickness of the membrane sustains a gradient,

$$z_1\left(u_1 kT \frac{dn_1}{dx} + z_1 e n_1 u_1 \frac{dV}{dx}\right) + z_2\left(u_2 kT \frac{dn_2}{dx} + z_2 e n_2 u_2 \frac{dV}{dx}\right) = 0$$

Solving this expression for the potential gradient, we obtain

$$\frac{dV}{dx} = -\frac{kT}{e} \frac{z_1 u_1 \, dn_1/dx + z_2 u_2 \, dn_2/dx}{z_1^2 n_1 u_1 + z_2^2 n_2 u_2}$$

From the condition of local neutrality of charge, $n_1 = n_2 = n$ in every dx. Then

$$\frac{dV}{dx} = -\frac{kT}{e} \frac{z_1 u_1 + z_2 u_2}{z_1^2 u_1 + z_2^2 u_2} \frac{dn}{n}$$

or

$$V'' - V' = -\frac{kT}{e} \frac{z_1 u_1 + z_2 u_2}{z_1^2 u_1 + z_2^2 u_2} \ln \frac{n''}{n'}$$

In our example the subscript 1 stands for Cl and 2 for K. Then

$$V'' - V' = -\frac{kT}{e} \frac{u_K - u_{Cl}}{u_K + u_{Cl}} \ln \frac{n''_{KCl}}{n'_{KCl}}$$

where $n_{KCl} = n_K = n_{Cl}$. This is called the diffusion potential. The mobilities are those within the membrane. If the mobilities of the two ions are equal, then the diffusion potential is zero. If the mobility of one of the ions is zero, then the diffusion potential reduces to the Nernst potential.

Both the Nernst and the diffusion potential result from the interplay of forces described by Fick's law and Ohm's law. Any other forces we may wish to consider will result in different expressions. In the Nernst relation one may consider that an arbitrarily small amount of charge crosses the membrane and sets up a potential which opposes further movement. Thus the concentration difference will be maintained indefinitely. In the diffusion relation, different mobilities of the ions would result in a separation of charge.

A potential is established which retards one ion and helps the other so that the two move together (on the average) with the same effective mobility. In this case, however, the concentration imbalance would eventually run down since both ions move.

The Nernst relation is an equilibrium potential which dissipates no energy and requires none to maintain it. The diffusion relation gives a steady-state potential. As long as the initial conditions exist, the potential does not change with time. Energy, however, is continually dissipated by the net movement of the ions.

No perfect membranes exist although many physical barriers approximate this condition. Any substance may be thought of as a perfect membrane for itself: K for a K membrane, Cu for a Cu membrane, etc. Water is a good example of a membrane permeable to both K and Cl. A diffusion potential will arise between freely mixing volumes of water containing different concentrations of KCl. In this case the K and the Cl each move through water. Thus the membrane is homogeneous. We might also have considered a heterogeneous membrane composed of patches of perfect K membranes and patches of perfect Cl membranes. This mosaic membrane may have the same macroscopic properties as the heterogeneous membrane, but it will have strikingly different microscopic properties.

PROBLEM. In water the K and the Cl ions move more or less equally well. Actual values for mobility give

$$\frac{u_{Cl}}{u_K} \simeq 1.03$$

Calculate the initial diffusion potential between saturated KCl and 4.5 mM KCl.

Suggested Reading. Comprehensive treatments of diffusion and physical electrochemistry are given in Butler (1951), Robinson and Stokes (1955), MacInnes (1961), and Moelwyn-Hughes (1961).

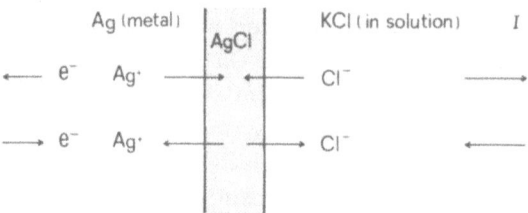

Figure 16.1. Diagram of a Ag–AgCl electrode. *e* represents an electron in the metal; if an electron leaves, a Ag ion from the metal combines with a Cl ion from solution to form the AgCl layer; if an electron enters, a Ag ion from the AgCl layer combines with it to form the metal, freeing a Cl ion in the solution. The net effect is seen by looking only at the electron and the Cl ion. By convention, current *I* is opposite to the flow of negative charge.

16. An Example of the Nernst Relation: The AgCl Electrode

The chemical AgCl may adhere to Ag to form a Ag–AgCl electrode. When this electrode is immersed in a salt solution that contains the Cl ion, it behaves approximately as a Cl electrode. Consider the reaction that occurs when a AgCl electrode is used to pass current. Two directions of current are shown in Figure 16.1. The symbol for the AgCl electrode will be as shown in Figure 16.2.

With respect to the solution, the AgCl admits or accepts only Cl. It is ideally a perfect Cl membrane. If a perfect Cl membrane separates two solutions containing the Cl ion, the potential difference between the solutions is given by the Nernst relationship:

$$V'' - V' = -\frac{kT}{ez_{Cl}} \ln \frac{n''_{Cl}}{n'_{Cl}}$$

In practice the potential difference is not measured across the AgCl but between two Ag wires that hold AgCl. This is illustrated in Figure 16.3. Each electrode is immersed in a salt solution of different concentration. The two solutions are joined by an ideal bridge that introduces no potential of its own and allows no ions to move (dotted line).

The potential measured in this hypothetical experiment has the op-

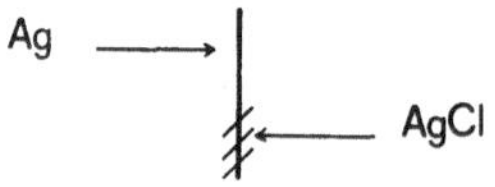

Ag ⟶ |

AgCl

Figure 16.2. Electrical symbol for a Ag–AgCl electrode.

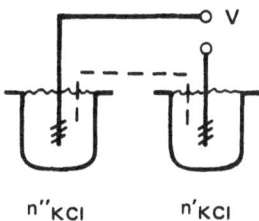

Figure 16.3. Two Ag–AgCl electrodes placed in solutions containing different concentrations of KCl and joined by an ideal bridge that allows no ions to move but still acts as an electrical connection (dashed line).

n''_{KCl} n'_{KCl}

posite sign of the Nernst potential, namely,

$$V'' - V' = + \frac{kT}{ez_{Cl}} \ln \frac{n''_{Cl}}{n'_{Cl}}$$

The sign is opposite to that in the previous equation because the potential difference is measured between the Ag wires holding the Cl membrane rather than across the Cl membrane itself.

PROBLEM. Derive the equation above by considering the Nernst potential at each boundary between V'' and V'.

Usually we consider V' as the reference (ground) potential and set it equal to zero. If the ratio of concentrations (n''/n') is about 10, then $V \simeq -60$ mV at room temperature $(z_{Cl} = -1)$.

In this idealized experiment, no current is flowing since V is measured with a perfect voltmeter. The AgCl on the electrodes is neither consumed nor formed and the setup is in equilibrium. In practice the AgCl electrode is used to pass steady current; the amount that may be passed before the AgCl on one electrode is consumed depends on the initial amount of AgCl laid down.

17. An Example of the Diffusion Potential: The Agar Bridge

Consider the experiment of Section 16 when the two beakers are joined by an agar/salt bridge. The agar will stop the bulk flow of solution, but ions in the bridge (prepared with n'_{KCl}) may move (see Figure 17.1). In this case, the potential between the two measuring points is given by the sum of the potentials discussed in Sections 15 and 16:

$$V'' - V' = - \frac{kT}{e} \ln \frac{n''_{Cl}}{n'_{Cl}} - \frac{kT}{e} \frac{u_K - u_{Cl}}{u_K + u_{Cl}} \ln \frac{n''_{Cl}}{n'_{Cl}}$$

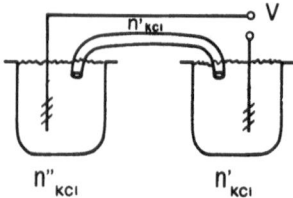

Figure 17.1. A diagram similar to Figure 16.3 except the solutions are now connected by an agar/salt bridge prepared with the KCl concentration of one of the solutions.

or

$$V = - \frac{kT}{e} \left(\frac{2u_{\mathrm{K}}}{u_{\mathrm{K}} + u_{\mathrm{Cl}}} \right) \ln \frac{n''_{\mathrm{Cl}}}{n'_{\mathrm{Cl}}}$$

PROBLEM. Consider an agar bridge separating two large reservoirs of KCl solutions: $n'_{\mathrm{KCl}} = 1\ M$ KCl and $n'_{\mathrm{KCl}} = 100$ mM. Let the cylindrical bridge have a cross-sectional area of 1 mm² and length of 10 cm. Calculate the initial resistance and the voltage seen by measuring device in Figure 17.1.

The entire setup in Figure 17.1 appears to the measuring instrument as a resistor and battery in series, as shown in Figure 17.2. Figure 17.2 is the equivalent circuit where R is the resistance of the bridge and salt solutions and E is the potential that is measured by the meter. Unlike the hypothetical situation shown in Figure 16.3, the initial conditions of Figure 17.1 shall change gradually. The two salts n'' and n' will diffuse through the bridge. Eventually the concentration of KCl will be uniform and the potential difference between the Ag wires will vanish.

18. Steady Current in Ionic Solutions

The Nernst and diffusion potentials result from constraints that restrict the net flow of current. They are not potentials that result from the passage of a net current through a resistor. The Nernst relation implies no net movement of charge. The diffusion relation implies the K and Cl ions

Figure 17.2. An equivalent circuit of the setup in Figure 17.1.

move equally in spite of their different mobilities. In both the Nernst and the diffusion potentials the net current is zero. Consider the case where a net current does flow.

The condition for the net flow of current through uniform area A is

$$eA \sum z_i \phi_i = I \ominus A$$

For example, consider a steady current in a solution of KCl. The condition above becomes

$$eA\left[kT(u_K - u_{Cl}) \frac{dn}{dx} + en(u_K + u_{Cl}) \frac{dV}{dx} \right] = I$$

where $n_K = n_{Cl} = n$. This implies local neutrality exists even in the presence of net current I. Rewriting the above expression gives

$$\frac{kT}{e}\left(\frac{u_K - u_{Cl}}{u_K + u_{Cl}} \right) \frac{dn}{n} + dV = I\left[\frac{1}{e^2(u_K + u_{Cl})n} \right] \frac{dx}{A}$$

Integration between two regions with steady-state concentrations n'' and n' results in

$$\frac{kT}{e}\left(\frac{u_K - u_{Cl}}{u_K + u_{Cl}} \right) \ln \frac{n''}{n'} + V = IR$$

where $V = V'' - V'$ and

$$R = \frac{1}{e^2(u_K + u_{Cl})A} \int_{'}^{''} \frac{dx}{n}$$

This is Ohm's law for direct current through a uniform solution of KCl separating the reservoirs n'' and n'. If A is not uniform in x it must be taken into the integral.

As an example, consider a hole of cross-sectional area A and length l separating two reservoirs of KCl at different concentrations (see Figure 18.1). The length l does not equal $\int_{'}^{''} dx$ since the solution concentration does not change abruptly between the reservoir and the hole. The integration is between points far from the hole where n and V are essentially constant in x.

Figure 18.1. Diagram of a uniform, watery hole of length l and area A separating two different concentrations of KCl.

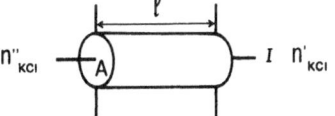

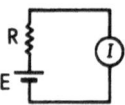

Figure 18.2. An equivalent circuit of the watery hole shown in Figure 18.1. An external branch (not shown in Figure 18.1) is introduced to measure the short circuit current I.

The equivalent circuit of Figure 18.1 is shown in Figure 18.2, where I is flowing in an external circuit through a current meter. In Figure 18.2 the voltage across the RE branch must be zero since the external circuit has zero resistance and can maintain no voltage difference. If $V = 0$, then

$$E = -\frac{kT}{e}\left(\frac{u_K - u_{Cl}}{u_K + u_{Cl}}\right)\ln\frac{n''}{n'}$$

When $I = 0$, the open circuit voltage $V = E$. We have ignored how contact is made between the current meter and the two solutions. If AgCl electrodes had been used, an additional term in E would result (see Section 17).

Above we treated K and Cl as moving together in water. Suppose instead they moved through physically separate, perfect K and Cl membranes. In this case we consider the two equations

$$eA_K\left(u_K kT\frac{dn_K}{dx} + en_K u_K\frac{dV}{dx}\right) = I_K$$

$$eA_{Cl}\left(-u_K kT\frac{dn_{Cl}}{dx} + en_{Cl}u_{Cl}\frac{dV}{dx}\right) = I_{Cl}$$

(18.1)

Now n_K and n_{Cl} are distinguished; we have assumed there are places where only one ion or the other may pass. Such a membrane is called a mosaic membrane. Equations (18.1) lead to

$$\frac{kT}{e}\ln\frac{n_K''}{n_K'} + V = I_K R_K$$

and

(18.2)

$$-\frac{kT}{e}\ln\frac{n_{Cl}''}{n_{Cl}'} + V = I_{Cl} R_{Cl}$$

where

$$R_K = \frac{1}{eu_K A_K}\int_{'}^{''}\frac{dx}{n_K}, \qquad R_{Cl} = \frac{1}{eu_{Cl} A_{Cl}}\int_{'}^{''}\frac{dx}{n_{Cl}}$$

$$E_K = -\frac{kT}{e}\ln\frac{n_K''}{n_K'}, \qquad E_{Cl} = \frac{kT}{e}\ln\frac{n_{Cl}''}{n_{Cl}'}$$

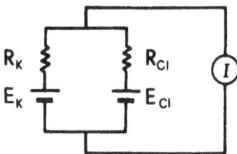

Figure 18.3. An equivalent circuit for a mosaic membrane with an external branch to measure the short circuit current *I*.

The equivalent membrane for the mosaic membrane with a parallel external pathway is shown in Figure 18.3. Mauro (1961) and Finkelstein and Mauro (1963) have given a detailed comparison of the mosaic membrane with different representations of the homogeneous membrane.

EXAMPLE. To illustrate the pitfalls in the equivalent circuit for the mosaic membrane, consider the open circuit across it. The open-circuit voltage is V when $I = 0$. From Equation (18.2), $I = I_K + I_{Cl} = 0$ gives

$$V = \frac{E_K R_{Cl} + E_{Cl} R_K}{R_K + R_{Cl}}$$

But this solution contradicts the principle that only K may pass through the K branch and only Cl through the Cl branch. If we should now assume $n_K = n_{Cl}$ (in spite of the separate pathways) then we may drop the subscript and write

$$E_K = -E_{Cl} = -\frac{kT}{e} \ln \frac{n''}{n'}$$

and

$$V = -\frac{kT}{e} \left(\frac{R_{Cl} - R_K}{R_K + R_{Cl}} \right) \ln \frac{n''}{n'}$$

$$= -\frac{kT}{e} \frac{u_K - u_{Cl}}{u_K + u_{Cl}} \ln \frac{n''}{n'}$$

as in the solution for the homogeneous membrane.

Such difficulties could have been anticipated from Equations (18.2), since there can be only one V across the membrane.

PROBLEM. The equivalent resistance of the mosaic membrane is the parallel combination of R_K and R_{Cl}, i.e.,

$$R = \frac{R_K R_{Cl}}{R_K + R_{Cl}}$$

Show that this is equivalent to the homogeneous case when we introduce the condition $n_K = n_{Cl}$.

PROBLEM. Assume $n_K = n_{Cl} = n$. From Equation (18.1) show that the concentration profile $n(x)$ is given by

$$n(x) = \left(\frac{n'' - n'}{x'' - x'}\right)x + n'$$

19. *The Integral Resistance*

In Section 18 equivalent circuits were drawn from flux equations in which the resistance appeared as an integral over the concentration profile $n(x)$. Here we derive an expression for resistance in which the concentration and the geometry are separate factors.

Consider an arbitrary hole separating two reservoirs n''_{KCl} and n'_{KCl}. Eventually the solutions will mix and the final concentration will be an arithmetic average of the initial concentrations. Consider the steady state, that is, infinitely large reservoirs where n'' and n' are constant. Let $A(x)$ be the cross-sectional area in the hole normal to the direction of current flow (see Figure 19.1). Fick's law of diffusion predicts the flux:

$$\phi = -D\frac{dn}{dx} \ominus \frac{\text{Number}}{\text{area} \cdot \text{time}}$$

This formula may be written in a form where the area appears explicitly:

$$\Phi = -DA\frac{dn}{dx} \ominus \frac{\text{Number}}{\text{time}}$$

Thus $\Phi = A\phi$ is the flux of particles through $A(x)$.

The resistance of the volume element $A(x)\,dx$ is

$$dR = \varrho\,\frac{dx}{A}$$

where ϱ is the resistivity of the solution in $A\,dx$. From Section 12, $\varrho = \varrho(n)$.

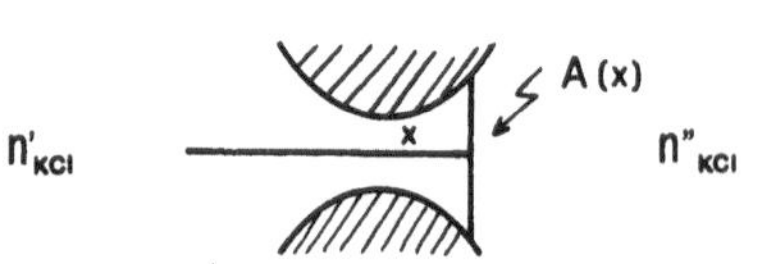

Figure 19.1. Diagram of a watery hole separating two different concentrations of KCl. The cross-sectional area of the hole, $A(x)$, varies with distance.

Since the two solutions n'' and n' will mix in the region of the hole, $n = n(x)$. Thus the resistivity changes with x because n does.

Combining the last two equations by eliminating the term dx/A results in the incremental resistance

$$dR = -\varrho \left(\frac{D}{\Phi} \right) dn$$

Since we consider only the steady state, $\Phi \neq \Phi(x)$. (If this were not true, particles would either build up or decrease somewhere.) Using Fick's law a second time gives

$$\frac{D}{\Phi} dn = -\frac{dx}{A} \qquad \text{or} \qquad \frac{D}{\Phi}(n'' - n') = -\int_{I}^{II} \frac{dx}{A}$$

Therefore

$$dR = \frac{\varrho \, dn}{n'' - n'} \int_{I}^{II} \frac{dx}{A}$$

or

$$R = \frac{\displaystyle\int_{I}^{II} \varrho \, dn}{n'' - n'} \int_{I}^{II} \frac{dx}{A}$$

This is the integral resistance for an arbitrary hole separating two solutions. The concentration and the geometry appear as separate factors.

If one makes the approximation (Section 11)

$$\varrho = \left(\frac{1000}{\varLambda} \right) \frac{1}{n} = \frac{1}{K_0 n}$$

where K_0 is constant, then

$$R = \frac{1}{K_0} \frac{\ln n''/n'}{n'' - n'} \int_{I}^{II} \frac{dx}{A} = \frac{1}{K_0 n_0} \int_{I}^{II} \frac{dx}{A}$$

Let n_0 be defined as the logarithmic average of n'' and n'. Consider the logarithmic average of x and y, viz.,

$$f(x, y) = \frac{x - y}{\ln x/y}$$

$f(x, y)$ is a number that lies between x and y, and has the property

$$f(x, y) = f(y, x)$$

Also, when $y = x$, $f(x, y) = x$. This may be shown from l'Hopital's rule:

$$\lim \frac{(d/dx)(x - y)}{(d/dx)(\ln x/y)} = \lim_{x \to y} \frac{1}{1/x} = x$$

The relations imply that n on either side of the hole may be interchanged with no change in integral resistance; if the concentrations are equal, then $n_0 = n$.

Table 19.1 gives $f(x, y)$ for a convenient range of x and y.

PROBLEM. Compare the logarithmic average of any two numbers with their arithmetic, geometric, and harmonic average.

PROBLEM.   For concentrations up to 0.1 M, $\Lambda(n)$ is given approximately by (see Section 12)

$$\Lambda(n) = \Lambda_0 - (a\Lambda_0 + b)n^{1/2}$$

where Λ_0, and a, and b are constants. Derive an expression for the integral resistance using this formula. Compare this integral resistance with that derived above for an arbitrary hole separating $n'' = 0.1$ M KCl and $n' = 0.01$ M KCl, assuming

$$\int_{'}^{''} \frac{dx}{A} = 1 \ \mu m^{-1}$$

PROBLEM. Show that u_{KCl} is the harmonic mean of u_K and u_{Cl}, i.e.,

$$u_{KCl} = \frac{2u_K u_{Cl}}{u_K + u_{Cl}}$$

Compare the integral resistance derived above with the integral resistance from Section 18 for arbitrary cross section, that is, compare

$$R = \frac{1}{e^2(u_K + u_{Cl})} \int_{'}^{''} \frac{dx}{An}$$

with

$$R = \frac{\int_{'}^{''} \varrho \ dn}{n'' - n'} \int_{'}^{''} \frac{dx}{A}$$

The two expressions converge for uniform cross section and the condition

$$n(x) = \left(\frac{n'' - n'}{x'' - x'} \right) x + n'$$

TABLE 19.1. The Value of $(x - y)/[\ln(x/y)]$ for x and y between 0.01 and 10,000 [a]

(x, y)	0.01000	0.03162	0.10000	0.31622	1.0000	3.1622	10.000	31.622	100.00	316.22	1000.0	3162.2	10000
0.0100	0.01000												
0.03162	0.01878	0.03162											
0.1000	0.03909	0.05939	0.10000										
0.31622	0.08866	0.12360	0.18781	0.31622									
1.0000	0.21498	0.28037	0.39087	0.59392	1.0000								
3.1622	0.54761	0.67981	0.88662	1.2360	1.8781	3.1622							
10.000	1.4462	1.7317	2.1498	2.8037	3.9087	5.9392	10.000						
31.622	3.9226	4.5733	5.4761	6.7981	8.8662	12.360	18.781	31.622					
100.00	10.856	12.405	14.462	17.137	21.498	28.037	39.087	59.392	100.00				
316.22	30.518	34.331	39.226	45.733	54.761	67.981	88.662	123.60	187.81	316.22			
1000.0	86.858	96.507	108.56	124.05	144.62	173.17	214.98	280.37	390.87	593.92	1000.0		
3162.2	249.70	274.67	305.18	343.31	392.26	457.33	547.61	679.81	886.62	1236.0	1878.1	3162.2	
10000.0	723.82	789.62	868.58	965.07	1085.6	1240.5	1446.2	1731.7	2149.8	2803.7	3908.7	5939.2	10000

[a] x and y are interchangeable.

Both expressions for R are a consequence of the simple form of Fick's law for a single salt. The second expression, having separated the geometric from the concentration profile, makes some calculations more convenient.

The value of the integral resistance depends on the concentration profile $n(x)$. If the ionic profile is altered by the passage of current through R, then R will depend on I; in this case, a plot of I versus V will not be a straight line. For the single salt considered above, $n(x)$ does not change with the passage of current. Thus the I–V curve given in Section 18

$$V + \frac{kT}{e}\left(\frac{u_K - u_{Cl}}{u_K + u_{Cl}}\right)\ln\frac{n''}{n'} = IR$$

is a straight line passing through the diffusion potential when $I = 0$. More complex cases will be considered below.

20. The Flux Equation and Potential Profile

The flux equation for an ion in a concentration and potential gradient is

$$-\phi = a\,\frac{dn}{dx} + b\,\frac{dV}{dx}$$

where $a = ukT$ and $b = zenu$. Rearranging this equation, we have

$$-\frac{\phi}{a} = \frac{dn}{dx} + \frac{b}{a}\,n\,\frac{dV}{dx} = \frac{d}{dx}\,(ne^{(b/a)V})e^{(-b/a)V}$$

Therefore

$$-\frac{\phi}{a}\,e^{(b/a)V}\,dx = d(ne^{(b/a)V})$$

In the steady state, the flux ϕ is not a function of x, i.e., nowhere does the concentration of the ion increase or decrease. Thus ϕ comes out of the integral and

$$-\frac{\phi}{a}\int_{'}^{''}e^{(b/a)V}\,dx = \int_{'}^{''}d(ne^{(b/a)V})$$

The integral on the right is in standard form. Writing $V = V(x)$ and substituting a and b into the expression gives

$$-\phi = ukT\,\frac{n''e^{(ze/kT)V''} - n'e^{(ze/kT)V'}}{\displaystyle\int_{'}^{''}e^{(ze/kT)V(x)}\,dx}$$

If this is regarded as the flux of one type of ion then ϕ, u, n, and z require the subscript i. Recall that the flux and the current are related through

$$I = zeA\phi$$

Thus the flux equation expresses the relationship

$$\text{flux} \propto \frac{\text{driving force}}{\text{barrier}}$$

PROBLEM. Assume that the potential profile is given by the linear equation

$$V(x) = \left(\frac{V'' - V'}{x'' - x'}\right)x + V'$$

where $x'' - x'$ is the thickness of the membrane. Evaluate the integral

$$\int_{'}^{''} e^{(ze/kT)V(x)}\, dx$$

Apply Kirchhoff's law for K and Cl to the membrane, namely,

$$\sum z_i\phi_i = 0$$

and show that

$$V'' - V' = -\frac{kT}{e}\ln\frac{u_K n''_K + u_{Cl}n'_{Cl}}{u_K n'_K + u_{Cl}n''_{Cl}}$$

This is called Goldman's equation (Goldman, 1943; Hodgkin and Katz, 1949). It has been applied to barriers that separate complex mixtures of electrolytes, including many types of biological membranes.

In general, the ionic profiles inside a membrane will not be linear; also the profiles may change with applied potential. Thus the integral resistance and the bias potential (e.g., the diffusion potential) may be expected to vary with the voltage across the membrane. However, even in relatively complex cases, the potential profile within the membrane may be approximately a linear function of x. One example is a homogeneous, fixed-charge membrane separating two solutions of equal concentrations of univalent electrolytes (Finkelstein and Mauro, 1963). It is for this reason that Goldman's equation has served membrane biology so well.

EXERCISE. Apply Patlak's (1960) derivation of Goldman's equation to a membrane separating two solutions of KCl at different concentrations. This derivation uses a rate of flux theory by Ussing.

PROBLEM. Discuss an equivalent circuit for Goldman's equation.

A general treatment of transport processes, equivalent circuits and electrical phenomena in membranes, with an historical viewpoint, may be found in a review by Teorell (1953).

21. An Example: The Glass Microelectrode

Glass tubing, drawn to a fine tip less than 1 μm in external diameter and filled with a salt solution, has been used to record from cells too small to be entered by other means (Ling and Gerard, 1949). The electrodes themselves are useful examples of ionic conductors to which the previous development may be applied.

Consider a glass microelectrode filled initially with n'' KCl and dipping into a solution of n' KCl. Eventually these solutions will mix and the final concentration will be the arithmetic average of n'' and n'. We are concerned here only with the initial state, or what amounts to the same thing, the condition that the total volumes of the n'' and n' are much larger than the volume in the region of the tip; in this case, the steady state is a fair approximation.

Assume the geometry of the electrode is a spherical cone of tip angle α and internal tip radius r (see Figure 21.1). Note that $A(x) = \pi r^2(x)$ and $r(x) = x \tan \alpha$. It is required to calculate the geometric term from Section 19. We assume that α is small; then $\tan \alpha \simeq \alpha$ and the contribution to the geometric term from the region outside the electrode is negligible compared to inside. With these assumptions

$$\int_{'}^{''} \frac{dx}{A} \simeq \frac{1}{\pi \alpha^2} \int_{x_r}^{\infty} \frac{dx}{x^2} = \frac{1}{\pi \alpha^2} \left(\frac{1}{x_r} \right) = \frac{1}{\pi \alpha r}$$

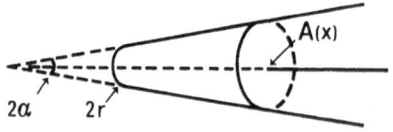

Figure 21.1. Diagram of the tip of a glass microelectrode. The cross-sectional area, $A(x)$, is a function of distance. α is the semiangle of the projected tip and r is the radius of the open tip.

Figure 21.2. The qualitative relationship between the resistance of a glass microelectrode, R, and the inverse of the external concentration, n'. The electrode was filled with concentration n'' and the two concentrations mix in the region of the tip.

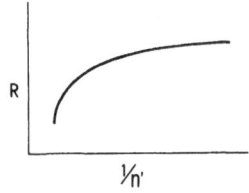

since $x_r = r/\tan \alpha \simeq r/\alpha$. From Section 19, the resistance of the glass microelectrode is given approximately by

$$R = \frac{1}{K_0} \frac{\ln n''/n'}{n'' - n'} \left(\frac{1}{\pi \alpha r} \right)$$

This formula assumes $\varrho \propto 1/n$; it ignores the geometric contribution outside the tip but does consider that the two solutions have mixed. Earlier treatments which assume that the electrode is entirely filled with the original filling solution n'', predict a linear dependence of R on $1/n''$ (Amatniek, 1958). Allowing the two solutions to mix predicts a linear dependence on $1/n_0$, where n_0 is the logarithmic average of n' and n'' (see Section 19).

Measurement of R for microelectrodes filled with n'' KCl and dipping into n' KCl, plotted against $(1/n')$ are markedly nonlinear and tend to saturate for small n' (see Figure 21.2). Such a relationship has been used by Lassen and Sten-Knudsen (1968; Figure 10) to study the resistance of red blood cells using microelectrodes. (Compare their Figure 10 with Figure 1 of Firth and DeFelice, 1971.)

Glass microelectrodes are considerably more complex than the treatment above would indicate (Firth and DeFelice, 1971). The dependence of R on $1/n_0$ deviates from a straight line for small resistors in which surface conduction along the wall of the glass plays a role. Also the diffusion potential (usually called the tip potential since it is greatly reduced when the tip is broken) is much larger than predicted by the relationship

$$V'' - V' = - \frac{kt}{e} \left(\frac{u_K - u_{Cl}}{u_K + u_{Cl}} \right) \ln \frac{n''}{n'}$$

if u_K and u_{Cl} are given their values in water. Because of the negative surface charge on the glass wall, there is a preferred pathway for the K ion; the effect of this pathway is to increase the diffusion potential beyond that expected for a watery hole. Finally the resistance of the microelectrode depends on the voltage across it, the more so for weak solutions. In general the I–V relationship is nearly linear for $n'' = n'$ (and strong solutions) and goes through the origin for $I = 0$. This relationship becomes highly

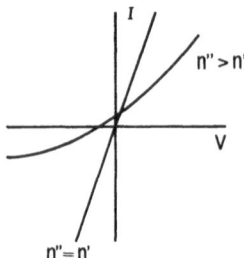

Figure 21.3. The qualitative steady-state I–V curves for two glass microelectrodes. The linear curve has equal ionic concentrations inside and out; the nonlinear curve has a greater concentration inside the electrode (far from the tip) than outside (far from the tip). Note the tip potential for $I = 0$.

curved for $n'' \neq n'$ (and weak solutions) and has increased tip potential (see Figure 21.3). Compare the above figure with Figure 2 of DeFelice and Firth (1971).

PROBLEM. Assume $n'' = n' = 3$ M KCl. Calculate R for a glass microelectrode when $r = 0.1$ μm with tip angle $3°$ (α must be in radians). Calculate the resistance when the electrode is dipped into 0.1 M KCl.

PROBLEM. In water $u_{Cl}/u_K \simeq 1.03$. Calculate the diffusion potential from a watery hole separating 3 M KCl and 0.1 M KCl. A typical tip potential from a microelectrode filled with 3 M KCl and dipping into 0.1 M KCl is -10 mV (the reference potential is considered to be the external bath).

Suggested reading. For a comprehensive treatment of electrodiffusion and the effect of surface charge densities such as that found on glass see Davies and Rideal (1961).

For additional information on the theory and practice of glass microelectrodes see the excellent review by Fatt (1961). Two text books dealing specifically with bioelectrodes are Geddes (1972) and Ferris (1974). Firth and DeFelice (1971) and DeFelice and Firth (1971) studied the resistance and the voltage noise of glass microelectrodes. Flasterstein (1966a,b) studied voltage noise from metal–electrolyte interfaces of the type used in electrophysiology. The early papers of Graham and Gerard (1946) and Ling and Gerard (1949) are also of interest.

22. The Effect of Pressure on Integral Resistance

The application of a pressure difference across an arbitrary hole separating two solutions n'' and n' will alter the concentration profile $n(x)$ and therefore the integral resistance. The number of particles in an elemental volume v is nv. The flux of these particles due to the volume flow is therefore

$$\Phi = n \frac{dv}{dt} \ominus \frac{\text{Number}}{\text{time}}$$

The total flux of particles due to diffusion and volume flow, writing $\dot{v}$ for dv/dt, is

$$\Phi = -DA \frac{dn}{dx} + n\dot{v}$$

The resistance of the elemental volume $A\,dx$ is

$$dR = \varrho \frac{dx}{A}$$

From the flux equation

$$\frac{dx}{A} = D \frac{dn}{n\dot{v} - \Phi} = \frac{D}{\dot{v}} \frac{dn}{n + a}$$

where $a = -\Phi/\dot{v}$. The expression for integral resistance with volume flow (convection) becomes

$$R = D \int_{'}^{''} \frac{\varrho\,dn}{n\dot{v} - \Phi} = \frac{D}{\dot{v}} \int_{'}^{''} \frac{\varrho\,dn}{n + a}$$

Make the simplifying assumption that $\varrho = 1/K_0 n$. Then

$$R = \frac{1}{K_0} \frac{D}{\dot{v}} \left[-\frac{1}{a} \ln\left(1 + \frac{a}{n}\right) \right]_{'}^{''},$$

$$= \frac{1}{K_0} \frac{D}{\dot{v}} \frac{1}{a} \ln \frac{1 + a/n'}{1 + a/n''}$$

From a previous result,

$$\int_{'}^{''} \frac{dx}{A} = \frac{D}{\dot{v}} \int_{'}^{''} \frac{dn}{n + a} = \frac{D}{\dot{v}} \ln \frac{n'' + a}{n' + a}$$

Letting

$$g = \frac{\dot{v}}{D} \int_{'}^{''} \frac{dx}{A}$$

and solving for a:

$$a = \frac{n'e^g - n''}{1 - e^g}$$

Substituting $D/\dot{v}$ and a into our last expression for R gives

$$R = \frac{1}{K_0} \frac{1 - e^{-g}}{g} \frac{\ln(n''e^{-g}/n')}{n''e^{-g} - n'} \int_{I}^{I''} \frac{dx}{A}$$

This is the integral resistance of an arbitrary hole separating two solutions that mix by diffusion and convection.

When there is no bulk flow of fluid ($\dot{v} = 0$), then $g = 0$ and R reduces to our previous result (Section 19). When $n'' = n'$, bulk flow cannot alter the concentration profile, and the dependence of R on g drops out of the expression. When $g \to \infty$,

$$R \to \frac{1}{K_0} \frac{1}{n'} \int_{I}^{I''} \frac{dx}{A}$$

i.e., the concentration profile shifts and the hole behaves as if it were filled with n'.

PROBLEM. Show that the total flux Φ due to diffusion and convection through an arbitrary pore is

$$\Phi = D \frac{1}{1 - e^{-g}} \frac{n''e^{-g} - n'}{\displaystyle\int_{I}^{I''} dx/A}$$

PROBLEM. Show that g is unitless.

The above theory has been applied to the glass microelectrode of Section 21 (Firth and DeFelice, 1971). For simple viscous flow, volume flow is related to pressure gradient by

$$\dot{v} \propto -\frac{dP}{dx}$$

where the minus sign indicates flow from a region of high pressure to low. By integrating the Poiseuille law of flow along a conical tube, it was found that

$$\dot{v} = -\frac{3\pi\alpha r^3}{8\eta}(P'' - P')$$

where P'' and P' are the pressures far from the tip. r and α are the electrode tip radius and semiangle (Figure 21.1) and η is the viscosity of the fluid.

With this result and the previous evaluation of the geometric term, viz.,

$$\int_{\iota}^{\iota\iota} \frac{dx}{A} \simeq \frac{1}{\pi\alpha r}$$

we have

$$g = -\frac{3}{8}\frac{r^2}{D\eta}(P'' - P') \equiv -GP$$

letting $P = P'' - P'$ and G be the constant of proportionality. Since g depends directly on the difference of pressure P between the inside and the outside of the microelectrode, the resistance R and the total flux Φ also depend on P.

PROBLEM.   Derive the units of viscosity η.

PROBLEM.   Show that as $P \to \infty$, the total flux from a glass microelectrode goes to

$$\Phi \to \frac{3\alpha\pi r^3}{8\eta} n'' P$$

Take $r = 0.1$ μm, $\alpha = 0.05$ rad, $D = 10^{-6}$ cm²/sec, and $\eta = 4 \times 10^{-2}$ poise, typical values for a KCl-filled microelectrode; show that at one atmosphere of pressure,

$$\Phi \simeq 1.5 \times 10^{-9} \text{ m} M/\text{sec}$$

Compare this with the iontophoretic flux obtained by applying 10 V across a 25-MΩ resistor.

An Analogy with the Goldman Equation (Section 20). The flux through a membrane separating the concentrations n'' and n' (assuming a uniform field within the membrane) is

$$\Phi = -\frac{D}{\delta}\frac{n''e^{\beta V} - n'}{e^{\beta V} - 1}$$

where δ is the membrane thickness, $\beta = ze/kT$ and $V = V'' - V'$. We have set $V' = 0$ as the reference potential (ground). It is of interest to note the similarity between this expression and the result of the first problem of this paragraph. The driving force is equivalent to g, which contains the pressure. However, in the hydrodynamic model it is the volume flow and not the pressure gradient that is kept constant.

23. Membrane Rectification and Reactance

The integral resistance of a membrane or hole separating two concentrations of salt solution depends on the concentration profile. If the profile is changed with the passage of current, the resistance is a function of the voltage across it; the I-V curve is not a straight line and the membrane is said to rectify. When changing from one voltage to another, the concentration profile may not change instantaneously. The resistance in this case is also time variant and the membrane is said to have a reactance.

A reactance due to a time-variant resistance has been called an apparent reactance by Teorell (1953) to distinguish it from the reactance due to electromagnetic fields.

The relationship between rectification and apparent reactance is easily demonstrated by a diagram introduced by A. Mauro (Figure 23.1). As an example, consider an outward rectifier, i.e., one whose I-V relationship is curved so that conductance is greater with outward (positive) current. The term "outward" is used since the side of the membrane that is at the reference potential (ground) is usually outside. An outward rectifier, e.g., a glass microelectrode, may have an I-V curve like that shown in Figure 23.1.

The current step in Figure 23.1 is forced on the resistance by an external circuit and is instantaneous. The voltage does not respond instantaneously because, we imagine, the ionic profile in the resistance takes time to readjust.

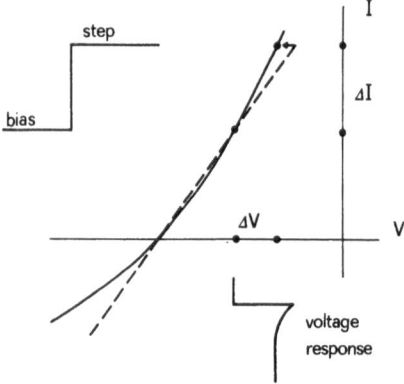

Figure 23.1. A nonlinear steady-state I-V relationship (solid curve) of a hypothetical ionic system. If the curve represents a microelectrode, V at $I = 0$ is the tip potential (Section 21). If the curve represents a perfect K membrane, V at $I = 0$ is the Nernst potential (Section 14). ΔI is a step change in current forced on the system by an external circuit. Note that the step occurs on top of a steady bias current. ΔV is the steady-state response of the system. The voltage first overshoots and then relaxes to the steady state as explained in the text.

The solid line in Figure 23.1 is the steady-state *I–V* curve, the dotted line is the instantaneous resistance (or chord resistance) at a particular voltage. The dots on the *I* and *V* axes show the final states; the current changes instantaneously because it is forced to do so by an external circuit (current clamp); the voltage, however, first overshoots and then relaxes to its final value. The ratio $\Delta V/\Delta I$ approaches the slope of the steady-state curve as the steps are made small and is called the slope resistance at any voltage.

An apparent reactance may be expected from any nonlinear resistor. For the microelectrode, the instantaneous voltage change results from the state of the ionic profile at the moment the current step is applied. The changing voltage is due to a change in ionic profile; eventually a new steady state is reached compatible with the new current. It is evident from the diagram that for zero bias current and small current steps the chord and slope resistance are equal and no reactance occurs.

The voltage response shown in Figure 23.1 is often wrongly called inductive since it mimics the response of a real inductor (Section 24) if the relaxation phase is exponential. The fundamental property of an inductor, however, is to resist changes in current; a capacitor resists changes in voltage; therefore, the relaxation phase in the voltage response may actually be described as capacitive. We shall return to this point in Section 80.

PROBLEM. Draw a Mauro diagram for a step current clamp of an outward rectifier in the lower left quadrant of the *I–V* curve. This response is called capacitive. Repeat the exercise when the voltage is forced to make a step change (voltage clamp) and the current is allowed to relax. Repeat the exercise for an inward rectifier (i.e., one with the opposite curvature to Figure 23.1).

Membrane rectification and membrane reactance due to nonlinear, time-variant resistors are general properties not restricted to a readjustment of a concentration profile. Any characteristic of a resistor that involves a relaxation time will create an apparent reactance; at sufficiently high frequencies, all resistors must display this property. The resistance of biological membranes is time and voltage dependent. In this case, however, the kinetics of the opening and closing of membrane channels are responsible for the apparent reactance (Sections 80–81). Membranes may also have real reactances due to the action of electromagnetic fields. The next two Sections review some of the properties of ideal reactances, real or apparent.

24. Inductance and Capacitance: Time Domain

Consider a purely inductive element represented by the symbol for a coiled wire. If a steady current is flowing through the coil, a static magnetic field is set up around it; any change in the current alters the field to oppose that change. This is known as Faraday's law and is expressed by assigning a voltage drop across the inductor proportional to the time rate of change of current through the inductor; thus

$$V = L \frac{di}{dt} \ominus V$$

In this expression, i is the current through the coil in units of amperes and L is the inductance of the coil in units of henrys.

Consider the circuit in Figure 24.1, where E is an applied voltage, R is a resistance, and V is the voltage across the inductor L. The lowest point on the diagram is taken as the reference potential (ground) to which E and V are referred. From Ohm's law, the current through R is

$$\frac{E - V}{R} = i \quad \text{or} \quad V = E - iR$$

From Faraday's law the voltage across the inductor is $L \, di/dt$; since these voltages must be the same, we equate them to arrive at the expression

$$L \frac{di}{dt} + iR = E$$

This equation may be rewritten as

$$\frac{d}{dt} (ie^{Rt/L}) e^{-Rt/L} = E/L$$

which can be verified by expansion. Consider the integral from some initial time $t = 0$ to time t; thus

$$\int_0^t d(ie^{t/\tau}) = \frac{1}{L} \int_0^t E e^{t/\tau} \, dt$$

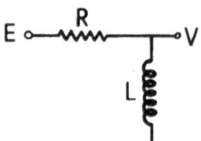

Figure 24.1. A resistor R and inductor L in series with respect to the input voltage E and in parallel with respect to the output voltage V.

where $\tau = L/R$. Perform the integration for the case that $E(t)$ is a step that changes instantaneously from zero to E at $t = 0$; then E comes out of the integral and

$$ie^{t/\tau} = \frac{\tau}{L} E(e^{t/\tau} - 1)$$

or

$$i(t) = \frac{E}{R} (1 - e^{-t/\tau})$$

With a step change in the driving potential the current through the inductor does not change at once but approaches the value E/R with a time constant L/R. The voltage across the inductor is derived from

$$V(t) = L \frac{di}{dt} = Ee^{-t/\tau}$$

$V(t)$ declines exponentially for a step change in $E(t)$.

The resistance across the inductor is the ratio of the voltage to the current, thus

$$R_L(t) = R \frac{e^{-t/\tau}}{1 - e^{-t/\tau}}$$

At $t = 0$, the resistance is infinite; at long times the resistance across the inductor approaches zero.

Consider a purely capacitive element, represented by the symbol for two conducting plates separated by a dielectric. In the presence of a constant voltage across the capacitor, an electric field is set up in the dielectric; any change in the potential alters this field and the charge on the condenser plates. This change is expressed by a capacitive current proportional to the time rate of change of the voltage across the capacitor, or

$$i = C \frac{dV}{dt}$$

Note that i does not represent charge moving through the capacitor but rather an apparent current due to a change in the charged state of the condenser plates. C is the capacitance in units of farads.

Let E be a voltage step applied through a resistor R to a capacitor C (see Figure 24.2). As before we may write $V = E - iR$; substituting for i gives

$$RC \frac{dV}{dt} + V = E$$

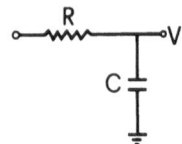

Figure 24.2. A voltage step E applied to a capacitor C through resistor R.

This equation may be solved by analogy with the previous expression for inductive current. The solution is

$$V(t) = E(1 - e^{-t/\tau})$$

where now $\tau = RC$. From the expression for capacitive current

$$i(t) = C\frac{dV}{dt} = \frac{E}{R}e^{-t/\tau}$$

In response to a step change in $E(t)$, the capacitive current declines exponentially.

The apparent resistance of the condenser is the ratio of $V(t)$ to $i(t)$, or

$$R_c(t) = R(e^{t/\tau} - 1)$$

When $t = 0$, the resistance is zero and at long times it becomes infinite.

Inductors and capacitors play comparable roles for current and voltage. If one attempts to change the current through an inductor abruptly, the new current is approached with some delay although the voltage across the inductor changes at once. An inductor opposes sudden changes in current. The converse is true for a capacitor, which tends to oppose sudden changes in voltage.

In Section 23 we saw that nonlinear, time-variant resistors may display either property.

The step response of an inductor and capacitor has been used as an example to simplify the integration. The formulas, however, apply to any input. For an inductor (Figure 24.1)

$$i(t) = \frac{e^{-t/\tau}}{L}\int_{-\infty}^{t} E(t)e^{t/\tau}\,dt$$

where $E(t)$ is the voltage applied through R and $i(t)$ is the current through L. For example, let $E(t)$ be a pure sinusoid of a given frequency $\omega_0 = 2\pi f_0$, where $1/f_0$ is the period of the wave. In this case, substituting

$$E(t) = E\sin\omega_0 t$$

into the above integration gives the response. An alternative approach to this problem exists when all circuit components are expressed in terms of frequency. This involves the Fourier transformation; the first example, given in the next section, will be the inductive and capacitive reactance.

PROBLEM.   Derive a general expression for $V(t)$ in terms of $E(t)$ for the circuit of Figure 24.2.

25. Fourier Transformation: The Delta Function

The Fourier transformation of any function of time $V(t)$ consists of multiplication of $V(t)$ by $e^{-i\omega t}$ and integration over all time. Here, the symbol i represents the complex number $\sqrt{-1}$, not to be confused with the same symbol used for current. A bar may be used to signify this transformation, thus

$$\bar{V}(f) = \int_{-\infty}^{\infty} V(t)e^{-i\omega t}\, dt$$

PROBLEM.   Substitute $E(t) = E_0 \sin \omega_0 t$ into the general expression for current through the inductor of Figure 24.1 and derive an expression for $i(t)$. What is the phase relationship between the input, $E(t)$, and the output, $V(t)$?

Notice that $\bar{V}$ is written as a function of f (f has units of hertz) rather than ω. We simply write $\omega = 2\pi f$ whenever it is convenient. Since we have integrated over all time, the transformed function depends only on f.

The above equation may be regarded as an integral equation in which $\bar{V}(f)$ is known and $V(t)$ is required. The solution of this integral equation is

$$V(t) = \int_{-\infty}^{\infty} \bar{V}(f)e^{i\omega t}\, df$$

This solution is called the inverse Fourier transform. To distinguish the two transformations, the former is called the minus-i transformation and the latter the plus-i transformation. A good summary of other notations and conventions is given in Bracewell (1965). We have adopted his system I (see Bracewell, 1965, Chap. 2).

The usefulness of the Fourier transform relies on this property: a pure sinusoid transforms into a delta function; information contained in an oscillatory function for all time is compressed into an impulse at the frequency of oscillation. This is expressed by writing

$$\delta(f) = \int_{-\infty}^{\infty} e^{-i\omega t}\, dt$$

where $e^{-i\omega t}$ is the general oscillatory function identically equal to

$$e^{-i\omega t} = \cos \omega t - i \sin \omega t$$

Also note that

$$e^{i\omega t} = \cos \omega t + i \sin \omega t$$

The delta function $\delta(f)$ is the Fourier transformation of $V(t) = 1$. The properties of the delta function are defined by the equations

$$\int_{-\infty}^{\infty} \delta(x - x_0)F(x)\, dx = F(x_0)$$

$$\int_{-\infty}^{\infty} \delta(x)\, dx = 1$$

where $F(x)$ is an arbitrary function.

Consider a specific sine wave described by the equation

$$V(t) = V_0 \sin \omega_0 t$$

By definition

$$\sin \omega_0 t = \frac{1}{2i} (e^{i\omega_0 t} - e^{-i\omega_0 t})$$

The Fourier transform of the sine wave of amplitude V_0 and frequency f_0 is

$$\bar{V}(f) = \frac{V_0}{2i} \int_{-\infty}^{\infty} (e^{i\omega_0 t} - e^{-i\omega_0 t})e^{-i\omega t}\, dt$$

$$= \frac{V_0}{2i} [\delta(f - f_0) - \delta(f + f_0)]$$

Similarly, the minus-i transformation of a cosine wave is

$$\bar{V}(f) = \frac{V_0}{2} [\delta(f - f_0) + \delta(f + f_0)]$$

Thus, in the frequency domain, a sine wave is represented by two imaginary delta functions symmetric about zero at the frequency of oscillation. The doublet is odd since the delta function on the negative frequency, $\delta(f + f_0)$, has the opposite sign to the delta function of the positive frequency axis, $\delta(f - f_0)$. A cosine wave has a similar representation on the real axis and is even.

PROBLEM. Sketch the functions $V(t) = \sin \omega_0 t, \cos \omega_0 t,$ and 1 in the time domain and their Fourier pairs (the minus-i transform) in the frequency domain.

26. Inductance and Capacitance: Frequency Domain

The basic formulas that describe the inductive and capacitive response in time are

$$V_L(t) = L \frac{d}{dt} [i_L(t)]$$

$$i_C(t) = C \frac{d}{dt} [V_C(t)]$$

Subscripts are used to distinguish the two devices. V_L refers to the voltage across the inductor and i_L refers to the current through the inductor. V_C and i_C have similar meanings for the capacitor. However, no physical charge moves through C; the current through the capacitor is an effective current induced by the electric field between opposite faces of the capacitor. An ideal capacitor is an open circuit (infinite resistance) for direct current. Similarly, the voltage across the inductor is an effective voltage induced by the magnetic field in the coiled wire. An ideal inductor has zero resistance for direct current. To describe the response of these elements in frequency, we take the Fourier transformation of both equations.

The derivative theorem of Fourier transforms states that

$$\int_{-\infty}^{\infty} \frac{dV}{dt} e^{-i\omega t} \, dt = i\omega \bar{V}$$

This may be verified by the identity

$$\frac{dV}{dt} e^{-i\omega t} = \frac{d}{dt} (V e^{-i\omega t}) + i\omega V e^{-i\omega t}$$

and the condition

$$V e^{-i\omega t} \Big|_{-\infty}^{\infty} = 0$$

The theorem may be applied by the two equations for $V_L(t)$ and $i_C(t)$ to give

$$\bar{V}_L(f) = i\omega L \bar{i}_L(f)$$

$$\bar{i}_C(f) = i\omega C \bar{V}_C(f)$$

The impedance of a device is defined as the ratio of voltage to current; thus

$$Z_L = i\omega L$$

$$Z_C = \frac{1}{i\omega C}$$

$Z(f)$ is referred to as the impedance of the inductor or capacitor.

As an example of the usefulness of this formulation reconsider the *RC* circuit of Section 24. In the frequency domain the output for any input may be written by inspection; thus

$$\frac{\bar{E} - \bar{V}_C}{R} = \frac{\bar{V}_C}{Z_C}$$

or

$$\bar{V}_C = \frac{\bar{E}}{1 + i\omega\tau}$$

where $\tau = RC$. Once the impedance of the device is formulated, the result is immediate.

To obtain the result in time, the reverse transformation (plus-*i* transformation) is performed. Let the driving force be

$$E(t) = E_0 \sin \omega_0 t$$

By a previous result (Section 25) we know that

$$\bar{E}(f) = \frac{E_0}{2i} [\delta(f - f_0) - \delta(f + f_0)]$$

To obtain $V(t)$, substitute $\bar{V}(f)$ into the plus-*i* transform:

$$V_C(t) = \int_{-\infty}^{\infty} \bar{V}_C(f) e^{i\omega t} \, df = \int_{-\infty}^{\infty} \frac{\bar{E}}{1 + i\omega\tau} e^{i\omega t} \, df$$

Recalling the properties of the delta function under the integral (Section 25),

$$V_C(t) = \frac{E_0}{2i} \left(\frac{e^{i\omega_0 t}}{1 + i\omega_0 t} - \frac{e^{-i\omega_0 t}}{1 - i\omega_0 t} \right)$$

$$= E_0 \frac{\sin \omega_0 t - \omega_0 \tau \cos \omega_0 t}{1 + \omega_0^2 \tau^2}$$

This solution for the response of a *RC* circuit to a sine wave shows that the output is a pure sinusoid, although decreased in amplitude and shifted in phase compared to the input. The solution is called the steady-state solution since it represents the response to an ever-present input.

PROBLEM. Show that the response of the *RC* circuit considered above to $E(t) = E_0 \cos \omega_0 t$ is

$$V_C(t) = E_0 \frac{\cos \omega_0 t + \omega_0 \tau \sin \omega_0 t}{1 + \omega_0^2 \tau^2}$$

Sketch the input and output on the same time base for $\omega_0 \gg \tau$, $\omega_0 = \tau$ and $\omega_0 \ll \tau$. Repeat for the sine input.

PROBLEM. The general expression for the response of the *RC* circuit of Section 24 is

$$V_C(t) = \frac{e^{-t/\tau}}{\tau} \int_{-\infty}^{t} E(t) e^{t/\tau} \, dt$$

Show by direct integration that this agrees with the above solutions for the sine and cosine input.

PROBLEM. Show that the current through the inductor in the *RL* circuit of Section 24, for $E(t) = E_0 \cos \omega_0 t$, is given by

$$\bar{i}(f) = \frac{\bar{E}}{R + Z_L} \quad \text{or} \quad i_L(t) = \frac{E_0}{R} \frac{\cos \omega_0 t + \omega_0 \tau \sin \omega_0 t}{1 + \omega_0^2 \tau^2}$$

Calculate the voltage drop $V_L(t)$ across L by taking the derivative of $i_L(t)$. Show that this agrees with the alternative derivation obtained from

$$\frac{\bar{E} - \bar{V}_L}{R} = \frac{\bar{V}_L}{Z_L}$$

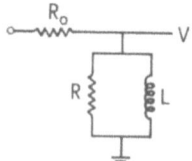

Figure 26.1. A parallel *RL* circuit. A step change in voltage E is applied through the resistor R_0.

PROBLEM. Consider the circuit shown in Figure 26.1. Show that the currents through the resistive and inductive branches in response to a step are

$$i_R(t) = \frac{E}{R + R_0}\, e^{-t/\tau} \quad \text{and} \quad i_L(t) = \frac{E}{R_0}\,(1 - e^{-t/\tau})$$

where $\tau = L(R + R_0)/RR_0$. The total current is therefore

$$i(t) = \frac{E}{R_0}\left(1 - \frac{R}{R + R_0}\, e^{-t/\tau}\right)$$

and the voltage

$$V(t) = \frac{ER}{R + R_0}\, e^{-t/\tau}$$

As a second example, consider a step input to the *RC* circuit of Section 24. The response to a step in the frequency domain requires the Fourier transformation of the step. The step function is called the Heaviside function, $H(t)$.

The first derivative of the Heaviside function is the delta function, thus

$$\delta(t) = \frac{d}{dt}\, H(t)$$

From the derivative theorem for Fourier transformation, we have

$$1 = i\omega H(f)$$

Notice that for both δ and H we eliminate the use of the bar to signify the Fourier-transformed quantity, and indicate the transform only by its function dependence. Thus, to within an arbitrary constant, the frequency representation of a step function is $1/i\omega$. (The constant may be taken as 0, 1/2, or 1, depending on the condition of the step at $t = 0$.)

Consider the step $E(t)$:

$$E(t) = 0, \qquad t < 0$$
$$E(t) = E_0, \qquad 0 \le t$$

The transform is taken as

$$\bar{E}(f) = E_0\left(\frac{1}{i\omega} + \delta(t)\right)$$

since $E = E_0$ at $t = 0$. The response of the RC circuit in the frequency domain is then

$$\bar{V}_C(f) = \frac{E_0}{i\omega}\frac{1}{1 + i\omega\tau} + \frac{E_0\,\delta(f)}{1 + i\omega\tau}$$

In the time domain, the response is the plus-i transformation, or

$$V_C(t) = E_0 \int_{-\infty}^{\infty} \frac{1}{i\omega}\frac{1}{1 + i\omega\tau}\, e^{i\omega t}\, df + E_0$$

To integrate the second term, we simply made use of the basic properties of the delta function. The first term may be integrated by using the converse of the derivative theorem discussed at the beginning of this section.

The converse of the derivative theorem states that

$$\int_{-\infty}^{\infty}\left[\int_{-\infty}^{t} V(t)\, dt\right] e^{-i\omega t}\, dt = \frac{1}{i\omega}\, \bar{V}(f)$$

or

$$\int_{-\infty}^{\infty} \frac{1}{i\omega}\, \bar{V}(f) e^{i\omega t}\, df = \int_{-\infty}^{t} V(t)\, dt$$

By this theorem the problem is reduced to finding the integral transform

$$\int_{-\infty}^{\infty} \frac{1}{1 + i\omega\tau}\, e^{i\omega t}\, df = \frac{1}{\tau}\, e^{-t/\tau}, \qquad 0 < t$$

[CF # I: 438, i.e., equation 438 of Table I in Campbell and Foster, 1948]. Since

$$\int_{-\infty}^{t} \frac{1}{\tau}\, e^{-t/\tau}\, dt = -e^{-t/\tau}$$

we have, finally,

$$V_C(t) = E_0(1 - e^{-t/\tau})$$

which is identical to our previous result (Section 24).

It is difficult to say which domain presents the easiest route to a solution. Once the impedance of circuit elements is defined, the steady-state solution to an input–output problem is often easiest in the frequency domain since the problem reduces to an application of Ohm's law and algebraic manipulation. For some purposes, a solution in the frequency domain is adequate; if the response in time is required, extensive tables of Fourier transformation exist to simplify the inversion.

PROBLEM.   Consider the step $E(t) = 0$, $t = 0$ and $E(t) = E_0$, $t \geq 0$, as input to the LR circuit of Figure 26.1. Describe the output in the frequency domain and obtain the response in time by the inverse Fourier transformation.

PROBLEM.   Discuss the units of $\delta(f)$ and $H(f)$, and of the step $\bar{E}(f) = E_0[\delta(f) + 1/i\omega]$.

PROBLEM.   Show that the impedance of an inductor L and resistor R in parallel is

$$Z = \frac{i\omega\tau R}{1 + i\omega\tau}, \qquad \tau = \frac{L}{R}$$

Show that the impedance of a capacitor C and resistor R in parallel is

$$Z = \frac{R}{1 + i\omega\tau}, \qquad \tau = RC$$

In these cases, Z is a complex number of the form $Z = a + ib$. Find expressions for a and b (the real and imaginary parts) of Z for both expressions above.

PROBLEM.   The conjugate of a complex number $Z = a + ib$ is Z^* $= a - ib$. Calculate $ZZ^* = a^2 + b^2$ for the two cases in the problem above, and plot the result for each as a function of frequency. Z^* is called the complex conjugate of Z. ZZ^* is called the modulus square, often written $|Z|^2$.

PROBLEM.   Find the square root of a complex number. $a + ib$ can be represented as a vector $(a^2 + b^2)^{1/2}e^{i\phi}$ in the complex plane where $b/a = \tan\phi$. Show that the square root is $(a^2 + b^2)^{1/4}e^{i\phi/2}$, where

$$(a^2 + b^2)^{1/4} = (A^2 + B^2)^{1/2} = A(1 + \tan^2\phi/2)^{1/2}$$

and

$$A = (a^2 + b^2)^{1/4}\cos\phi/2$$
$$B = (a^2 + b^2)^{1/4}\sin\phi/2$$

Plot $a + ib$ and its square root $A + iB$ in the complex plane, showing the magnitudes a, b and A, B on the real and imaginary axis.

27. *Equivalent Circuits of Membranes*

Three circuit elements have been discussed so far: resistance (R), capacitance (C), and inductance (L). These may be related to the material and the geometry of the device in which they originate. In membranes of uniform thickness, the material property and the thickness may be combined into a single parameter.

The resistance of an object of length l, cross-sectional area A, and resistivity ϱ is

$$R = \varrho \, \frac{l}{A}$$

ϱ in units of Ω cm depends only on the material (e.g., its composition, temperature, etc.). The factor l/A describes the contribution of shape to resistance. For good conductors, e.g., silver, $\varrho \simeq 10^{-6}$ Ω cm. For poor conductors, like glass, $\varrho \simeq 10^{14}$ Ω cm. 1 M NaCl has a resistivity of about 10 Ω cm (Section 11).

For membranes of uniform thickness l, the product ϱl may be combined into a single term called the specific resistance of the membrane:

$$R_m = \varrho l \,\ominus\, \Omega \text{ cm}^2$$

A membrane with specific resistance $R_m = 1000$ Ω cm^2 and 100 Å thickness would have

$$\varrho = \frac{1000 \ \Omega \text{ cm}^2}{10^{-6} \text{ cm}} = 10^9 \ \Omega \text{ cm}$$

Compared with physiological salt solutions, such a membrane is a poor conductor.

In homogeneous membranes, ϱ represents the resistivity of the membrane material. In heterogeneous membranes, ϱ represents a composite resistivity. For example, consider a membrane composed of a fatty material dotted with small conducting pores. For the fatty material, $\varrho \gg 10^9$ Ω cm. For the conducting pores, $\varrho \ll 10^9$ Ω cm. These could result in the value 10^9 Ω cm, or 1000 Ω cm^2 for a 100-Å membrane, depending on the number and size of the pores.

PROBLEM. Assume two parallel pathways in a membrane 100 Å thick, with $\varrho_1 = 10$ Ω cm, $\varrho_2 = 10^{14}$ Ω cm, and $R_m = 1000$ Ω cm^2. Estimate the ratio of areas A_1/A_2.

The resistivity and the geometric term are usually constant; however, in some devices these may depend on the current through the membrane or the voltage across it. For example, in a glass microelectrode, $\varrho = \varrho(n)$ may change because the concentration profile changes. In a heterogeneous membrane, the number of conducting pores may depend on the voltage across the membrane.

If the change in R is not instantaneous, an apparent capacitance or inductance may result. In addition, devices may also have a real capacitance or inductance, real in the sense that the property of capacitance or inductance results from an electric or magnetic field.

The capacitance of two conducting surfaces of area A, separated by distance l, is

$$C = \varepsilon \frac{A}{l}$$

where ε, having the units of F/cm, is the dielectric constant of the medium between the surfaces. In empty space $\varepsilon_0 = 8.854 \times 10^{-14}$ F/cm. For common dielectrics, $\varepsilon = K\varepsilon_0$, where K ranges between 1 in a vacuum and about 10 in oil. An ideal capacitor stores energy in a purely electrostatic field between its conducting surfaces. A current through an ideal capacitor does not represent the movement of free charge, as in a resistor, but rather a charge induced between conductors. Only the electric field changes; no magnetic field is associated with this current, nor does it generate heat through dissipation.

For membranes of uniform thickness, one defines the specific capacitance as

$$C_m = \varepsilon/l \ominus \text{F/cm}^2$$

For example, a 100-Å membrane with $C_m = 1$ μF/cm² has

$$\varepsilon = 10^{-6} \text{ F/cm}^2 \times 10^{-6} \text{ cm} = 10^{-12} \text{ F/cm}$$

Thus $K = 11.3$. Such a membrane is a good dielectric.

In a heterogeneous membrane composed of a fatty material dotted with small conducting pores, the capacitance is due to the larger area. The pores may be regarded as having no capacitance; since parallel capacitors add, ε is due primarily to the fatty material.

PROBLEM. Show that

$$\varepsilon \ominus \frac{\text{charge/area}}{\text{force/charge}}$$

The above model of a heterogeneous membrane gives a picture of a leaky dielectric in parallel with ionic channels; the circuit diagram is given

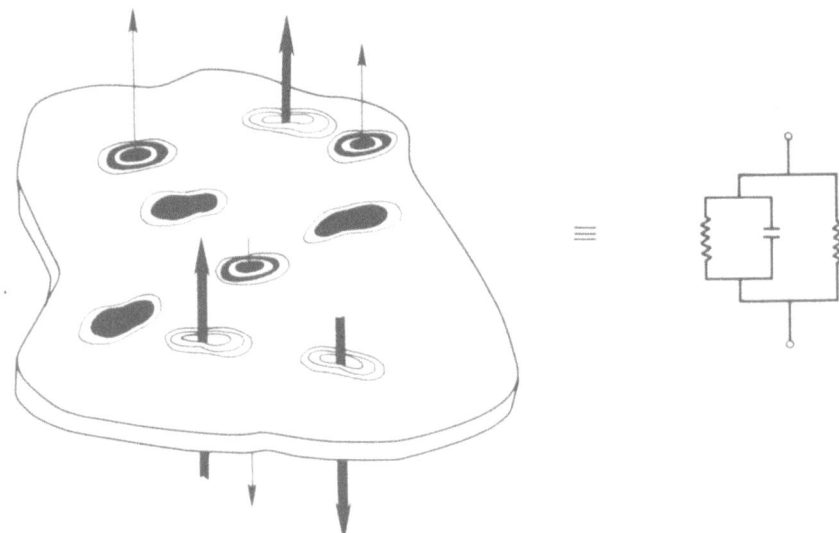

Figure 27.1. Diagram of a heterogeneous membrane and its equivalent circuit. The various patches represent different pathways through which specific ions flow (arrows). These are imbedded in a lipid matrix, represented in the equivalent circuit as parallel resistor/capacitor network. The purely resistive branch of the equivalent circuit represents the composite resistance of all the ionic pathways; this branch also contains a battery (not shown) that would stand for the weighted sum of the individual Nernst potentials of the ionic pathways (Section 30).

in Figure 27.1. Note that Figure 27.1 reduces to the familiar RC network since the two resistors are in parallel. If the ionic channels are time-variant resistors, capacitive or inductive elements may be included in the circuit diagram (see Section 81).

Biological membranes do not appear to generate the property of real inductance. (For an alternative view, see Leiberstein, 1973.) When inductors are used to represent time-variant resistors, the specific membrane inductance is defined as

$$L_m \ominus \text{H cm}^2$$

Inductors in parallel add like resistors, not like capacitors.

28. Equivalent Circuits of Cells

Consider a cylindrical cell with three distinct homogeneous regions: an inside (double prime), an outside (single prime) and a membrane (m). All three regions resist the flow of electical current; because of the geometry

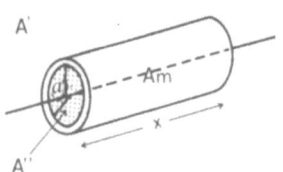

Figure 28.1. Diagram of a cylindrical cell. The inside, the outside, and the membrane are designated by a prime, a double prime, and the letter m, respectively. A'' is the cross-sectional area of the cell and A' is an unspecified cross-sectional area of the space around the cell. a is the cell radius and x is the distance along the long axis of the cell.

of the cell and the directions of current flow, these resistances are usually expressed in different units.

The three regions of the cell are illustrated in Figure 28.1. Suppose the cylindrical cell has a radius $a = 0.1$ mm and a length of 1 mm. Let $R_m = 1000\ \Omega$ cm². The resistance to current flow perpendicular to $A_m = 2\pi a x$ is

$$R = \frac{R_m}{2\pi a x} = 0.16\ \text{M}\Omega$$

In order to describe the membrane resistance encountered per unit length of cell, we define the quantity

$$r_m = Rx = \frac{R_m}{2\pi a} \ominus \Omega\,\text{cm}$$

For the example above, $r_m = 0.016$ MΩ cm; a 1-mm length of this cell has a membrane resistance of 0.16 MΩ as calculated. Note that r_m contains the geometry of the cell through $2\pi a$:

$$2\pi a = \frac{\text{membrane area}}{\text{length of cylindrical cell}}$$

r_m also contains the membrane material through R_m. Although r_m has the units of resistivity, it is not a resistivity; the cm in the units of r_m refers to a different direction than it does in ϱ.

Consider the resistance to current flow parallel to the long axis of the cylindrical cell. The area presented to this current is $A'' = \pi a^2$. The resistance to current flow perpendicular to A'' is $\varrho'' x / A''$. ϱ'' is a property of the cytoplasm and x is the cell length. The resistance to the longitudinal current, encountered per unit length of cell interior, is

$$r'' = \frac{\varrho'' x / A}{x} = \frac{\varrho''}{\pi a^2} \ominus \frac{\Omega}{\text{cm}}$$

A reasonable value for ϱ'' is 150 Ω cm, corresponding roughly to a 50 mM

solution of KCl (Section 11). Keeping to the same example,

$$r'' = \frac{150 \ \Omega \ cm}{\pi(0.01 \ cm)^2} = 477 \ \frac{K\Omega}{cm}$$

A current flowing along the long axis of this cell encounters 477 KΩ resistance for each centimeter length of cell.

Finally, consider the external resistance to current flow along the long axis of the cell. Whatever the external area may be, denote it by A'. Its value will depend on the external environment, the proximity of other cells, etc. By analogy with r'', define

$$r' = \frac{\varrho'}{A'} \ominus \frac{\Omega}{cm}$$

Normally ϱ' and ϱ'' are of the same order of magnitude. If $A' \ll A''$, e.g., if a cell is in a volume much greater than its own, then

$$r' \ll r''$$

This often leads to the approximation $r' = 0$, compared to r''. If cells form a tightly packed tissue, rules derived from this approximation are not expected to hold.

Just as R_m and r_m were described for the resistive part of the membrane, we now describe C_m and c_m for the capacitive part. For a cylindrical cell of length 1 mm and radius 0.1 mm, and $C_m = 1 \ \mu F/cm^2$,

$$C = C_m 2\pi a x = 6.28 \ nF$$

To describe the membrane capacitance encountered per unit length of cell, we define

$$c_m = \frac{C}{x} = 2\pi a C_m \ominus \frac{F}{cm}$$

For the above example, $c_m = 62.8$ nF/cm. Thus a 1-cm length of this cell has 6.28 nF membrane capacitance as calculated.

Note that R and C, R_m and C_m, and r_m and c_m depend oppositely on geometry. Therefore $RC = R_m C_m = r_m c_m$ are independent of membrane area or cell length.

The internal and external media have insignificant capacitance; they are inherently far better conductors than the cell membrane; capacitive effects in the internal and external media are only observable at frequencies greater than 100 kHz.

PROBLEM.   Derive the units for the membrane inductance, l_m, encountered per unit length of axon. Are there formulas that relate l_m to axon
geometry (as there are for r_m and c_m)?

29. Cable Equation: Passive Properties

Consider the cylindrical cell of Figure 28.1, represented by the lumped
circuit diagram in Figure 29.1. Recall that r' has the units of r'' which has
the units of Ω/cm, r_m has the units of $\Omega \cdot cm$ and c_m has the units of F/cm.
The unit cm implies length in the direction x of the long axis of the cylinder.
Applying Ohm's law to one segment of Figure 29.1 gives

$$dV'' = -i''(r''dx)$$
$$dV' = -i'(r'dx)$$

where dV is the voltage drop in distance dx. The negative sign expresses the
convention that current flows to the right for $dV/dx < 0$, as indicated in
Figure 29.2. Combining these equations we obtain

$$\frac{d}{dx}(V'' - V') = \frac{dV}{dx} = i'r' - i''r''$$

In words, the change in transmembrane voltage $(V'' - V')$ with x is equal
to the difference in ir drops inside and outside the cell.

Since we are dealing with a closed circuit, it is required that

$$\frac{di'}{dx} = -\frac{di''}{dx} = i_m$$

The current gained on the outside must equal that lost from the inside,
per unit length. Since these two currents are joined only though the
membrane, they must also equal the current per unit length, i_m, passing there.
Note that i' has the units of i'' which is in amperes, but that i_m has the units
A/cm.

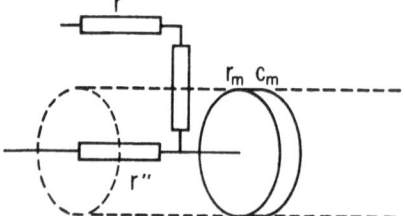

Figure 29.1. The equivalent circuit of
the cell in Figure 28.1. The cell membrane
is represented by a resistor (all ionic
pathways, plus the leakage resistance, in
parallel) and a capacitor (the lipid substrate) in parallel.

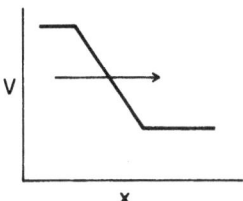

Figure 29.2. Voltage as a function of distance along a cylindrical cell. By convention, positive current flows down the voltage gradient.

The last two equations may be used to eliminate all but membrane currents, thus

$$\frac{d}{dx}\frac{dV}{dx} = r'\frac{di'}{dx} - r''\frac{di''}{dx} = i_m(r' + r'')$$

If the membrane is modeled by the $r_m c_m$ circuit, then

$$i_m = c_m \frac{dV}{dt} + \frac{V}{r_m}$$

and

$$\frac{d^2V}{dx^2} = (r' + r'')\left(c_m \frac{dV}{dt} + \frac{V}{r_m}\right)$$

This can be written in the convenient form

$$\lambda^2 \frac{d^2V}{dx^2} = \tau \frac{dV}{dt} + V$$

where

$$\lambda^2 = \frac{r_m}{r' + r''} \ominus \text{cm}^2$$

and

$$\tau = r_m c_m \ominus \text{sec}$$

The equation above is called the cable equation. The solution of the cable equation, $V(x, t)$, is a function of both distance and time. The equation is reasonably difficult to solve in the time domain. However, if we take the Fourier transformation (Section 25) of the cable equation, the first time derivative may be replaced by $i\omega$ (Section 26) and a solution in the frequency domain is easily obtained. Using the bar to denote the Fourier transform,

$$\lambda^2 \frac{d^2\bar{V}}{dx^2} = (1 + i\omega\tau)\bar{V}$$

By inspection, a solution to this equation is

$$\bar{V}(x,\,\omega) = V_0 \exp\left[-\frac{x}{\lambda}(1 + i\omega\tau)^{1/2}\right]$$

A dc potential ($\omega = 0$) decays exponentially along the cell with length constant λ.

From a previous result, $i' = -i''$ and therefore

$$\frac{d\bar{V}}{dx} = -i''(r' + r'')$$

From the solution of the cable equation, we also know that

$$\frac{d\bar{V}}{dx} = V_0\left[-\frac{1}{\lambda}(1 + i\omega\tau)^{1/2}\right]\exp\left[-\frac{x}{\lambda}(1 + i\omega\tau)^{1/2}\right]$$

Eliminating $d\bar{V}/dx$ from the last two equations gives

$$\frac{\bar{V}}{\bar{\imath}''} = \frac{\lambda(r' + r'')}{(1 + i\omega\tau)^{1/2}}$$

The ratio $\bar{V}/\bar{\imath}''$ is called the input impedance Z of the cable (see Figure 29.3). Define $Z_m = r_m/(1 + i\omega\tau)$; the final expression for the input impedance may be written

$$Z = [Z_m(r' + r'')]^{1/2}$$

This result is valid for the infinite cable. It also holds when r' and r'' are replaced by complex networks z' and z''.

PROBLEM.   Take the Fourier transform of

$$\frac{d^2V}{dx^2} = i_m(r' + r'')$$

and show that

$$Z_m = \frac{\bar{V}}{\bar{\imath}_m} = \frac{r_m}{1 + i\omega\tau}$$

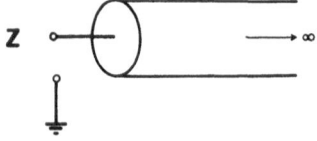

Z

Figure 29.3. The impedance Z of an infinitely long cable.

PROBLEM. Let the cell radius be $a = 10$ μm, $\varrho'' = 150$ Ω cm and $R_m = 1000$ Ω cm². Show that $\lambda \simeq 0.5$ mm and $\tau \simeq 2$ msec, and that the dc input resistance of the cell is $Z \simeq 3$ MΩ. Note that Z is the geometric mean of membrane and longitudinal cell resistance.

PROBLEM. For a cylindrical cell in a large external volume, show that

$$\lambda \propto a^{1/2}$$
$$\tau \neq \tau(a)$$
$$Z \propto a^{-3/2}$$

where a is the cell radius.

The previous treatment applies to cylindrical cells uniformly exposed to the exterior. This is approximately the situation in unmyelinated nerve. In myelinated nerve cells only nodes of membrane are exposed at regular intervals along the length of the cell; this has a marked effect on the equivalent circuit.

In myelinated nerves, the internodal distance L is approximately proportional to the radius of the nerve axon:

$$L \propto a$$

That is, for larger diameter cells, the nodes are proportionally farther apart. This is illustrated in Figure 29.4, in which only the nodes (exposed membrane) are shown. Also indicated in Figure 29.4 is the nodal width d, which is approximately constant in both large and small cells. This leads to the following result for myelinated nerves:

$$\frac{\text{membrane area}}{\text{length of cylindrical cell}} = \frac{2\pi a d}{L} = \frac{2\pi(2a)d}{2L} = \text{const}$$

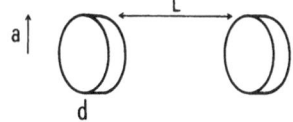

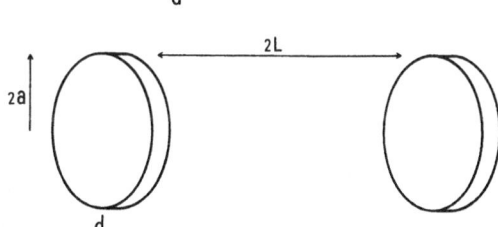

Figure 29.4. An idealization of the scaling of myelinated nerve. L is the internodal length, a is the axon radius and d is the width of the node.

If $L \propto a$ and if d does not vary with cell size, the exposed area per unit length of myelinated nerve is nearly independent of cell radius. Recall that in unmyelinated nerve, the same ratio is $2\pi a$.

PROBLEM.   For a cell with $A/x = $ const, in a large external volume, show that the equivalent circuit implied by the cable equation has

$$\lambda \propto a$$

$$\tau \neq \tau(a)$$

$$Z \propto a^{-1}$$

PROBLEM.   Calculate the membrane resistance of a node for $R_m = 1000 \ \Omega \ \text{cm}^2$, $a = 10 \ \mu\text{m}$, and $d = 2 \ \mu\text{m}$.

PROBLEM.   Show that $A/x = $ const also holds for myelinated nerves that are constructed with constant internodal distance, but $d \to d/2$ as $a \to 2a$.

30. Equivalent Circuits and Active Membranes

The remainder of this work will deal primarily with the subthreshold behavior of excitable membranes. The concepts introduced in Sections 11–29 are used frequently in subsequent chapters. Here we draw attention to the formal connection between excitability and the equivalent circuits that will be considered.

From passive cable theory for the cylindrical cell

$$\frac{d^2V}{dx^2} = (r' + r'')i_m$$

where

$$i_m = c_m \frac{dV}{dt} + \frac{V}{r_m}$$

The fatty material of the membrane is represented by the capacitor c_m; the leakage pathway through this dielectric is represented by r_m.

For the heterogeneous membrane (the fatty material dotted with small conducting pores), r_m is modeled by ionic pathways in parallel. In

Figure 30.1. The qualitative dependence of the steady-state potassium
current on transmembrane voltage.
The open circle on the V axis is
imagined to move (arrows); the filled
circle on the I-V curve will follow
the curve only for very slow movements. Rapid movements are off
the line and instantaneous movements
follow the chord—a straight line
drawn between the filled circle and
V_K.

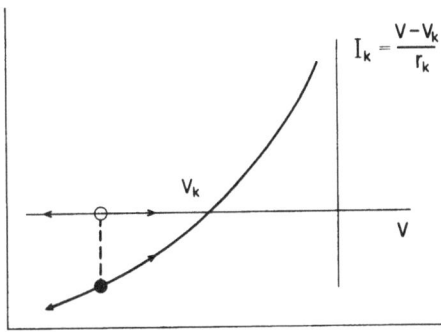

the Hodgkin and Huxley (1952) model for the giant axon of the squid

$$\frac{V}{r_m} = \frac{V - V_K}{r_K} + \frac{V - V_{Na}}{r_{Na}} + \frac{V - V_L}{r_L}$$

r_K and r_{Na} represent specific ionic pathways for these ions. V_K and V_{Na} are the equilibrium or reversal potentials for these pathways. For example, if the membrane potential $V = V_{Na}$, no current flows through the Na pathway. The third term represents a general leakage pathway which may pass more than one type of ion.

In this model, excitability is described by letting r_K and r_{Na} depend on V. For example, the K system has the qualitative behavior shown in Figure 30.1. As the membrane depolarizes conductance ($1/r_K$) increases. Some sort of recruiting of K pathways may be envisioned, or each K pore may be thought of as becoming more effective. We shall consider this question in detail when subthreshold phenomena are discussed in later chapters.

In Figure 30.1 the curve describes the steady-state behavior of the K system. If the open dot moves, e.g., sinusoidally, the closed dot will only follow the curve for very low frequencies. This time dependence accounts for the apparent impedance of excitable membranes, as previously explained (Section 23). When the perturbations are small, the impedance may be described by linear circuit theory (Section 81).

To describe the active propagation of signals along such a cable, it is convenient to define

$$I_m = \frac{i_m}{2\pi a} \ominus \frac{A}{cm^2}$$

I_m is the amount of current which flows through the membrane per unit area of a cylindrical cell of radius a. Let $r'' \gg r'$. Then

$$i_m \simeq \frac{1}{r''} \frac{d^2V}{dx^2}$$

and

$$I_m = \frac{a}{2\varrho''} \frac{d^2V}{dx^2}$$

(See Section 28 for definitions of r'' and ϱ''.)

If the signal is carried along the cable without loss and at uniform velocity θ, then

$$x = \theta t$$

Making this substitution,

$$I_m = \frac{a}{2\varrho''\theta^2} \frac{d^2V}{dt^2}$$

Whatever the active properties of the membrane, the ratio

$$K = \frac{1}{I_m} \frac{d^2V}{dt^2}$$

is constant. Then

$$\theta^2 = \frac{a}{2\varrho''} K$$

or

$$\theta \propto a^{1/2}$$

This is a well-known result for unmyelinated axons; it says that the velocity of propagation is proportional to the square root of the axon radius. The derivation, which assumes only the equivalent circuit and uniform propagation, is due to Hodgkin (1954).

PROBLEM. Show that for myelinated nerves, with $A/x = \text{const}$ (see Section 29)

$$\theta \propto a$$

In a limited-size nerve trunk, more information may flow if the nerves are myelinated since smaller-sized myelinated fibers may have propagation velocities equal to larger-sized unmyelinated fibers.

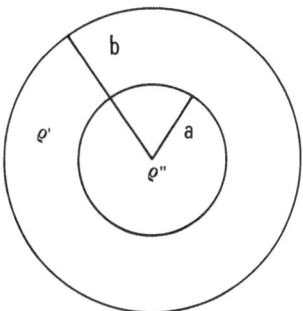

Figure 30.2. Diagram of an unmyelinated axon of radius a and internal resistivity ϱ''; current flow in extracellular space is limited by a cylinder of radius b and extracellular resistivity ϱ'.

PROBLEM. Consider an unmyelinated axon of radius a, enclosed in a cylindrical volume of radius b (see Figure 30.2). Show that the velocity of uniform propagation is given by

$$\theta^2 = \frac{a(b^2 - a^2)}{2[(b^2 - a^2)\varrho'' + a^2\varrho']} K$$

Consider the limiting cases for the values of a and b. Discuss the same problem for a myelinated nerve.

PROBLEM. Show that the units of K^{-1} are $C_m \cdot$ time. Discuss the assumption that $K = $ const and how this is related to uniform propagation.

Suggested Reading. In the general area of excitability see Hodgkin (1951), Hodgkin and Huxley (1952), FitzHugh (1969), Ehrenstein and Lecar (1972), and Stämpfli and Hille (1976). For an elementary treatment see Katz (1966). Comprehensive mathematical treatments on all aspects of cellular electrophysiology are found in the textbooks by Plonsey (1969), Cole (1968), Leibovic (1972, part one), Jack, Noble, and Tsien (1975), and in Brookhart and Mountcastle (1977, Chaps. 3–7).

The equivalent circuit of cells of other geometries, or of groups of cells forming a tissue, is beyond the scope of this book. *Suggested reading* in this area includes Fatt (1964), Falk and Fatt (1964), Eisenberg and Johnson (1970), Schneider (1970), Hodgkin and Nagajima (1972), Adrian and Almers (1973, 1976), Valdiosera *et al.* (1974a,b), Schoenberg *et al.* (1975), Eisenberg *et al.* (1977, 1979), and Mathias *et al.* (1979).

3

Basic Circuit Theory

31. Voltage and Current Sources

An emf is a device that always maintains the same voltage across its terminals regardless of the current passing through it. A steady emf is represented by the symbol shown in Figure 31.1, where the long bar indicates positive voltage $(+)$ and the shorter bar negative voltage $(-)$. The term "dc" is used to indicate a steady or average property; thus the symbol above represents a dc-emf.

A fluctuating emf is represented by the symbol in Figure 31.2 and is called an ac-emf. Since this emf alternates from $(+)$ to $(-)$ a sign convention is not required. When the fluctuation is random the symbol represents a noise source.

Consider the dc circuit corresponding to two parallel pathways (Figure 31.3). The reference potential (ground) is usually drawn lowest in the circuit, even when the symbol for ground is not used. To analyze the circuit, select an initial direction for the current i—say clockwise—then

$$E_1 - E_2 = i(r_1 + r_2)$$

A clockwise current is the normal direction for E_1, since an emf attempts to drive the current from $(+)$ to $(-)$ through the circuit. If $E_2 > E_1$, then $i < 0$ and the actual current will be counterclockwise.

E

Figure 31.1. Symbol for a steady emf (an ideal battery).

85

e

Figure 31.2. Symbol for a fluctuating emf, e.g., a noise source.

It is required to calculate the open circuit voltage (V) across the parallel circuit with respect to ground. If V is initially thought of as positive, then

$$V = E_1 - ir_1$$

for the left branch of the circuit, and

$$V = ir_2 + E_2$$

for the right branch. V has the same sign as E_1, but the opposite sign to ir_1 since i is flowing toward V through r_1. Either of these equations may be used to eliminate i from the initial expression. For example, using $V = E_1 - ir_1$ gives

$$E_1 - E_2 = \left(\frac{E_1 - V}{r_1}\right)(r_1 + r_2)$$

which may be rearranged to give

$$V = \frac{E_1 r_2 + E_2 r_1}{r_1 + r_2}$$

PROBLEM. Show that $V = ir_2 + E_2$ results in the same expression for V given above. Draw the parallel circuit with E_2 in the reverse direction (positive side down) and show that

$$V = \frac{E_1 r_2 - E_2 r_1}{r_1 + r_2}$$

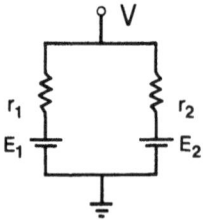

Figure 31.3. A network consisting of two parallel pathways, each containing a battery and a resistance. The resistors may be considered as the internal resistance of real batteries or as separate components outside ideal batteries with no internal resistance.

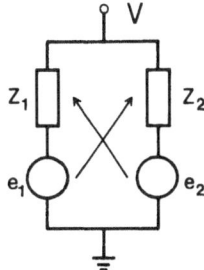

Figure 31.4. A general parallel network to illustrate the cross-multiply rule for open circuit voltage.

   The above result suggests a simple maneuver for calculating the open-circuit voltage for any parallel network: to find the open-circuit voltage, cross-multiply the voltage source in one branch with the resistance in the the other, add the two products, and divide by the sum of the circuit resistance.

   This rule also applies when the resistances are complex impedances and the sources are alternating. In this case it is the Fourier transformation of the ac-emf that is considered. The Fourier transform of $e(t)$ is

$$\bar{e}(f) = \int_{-\infty}^{\infty} e(t)e^{-i\omega t}\, dt$$

Let an open box stand for a general circuit element; the rule states, for the parallel circuit shown in Figure 31.4, that the open-circuit voltage is given by

$$\bar{V}(f) = \frac{\bar{e}_1 z_2 + \bar{e}_2 z_1}{z_1 + z_2}$$

   The bar signifies Fourier transformation. It may be convenient to omit the bars if it is understood that one is working entirely in the frequency domain (Figure 31.4).

   Once the open-circuit voltage of a parallel network is calculated, the network may be replaced by a single equivalent branch, as, for example, in Figure 31.5.

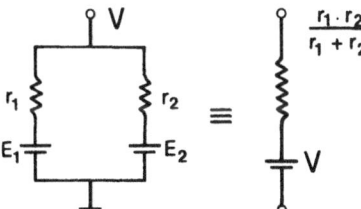

Figure 31.5. A parallel network and its electrical equivalent.

Consider now the closed-circuit current for Figure 31.4. The closed-circuit current is the current that would flow in an external pathway placed across the parallel network. From Figure 31.5 this is readily seen to be

$$I = \frac{V}{r_1 r_2 / (r_1 + r_2)} = \frac{E_1 r_2 + E_2 r_1}{r_1 r_2}$$

This result suggests a maneuver similar to that for obtaining the open-circuit voltage: to find the closed-circuit current, cross-multiply the voltage source in one branch with the resistance in the other, add the two products, and divide by the product of the circuit resistance. For the circuit shown in Figure 31.6 the current in the external branch is given in the frequency domain by

$$\bar{I}(f) = \frac{\bar{e}_1 z_2 + \bar{e}_2 z_1}{z_1 z_2}$$

The ratio of the open-circuit voltage to the closed-circuit current is the network impedance:

$$Z = \frac{\bar{V}}{\bar{I}} = \frac{z_1 z_2}{z_1 + z_2}$$

PROBLEM.   Consider a network with three parallel resistive branches, each having a voltage source. Show that the open-circuit voltage is

$$V = \frac{E_1 r_2 r_3 + E_2 r_1 r_3 + E_3 r_1 r_2}{r_1 r_2 + r_1 r_3 + r_2 r_3}$$

Derive the expression for closed-circuit current and check this result against the network impedance.

The conductance is defined as one over resistance. Similarly, the admittance of a circuit is defined as $1/Z$.

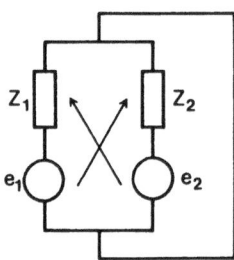

Figure 31.6. A general parallel network to illustrate the cross-multiply rule for the closed-circuit rule.

Figure 31.7. Symbol for an ideal current source.

PROBLEM. Derive an expression for the open-circuit voltage of a two-branch and a three-branch network (similar to those considered above) in terms of conductances rather than resistances.

PROBLEM. Devise a general rule using summation signs for the voltage across a circuit with N branches in parallel.

It is often convenient to replace a voltage source by an equivalent current source. A current source is a device that always provides the same current regardless of the voltage across it. Let the symbol shown in Figure 31.7 designate such a source (either ac, or dc, or both). Unlike the voltage source, which expresses itself even as an isolated branch, a current source must be connected to an external circuit to provide a net current. The correspondence between a voltage source and a current source is as shown in Figure 31.8, where $i = e/r$.

PROBLEM. Prove the equivalence $i = e/r$ described above by attaching each circuit to an external resistor R and calculating the current through R and the voltage across it.

PROBLEM. Show that a parallel RC circuit, with noise source $e(t)$ in series with R, has an open-circuit voltage and a closed-circuit current given by

$$\bar{V}(f) = \frac{\bar{e}}{1 + i\omega\tau} \quad \text{and} \quad \bar{I}(f) = \frac{\bar{e}}{R} = \bar{i}(f)$$

where $\tau = RC$.

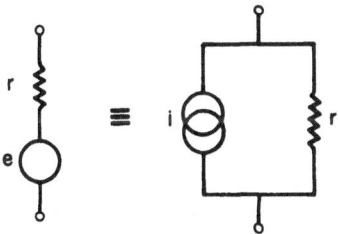

Figure 31.8. The equivalence of a voltage source and a current source. The circuit on the right may always replace the one on the left if $i = e/r$.

32. *Frequency Composition of Signals*

We have seen that for purposes of analysis it is often convenient to consider circuit elements and voltage or current sources in the frequency domain. Here we give examples of the frequency composition of some common sources.

The frequency composition of a signal is defined as the modulus (absolute value) of its Fourier transform. Let $e(t)$ be an arbitrary signal, and

$$\bar{e}(f) = a + ib$$

be its Fourier transform. In general, $\bar{e}$ has both real (a) and imaginary (b) parts. The frequency composition is then

$$|\bar{e}(f)| = (a^2 + b^2)^{1/2}$$

When $e(t)$ is a voltage, $|\bar{e}|$ has the units of V/Hz.

A few examples are given below.

a. *Rectangular Pulse*

Let $e(t)$ be a pulse of height e_0 and duration T (Figure 32.1). Since the signal is zero everywhere except for t between 0 and T,

$$\bar{e}(f) = \int_{-\infty}^{\infty} e(t)e^{-i\omega t}\, dt = e_0 \int_0^T e^{-i\omega t}\, dt$$

$$= \frac{e_0}{i\omega}(1 - e^{-i\omega T})$$

From the identity $e^{-i\theta} = \cos\theta - i\sin\theta$

$$\bar{e}(f) = \frac{e_0}{\omega}[\sin\omega T - i(1 - \cos\omega T)]$$

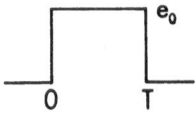

Figure 32.1. Pulse of height e_0 and duration T.

Figure 32.2. A sequence of m pulses, each of duration T_1. The pulses arrive regularly at intervals $t = 0$, T_2, $2T_2$, etc., $T_2 > T_1$.

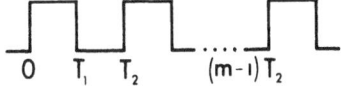

The frequency composition of a rectangular pulse is therefore

$$| \bar{e}(f) | = \frac{e_0}{\omega} [\sin^2 \omega T + (1 - \cos \omega T)^2]^{1/2}$$

$$= e_0 \frac{2^{1/2}}{\omega} (1 - \cos \omega T)^{1/2}$$

The last step follows from $\cos^2 \theta + \sin^2 \theta = 1$.

PROBLEM.   Plot $| \bar{e} |$ versus ω for the rectangular pulse. (Note that $| \bar{e} |$ is always positive.) Show that another form for $| \bar{e} |$ is

$$\left| \frac{2e_0}{\omega} \sin \left(\frac{\omega T}{2} \right) \right|$$

The maxima of this function are given by the condition

$$\frac{\omega T}{2} = \tan \frac{\omega T}{2}$$

and the value of $| \bar{e} |$ at $\omega = 0$ is $e_0 T$. Investigate this function as T becomes large.

PROBLEM (from Goldman, 1948, p. 65).   Find the frequency distribution of m pulses of length T_1 and repetition period T_2 as shown in Figure 32.2. Draw $| \bar{e}(f) |$ for $m = 10$ and compare this result with the above problem.

b. Cosine Wave (from Goldman, 1948, p. 56)

Let $e(t)$ be the cosine function between $t = T_1$ and T_2 (Figure 32.3). The Fourier transform is

$$\bar{e}(f) = e_0 \int_{T_1}^{T_2} \cos \omega_0 t e^{-i\omega t} \, dt$$

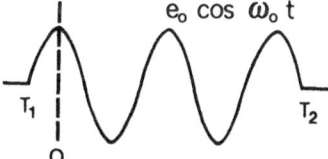

Figure 32.3. A cosine function that is zero everywhere except between T_1 and T_2. The amplitude of the wave is e_0 and its period is $2\pi/\omega_0$.

This integration may be divided into four parts:

$$\int_{T_1}^{T_2} \cos \omega_0 t \cos \omega t \, dt = + \left[\frac{\sin(\omega + \omega_0)t}{2(\omega + \omega_0)} + \frac{\sin(\omega - \omega_0)t}{2(\omega - \omega_0)} \right]_{T_1}^{T_2}, \qquad \omega \neq \omega_0$$

$$\int_{T_1}^{T_2} \cos \omega_0 t \sin \omega t \, dt = - \left[\frac{\cos(\omega + \omega_0)t}{2(\omega + \omega_0)} + \frac{\cos(\omega - \omega_0)t}{2(\omega - \omega_0)} \right]_{T_1}^{T_2}, \qquad \omega \neq \omega_0$$

$$\int_{T_1}^{T_2} \cos^2 \omega_0 t \, dt = \left(\frac{t}{2} - \frac{\sin 2\omega_0 t}{4\omega_0} \right)_{T_1}^{T_2}, \qquad \omega = \omega_0$$

$$\int_{T_1}^{T_2} \cos \omega_0 t \sin \omega t \, dt = - \frac{1}{4\omega_0} (\cos 2\omega_0 t)_{T_1}^{T_2}, \qquad \omega = \omega_0$$

The sum of these four expressions (times e_0) gives the Fourier transform of $e_0 \cos \omega_0 t$ for arbitrary length. This sum is real; its absolute value is the frequency composition of the cosine wave.

PROBLEM.   Plot $|\bar{e}(f)|$ for the cosine wave with $T_1 = 0$ and $T_2 = \pi/2\omega_0, \pi/\omega_0, 2\pi/\omega_0$, and $4\pi/\omega_0$.

c. Constant

Let $e(t) = e_0$ for all time. Then

$$\bar{e}(f) = e_0 \int_{-\infty}^{\infty} e^{-i\omega t} \, dt$$

$$= e_0 \, \delta(f)$$

Since this is real, the frequency composition is simply the absolute value:

$$|\bar{e}(f)| = |e_0| \, \delta(f)$$

PROBLEM.   Compare the solution to (c) with the solution to (a) as the rectangular pulse (T) becomes large.

d. Cosine

Let $e(t) = e_0 \cos \omega_0 t$ for all time. Then

$$\bar{e}(f) = e_0 \int_{-\infty}^{\infty} \cos \omega_0 t \, e^{-i\omega t} \, dt$$

$$= \frac{e_0}{2} [\delta(f - f_0) + \delta(f + f_0)]$$

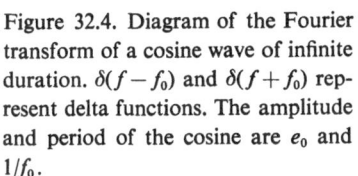

Figure 32.4. Diagram of the Fourier
transform of a cosine wave of infinite
duration. $\delta(f - f_0)$ and $\delta(f + f_0)$ represent delta functions. The amplitude
and period of the cosine are e_0 and
$1/f_0$.

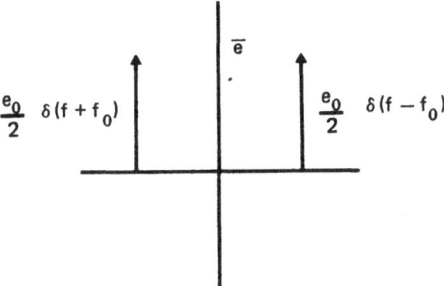

(see Section 25.) The pictorial representation of this function, for e_0 positive,
is shown in Figure 32.4, where the arrows represent delta functions. This
function is real; the frequency composition of the cosine is just its absolute
value:

$$| \bar{e}(f) | = \frac{| e_0 |}{2} [\delta(f - f_0) + \delta(f + f_0)]$$

PROBLEM. Compare the solution to (d) with the solution to (b) as
the wave increases in length.

PROBLEM. What is the frequency composition of the sine wave
$e(t) = e_0 \sin \omega_0 t$ for all time?

A useful pictorial dictionary of Fourier transforms is given in Chapter
18 of Bracewell (1965). In this dictionary, the notation

$$F(s) = \int_{\infty}^{\infty} f(x) e^{-i2\pi xs} \, dx$$

is used. $F(s)$ is the Fourier transform of $f(x)$. The variables x and s are
regarded as unitless. For example, Bracewell gives the pair:

$$\cos \pi x, \qquad \tfrac{1}{2}[\delta(x + \tfrac{1}{2}) + \delta(x - \tfrac{1}{2})]$$

in place of the pair

$$\cos(2\pi f_0)t, \qquad \tfrac{1}{2}[\delta(f + f_0) + \delta(f - f_0)]$$

It is useful to browse through this dictionary and to keep the more common waveforms and their transformations in mind.

PROBLEM. Prove that $| \bar{e}(f) | = | \bar{e}(-f) |$, i.e., that the frequency
composition is a symmetric function of frequency.

33. The Mean and the Variance

In the previous section, the frequency composition of specific analytic functions was considered. Before turning to noise signals, it is necessary to define the mean and the variance of an arbitrary signal.

As a first example consider the function

$$x(t) = A + B \sin \omega t$$

between $t = 0$ and T. Here ω is the angular frequency $2\pi f$ and T is called the period. Thus ω has the units of rad/sec and f has the units of $1/\text{sec} = \text{Hz}$ (Figure 33.1).

The mean value of $x(t)$ is defined as

$$\langle x(t) \rangle = \frac{1}{T} \int_0^T x(t)\, dt$$

Substitution gives

$$\langle x(t) \rangle = \frac{1}{T} \int_0^T (A + B \sin \omega t)\, dt$$

$$= \frac{1}{T} \int_0^T A\, dt + \frac{1}{\omega T} \int_0^T B \sin \omega t\, d(\omega t)$$

$$= A + \frac{B}{T\omega} [- \cos \omega t]_0^T$$

The last term is zero. Hence

$$\langle x(t) \rangle = A$$

as is obvious from Figure 33.1 since the signal spends as much time above as below the value A.

The mean value of the square of $x(t)$ is defined as

$$\langle x^2(t) \rangle = \frac{1}{T} \int_0^T x^2(t)\, dt$$

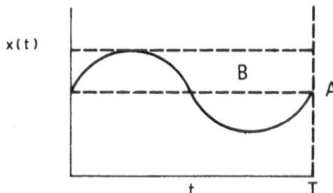

Figure 33.1. The function $A + B \sin \omega t$ between $t = 0$ and $t = T$. $T = 2\pi/\omega$ is the period of the sine wave; $f = 1/T$ is the frequency of the sine wave.

Substitution gives

$$\langle x^2(t) \rangle = \frac{1}{T} \int_0^T (A + B \sin \omega t)^2 \, dt$$

$$= \frac{1}{T} \int_0^T (A^2 + B^2 \sin^2 \omega t + 2AB \sin \omega t) \, dt$$

By the previous argument, the last term is zero.
Evaluating the middle term gives

$$\frac{B^2}{\omega T} \int_0^T \sin^2 \omega t \, d(\omega t) = \frac{B^2}{\omega T} \left(\frac{\omega t}{2} - \frac{1}{2} \cos \omega t \sin \omega t \right)_0^T$$

$$= \frac{B^2}{\omega T} [\pi - 0] = \frac{B^2}{2}$$

Thus

$$\langle x^2(t) \rangle = A^2 + \tfrac{1}{2} B^2$$

or

$$\langle x^2(t) \rangle - A^2 = \tfrac{1}{2} B^2$$

This is called the variance of $x(t)$; in words, the variance is the mean of the square minus the square of the mean. The variance of an arbitrary signal $x(t)$ is given the symbol σ^2 and is defined as

$$\sigma^2 = \langle x^2(t) \rangle - \langle x(t) \rangle^2$$

PROBLEM. Show that an equivalent expression for the variance is

$$\sigma^2 = \langle (x - \langle x \rangle)^2 \rangle$$

The variance and the mean are essential properties of a noise signal for which no simple analytic function exists.

PROBLEM. The root-mean-square (rms) of a signal is the square root of the variance, or σ. Consider a sine wave; show that the peak-to-peak (pp) value of the wave is related to the rms by

$$\text{pp} = 2(2^{1/2})\text{rms} \simeq 2.8 \text{ rms}$$

For example, a sine wave coursing $\pm 1\tfrac{1}{2}$ V around zero has an rms of about 3 V/2.8 $\simeq$ 1 V. This rule also applies to noise signals; the peak-to-peak value of noise, however, is less well defined. An expected pp value of noise may be defined from the rms value; this pp value corresponds roughly to the thickness of a noise trace measured, say, on a storage oscilloscope.

34. *Spectral Density and Rayleigh's Theorem*

We have seen that an arbitrary signal in time, $e(t)$, may be represented by its transform in the frequency domain, $\bar{e}(f)$. The two representations are equivalent and contain the same information although in different forms.

The frequency composition of $e(t)$ is a symmetric function. This means that $|\bar{e}(f)|$ on the negative frequency axis is the mirror image of $|\bar{e}(f)|$ on the positive frequency axis, i.e.,

$$|\bar{e}(-f)| = |\bar{e}(f)|$$

There are two fundamental properties of the frequency composition of a signal. First, the function $e(t)$ need not be represented by a simple analytic function (as in Section 32) in order to define its frequency composition. Second, although $e(t)$ may be nonzero only during a finite time interval, its frequency composition extends to all values of the positive and negative frequency axis.

Consider an arbitrary wave form for which no explicit elementary function exists. Although we may regard the region $0 < t < T$ as our signal, its representation in the frequency domain extends to infinity. Thus, for the signal shown in Figure 34.1, we may have the frequency composition shown in Figure 34.2. Only the positive frequency axis is shown since $|\bar{e}(f)|$ is a symmetric function. Although $e(t)$ may not be described conveniently, its frequency composition can be measured and used to describe the signal. This is the basis of noise analysis in the frequency domain.

To define the spectral density we introduce Rayleigh's theorem. An excellent discussion of this theorem is given in Bracewell (1965). Rayleigh's theorem states that the area under the square modulus of a function is equal to the area under the square modulus of its Fourier transform. In

Figure 34.1. An arbitrary wave (e.g., from a noise source) that is zero everywhere except between 0 and T.

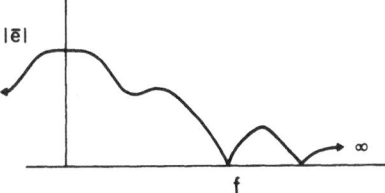

Figure 34.2. The frequency composition
of an arbitrary wave such as that shown in
Figure 34.1. Although the signal in the
time domain is finite, its frequency composition extends to $\pm\infty$.

symbols,

$$\int_{-\infty}^{\infty} |e(t)|^2 \, dt = \int_{-\infty}^{\infty} |\bar{e}(f)|^2 \, df$$

Rayleigh's theorem is easily verified for a specific case. For example,
consider the rectangular pulse of height e_0 and duration T (Section 32).
The Fourier transform of the rectangular pulse is

$$\frac{2e_0}{\omega} \sin\left(\frac{\omega T}{2}\right)$$

Let $x = \omega T/2$. Then

$$\int_{-\infty}^{\infty} \frac{4e_0^2}{\omega^2} \sin^2\left(\frac{\omega T}{2}\right) dt = \frac{e_0^2 T}{\pi} \int_{-\infty}^{\infty} \frac{\sin^2 x}{x^2} \, dx = e_0^2 T$$

since the integral in x is equal to π. The quantity $e_0^2 T$ is also the area under
the square of the rectangular pulse. The absolute value signs have been
omitted since evidently

$$\left|\frac{\sin x}{x}\right|^2 = \frac{\sin^2 x}{x^2}$$

For an arbitrary, real signal $e(t)$, which exists only between $t = 0$ and T,
Rayleigh's theorem states that

$$\int_0^T e^2(t) \, dt = \int_{-\infty}^{\infty} |\bar{e}(f)|^2 \, dt$$

Define the average value of $e^2(t)$ as

$$\frac{1}{T} \int_0^T e^2(t) \, dt$$

Note that $\langle e^2(t) \rangle$ is the variance of $e(t)$ if $\langle e(t) \rangle = 0$ (Section 33). Since
$|\bar{e}(f)|$ is an even function of frequency, the integral may be expressed
over all frequencies as twice the integral over the positive frequency axis.

Figure 34.3. An arbitrary wave (e.g., from a noise source) extending to all time. The time interval T represents a finite sample of the perpetual wave.

Using the symbol σ^2 for the variance, Rayleigh's theorem states that

$$\sigma^2 = \frac{1}{T} \int_0^T e^2(t)\, dt = \frac{2}{T} \int_0^\infty \mid \bar{e}(f) \mid^2 df$$

The integrand on the right-hand side is called the spectral density* $S(f)$:

$$\hat{S}(f) = \frac{2}{T} \mid \bar{e}(f) \mid^2$$

The last two equations are written only for positive f.

The caret notation for $S(f)$ indicates a spectral density for finite time. The spectral density for a process continuing for all time is

$$S(f) = \lim_{T \to \infty} \hat{S}(f)$$

One regards $\hat{S}(f)$ as a representative of some perpetual signal that will approach the spectral density of the perpetual signal if a large enough sample is taken (Figure 34.3).

If $e(t)$ has the units of volts, then $\bar{e}(f)$ has the units of V sec and $S(f)$ the units of V^2 sec. Usually the spectrum is represented as a frequency density:

$$S(f) \ominus V^2/Hz$$

The meaning of such a density may be clarified by an analogy with weight.

---

* In some books (for example, Bendat and Piersol, 1971) the symbol $S(f)$ is reserved for the double-sided spectral density, and $G(f)$ for the positive frequency spectrum; see for example their page 77. Rice (1944) uses $S(f)$ to stand for the Fourier transform. This equation describes the basic relationship between the Fourier transform and the spectral density. *Suggested reading.* For an introduction to the fast Fourier transform (FFT) used in digital computers see Cochran and Cooley (1967). Other useful books are by Bendat (1958) and Bendat and Piersol (1966, 1971). An excellent mathematical overview is given by Hochstadt (1973) and by Bracewell (1965). Earlier books include Byerly (1893), Carslaw (1930), Goldman (1948, 1949) and Lighthill (1959).

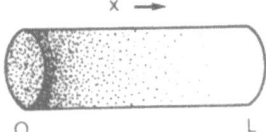

Figure 34.4. Diagram of a rod with nonuniform weight
distribution. The density of the rod is a function of *x*.

Consider a rod with a nonuniform weight distribution being heavier at one
end than the other (Figure 34.4). Let $\varrho(x)$ be the distribution of the weight/
length (Figure 34.5). The weight of the rod in any element of length *dx*
is $\varrho \, dx$, and total weight is the sum of all the elementary weights:

$$W = \int_0^L \varrho(x) \, dx$$

Similarly the variance of a signal in a narrow frequency band *df* is $S(f) \, df$,
and the total variance is the sum of these:

$$\sigma^2 = \int_0^\infty S(f) \, df$$

If the signal is deterministic and is defined by an analytic expression
(such as cos ωt) the spectral density and the variance are absolutely determined. If the signal is probabilistic, e.g., a random noise source, the
spectral density and the variance will be different for each time interval *T*.
However, if the random signal is stationary, the different estimates for spectral density and variance will tend toward a mean value. In general, the
symbols σ^2 and $S(f)$ will be used to refer to the mean properties of noise
signals.

35. Spectral Density and Source Impedance

An internal signal seen at an external point may appear distorted as a
consequence of the impedance of the source (Section 31). For example, an
internal rectangular voltage pulse from an *RC* circuit is distorted when

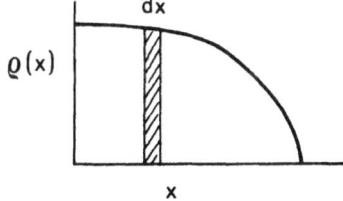

Figure 34.5. The density of the rod in Figure
34.4 as a function of length along the rod.
The weight of the rod in *dx* is $\varrho(x) \, dx$.

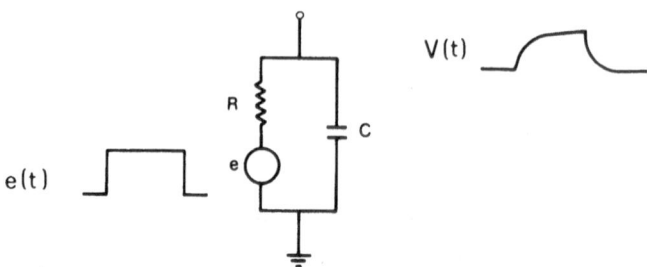

Figure 35.1. A parallel *RC* network with an emf in the resistive branch. If the emf is a rectangular pulse *e(t)*, the voltage measured across the parallel circuit has the qualitative shape *V(t)*.

observed across the parallel network (Figure 35.1). The problem illustrated in Figure 35.1 was treated in Section 24 as an input–output relationship in the time domain and in Section 26 in the frequency domain.

In the preceding section, a noise source *e(t)* was described by its spectral density $S(f)$. For the parallel *RC* circuit, the output voltage in the frequency domain is

$$\bar{V}(f) = \frac{\bar{e}(f)}{1 + i\omega\tau}, \qquad \tau = RC$$

Taking the square modulus of both sides of this equation gives

$$|\bar{V}(f)|^2 = \frac{|\bar{e}(f)|^2}{1 + \omega^2\tau^2}$$

Multiplying both sides by $2/T$ gives

$$\frac{2}{T}|\bar{V}(f)|^2 = \frac{(2/T)|\bar{e}(f)|^2}{1 + \omega^2\tau^2}$$

In the limit of large *T*,

$$S_V(f) = \frac{S_e(f)}{1 + \omega^2\tau^2}$$

The subscript in $S_V(f)$ indicates that $S(f)$ is the spectral density of the signal *V(t)* observed at the output. This is called the voltage spectral density. Similarly, the subscript in $S_e(f)$ indicates the voltage spectral density of the internal source of noise, *e(t)*. Note that S_V has the units of S_e, namely, V²/Hz.

In a voltage measurement, the spectral shape of the internal noise source is distorted by the impedance of the source. For example, if $S_e(f)$

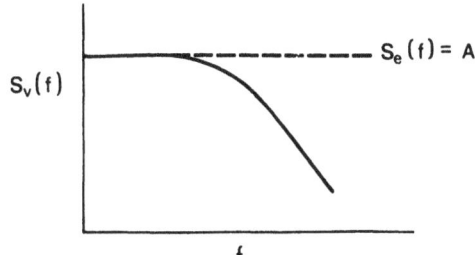

$S_v(f)$

$S_e(f) = A$

f

Figure 35.2. The voltage spectral density observed across a parallel circuit like that shown in Figure 35.1 if the emf in R is a white noise source (flat spectral density for all frequencies).

$= A$ is a white noise source (no frequency dependence) then S_V has the form shown in Figure 35.2.

If current I is allowed to flow in an external branch the time domain current is $I(t) = e(t)/R$ and the frequency domain current is

$$\bar{I}(f) = \frac{\bar{e}(f)}{R}$$

This is illustrated in Figure 35.3. In this case the measured current spectral density is

$$S_I(f) = \frac{S_e(f)}{R^2}$$

Note that S_I has the units of A²/Hz.

If the voltage noise source $e(t)$ is replaced by a current noise source $i(t)$ such that

$$i(t) = \frac{e(t)}{R}$$

then

$$S_I(f) = S_i(f)$$

where $S_i(f)$ is the spectral density of $i(t)$.

The output current noise spectral density in Figure 35.3 is related directly to the internal noise source; the shape of the spectral density is not altered by the source impedance (Figure 35.4).

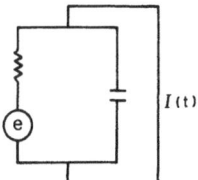

$I(t)$

Figure 35.3. A parallel network similar to Figure 35.1 except that current is allowed to flow through an external short.

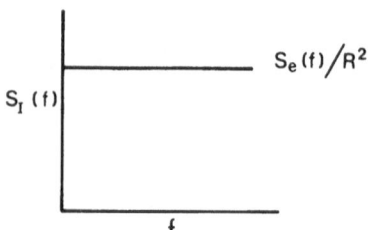

Figure 35.4. The current spectral density observed in the external short of Figure 35.3 if the emf in R is a white noise source.

PROBLEM.   Consider the RC circuit above with $\tau = 1$ msec. Let

$$S_e(f) = \frac{K\theta}{1 + \omega^2\theta^2}$$

where $K = 1$ V² and θ is the time constant of the internal noise source $e(t)$. Plot $S_V(f)$ when $\theta = 0.1\ \tau$, $\theta = \tau$, and $\theta = 10\ \tau$.

PROBLEM.   Consider a white noise source, $S_e(f) = \text{const}$, in the two circuits shown in Figure 35.5. Calculate the open-circuit voltage noise and closed-circuit current noise spectra in each case. Show that the ratio S_V/S_I is the modulus square of the source impedance.

The formal relationship between the spectral density and the frequency composition of signals enables one to apply the rules developed in Section 31 to noise analysis. For example, the voltage across the parallel $z_1 z_2$ circuit is

$$\bar{V}(f) = \frac{\bar{e}_1 z_2 + \bar{e}_2 z_1}{z_1 + z_2}$$

and the voltage spectral density is given by

$$S_V(f) = \frac{2}{T}|\bar{V}(f)|^2 = \frac{2}{T}\frac{|\bar{e}_1 z_2 + \bar{e}_2 z_1|^2}{|z_1 + z_2|^2}$$

in the limit of large T.

Figure 35.5. A parallel and a series RL circuit. In each case, the resistor has a noise source e associated with it.

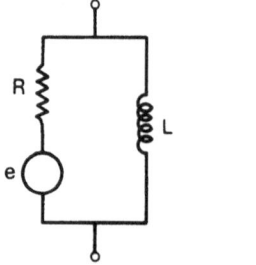

PROBLEM. Show that if $x = a_1 + ib_1$ and $y = a_2 + ib_2$ are complex numbers, then

$$|x + y|^2 = |x|^2 + |y|^2 + 2(a_1a_2 + b_1b_2)$$

The result of the above problem implies that the voltage spectral density across a parallel z_1z_2 circuit is given by

$$S_V(f) = \frac{2}{T} \frac{|\bar{e}_1z_2|^2 + |\bar{e}_2z_1|^2 + \text{cross terms}}{|z_1 + z_2|^2}$$

Since $|xy|^2 = |x|^2 |y|^2$,

$$S_V(f) = \frac{S_{e_1}|z_2|^2 + S_{e_2}|z_1|^2}{|z_1 + z_2|^2} + \text{cross terms}$$

$S_{e_1}(f)$ and $S_{e_2}(f)$ are the voltage spectral densities of the sources $e_1(t)$ and $e_2(t)$ if one could measure them directly. We shall show below that if e_1 and e_2 are independent random noise sources, the cross terms will approach zero.

As an example let $z_1 = z_2 = r$ be purely resistive elements; consider the qualitative summation shown in Figure 35.6. The factor $1/2$ comes from voltage division of each source by the parallel rr network. Thus

$$V(t) = \tfrac{1}{2}e_1(t) + \tfrac{1}{2}e_2(t)$$

Consider the variance of this sum. By definition

$$\sigma_V{}^2 = \frac{1}{T} \int_0^T V^2(t)\, dt = \frac{1}{T} \int_0^T \frac{1}{4}(e_1{}^2 + e_2{}^2 + 2e_1e_2)\, dt$$

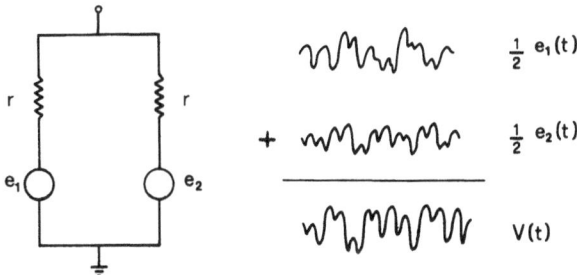

Figure 35.6. Two resistors in parallel, each associated with a noise source *e*. Since the two resistors are equal, half of each noise source appears across the parallel network and add to give $V(t)$.

If $e_1(t)$ and $e_2(t)$ are random, independent sources,

$$\int_0^T e_1 e_2 \, dt \to 0$$

for sufficiently long T. This may be verified by considering the product of two series of random numbers of equally likely positive and negative values. The product is equally distributed about zero, with an average value of zero. Finally

$$\sigma_V{}^2 = \tfrac{1}{4}\sigma_{e_1}^2 + \tfrac{1}{4}\sigma_{e_2}^2$$

Therefore, in the frequency domain,

$$S_V(f) = \tfrac{1}{4}S_{e_1}(f) + \tfrac{1}{4}S_{e_2}(f)$$

If $e_1(t)$ and $e_2(t)$ have the same average spectral densities (though their instantaneous values are different), then

$$S_{e_1}(f) = S_{e_2}(f) = S_e(f)$$
$$S_V(f) = \tfrac{1}{2}S_e(f)$$

PROBLEM. Consider the circuit in Figure 35.7 where $e_1(t)$ and $e_2(t)$ are random, independent noise sources. Show that variance and spectral density of the open-circuit voltage are given by

$$\sigma_V{}^2 = \sigma_{e_1}^2 + \sigma_{e_1}^2$$

and

$$S_V(f) = S_{e_1}(f) + S_{e_2}(f)$$

In general, if $e_1(t)$ and $e_2(t)$ are random noise sources in a parallel $z_1 z_2$ circuit, then the spectral density of the open-circuit voltage is given by

$$S_V(f) = \frac{S_{e_1}\,|\,z_2\,|^2 + S_{e_2}\,|\,z_1\,|^2}{|\,z_1 + z_2\,|^2}$$

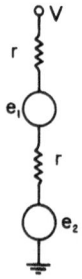

Figure 35.7. Two resistors and their associated noise sources in series.

PROBLEM. In the examples above, the noise sources $e(t)$ had no dc component. Consider the case when each source is of the form $E = E_0 + e(t)$, where E_0 is a steady voltage.

PROBLEM. Show that if $e_1(t)$ and $e_2(t)$ are random noise sources in a parallel z_1z_2 circuit, then the current spectral density observed in an external branch is

$$S_I(f) = \frac{S_{e_1} |z_2|^2 + S_{e_2} |z_1|^2}{|z_1z_2|^2}$$

It is evident from the last problem that for the parallel z_1z_2 network

$$\frac{S_V(f)}{S_I(f)} = \frac{|z_1z_2|^2}{|z_1 + z_2|^2}$$

Let $Z = z_1z_2/(z_1 + z_2)$ be the circuit impedance at the measurement point V. Then

$$S_V(f) = S_I(f) |Z|^2$$

This is true for an arbitrary circuit when V implies the open-circuit voltage and I the closed-circuit current. The result is also evident from Ohm's law, written in the frequency domain, viz.,

$$\bar{V} = \bar{I}Z$$

which is essentially a definition of impedance.

PROBLEM. Let two noise sources have rms values of 0.3 V and 0.5 V. What is the expected peak-to-peak value of the noise obtained by the sum of these two sources?

36. Examples

In the preceding section we derived the general relationship between voltage noise $S_V(f)$, current noise $S_I(f)$, and impedance Z. Here we catalog some circuits frequently encountered.

The results are obtained conveniently by reducing the circuits to simpler equivalents. For example, consider the circuit in Figure 36.1, where

$$\bar{V} = \frac{\bar{e}(i\omega L)}{R + i\omega L}, \qquad Z = \frac{i\omega LR}{R + i\omega L}$$

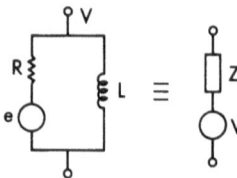

Figure 36.1. A parallel *RL* circuit and its electrical equivalent.

It is convenient to define $\tau = L/R$. The circuit on the right in Figure 36.1 will always replace the one on the left if V is taken as the open-circuit voltage and Z the circuit impedance between the two measurement points. $\bar{V}$ is readily obtained by the cross-product rule (Section 31), and Z from the product over the sum of the impedances of each branch. Z is free from all voltage sources, which are now collected in V.

a. Single Noise Source (Figure 36.2)

Since the open-circuit voltage in Figure 36.2 is $V = e$, the voltage noise at the output is simply

$$S_V(f) = S_e(f)$$

independent of Z.

Consider a circuit composed of two components, one of which contains a noise source (Figure 36.3). The top component in Figure 36.3 contains the noise source and reduces to the equivalent source ($\tau_1 = R_1 C_1$),

$$\bar{e}_1 = \frac{\bar{e}}{1 + i\omega\tau_1}$$

in series with the impedance Z_1, where

$$Z_1 = \frac{R_1}{1 + i\omega\tau}$$

The total impedance of the network is ($\tau_2 = R_2 C_2$)

$$Z = \frac{R_1}{1 + i\omega\tau_1} + \frac{R_2}{1 + i\omega\tau_2}$$

Figure 36.2. General equivalent circuit of a voltage noise source.

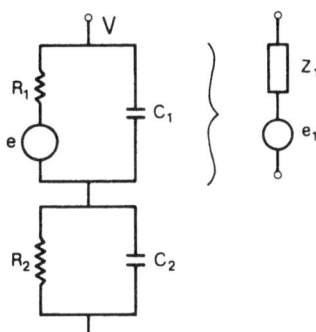

Figure 36.3. A parallel *RC* circuit that contains a noise source *e* in series with a similar but noiseless circuit. The electrical equivalent of the noisy circuit is shown to the right of the curly bracket.

The voltage noise at the output of the two component circuits of Figure 36.3 is therefore

$$S_V(f) = \frac{S_e(f)}{1 + \omega^2 \tau_1^2}$$

which is independent of τ_2. Thus, the voltage noise spectral density from a source is not distorted by access to that source if the access impedance contains no noise sources.

The current noise in an external loop connecting the point V and ground in the above example (Figure 36.3) is given by

$$S_I = \frac{S_V}{|Z|^2} = \frac{S_e(f)}{1 + \omega^2 \tau_1^2} \, |Z|^{-2}$$

which depends on both τ_1 and τ_2. In general, current noise from a source is distorted by the access impedance.

PROBLEM.   Calculate $|Z|^2$ for the network in the above example (Figure 36.3). Plot S_I when $\tau_1 = 10\tau_2$, $\tau_1 = \tau_2$, and $10\tau_1 = \tau_2$.

PROBLEM.   Consider the circuit shown in Figure 36.4. Assume that

$$S_e(f) = \frac{K\theta}{1 + \omega^2 \theta^2}$$

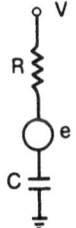

Figure 36.4. A resistor with associated noise source in series with a capacitor.

and calculate the open-circuit voltage noise, closed-circuit current noise, and impedance. Plot $S_I(f)$ on log–log paper for $K = 1$ V², $C = 1$ μF. and $\theta = 1$ msec, for the three cases $R = 1$ MΩ, 1 kΩ, and 1 Ω.

b. Parallel–Series Circuits

Consider the network shown in Figure 36.5. By the cross-product rule (Section 31), the open-circuit voltage in the frequency domain is

$$\bar{V} = \frac{(\bar{e}_1 + \bar{e}_3)(z_2 + z_4) + (\bar{e}_2 + \bar{e}_4)(z_1 + z_3)}{z_1 + z_2 + z_3 + z_4}$$

The circuit impedance is

$$Z = \frac{(z_1 + z_3)(z_2 + z_4)}{z_1 + z_2 + z_3 + z_4}$$

Let the sources $e(t)$ represent random voltage noise sources. $S_V(f)$ and $S_I(f)$ may be obtained by the previously developed rules, e.g.,

$$S_V = \frac{(S_{e_1} + S_{e_3})\,|\,z_2 + z_4\,|^2 + (S_{e_2} + S_{e_4})\,|\,z_1 + z_3\,|^2}{|\,z_1 + z_2 + z_3 + z_4\,|^2}$$

All cross products have been set equal to zero.

Consider the special case that all four noise sources have the same average properties, and the impedances are pure resistances of equal value, i.e.,

$$S_{e_1} = S_{e_2} = S_{e_3} = S_{e_4} = S_e$$

$$z_1 = z_2 = z_3 = z_4 = r$$

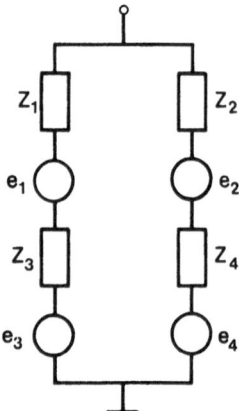

Figure 36.5. A parallel–series combination of four general equivalents (Z and e in series) of voltage noise sources.

Then

$$S_V(f) = \frac{(S_e + S_e)(r + r)^2 + (S_e + S_e)(r + r)^2}{(4r)^2}$$

$$= S_e(f)$$

and

$$Z = \frac{(r + r)(r + r)}{4r} = r$$

The open-circuit voltage noise of the array, namely, $S_V(f)$, is equal to the voltage noise of an individual source in the array. The result assumes that each source is independent of the others.

PROBLEM. Calculate the closed-circuit current noise $S_I(f)$ for case (b) above (Figure 36.5). Replace each voltage noise source with an equivalent current noise $i(t)$. Show that if the spectral densities of each noise source are equal, on the average, then

$$S_I(f) = S_i(f)$$

where $S_i(f)$ is the spectral density of $i(t)$.

PROBLEM. Show that the open-circuit voltage noise, or closed-circuit current noise, is unaltered if the four-component circuit (Figure 36.5) is short-circuited at its midpoint.

PROBLEM. Let each source in Figure 36.5 be instantaneously correlated, i.e., the four sources $e(t)$ superposed in time. Show that in this case

$$S_V(f) = 2S_e(f)$$

The results are extended easily to a 3×3, 4×4, ..., array. If the noise sources are arranged so that the array impedance is equal to the impedance of an individual source, the spectral density of the array will equal the spectral density of an individual source if the noise sources are independent.

c. Continuous, Homogeneous Noise Sources

Consider a volume conductor of length l and cross-sectional area A. Imagine the resistor R to be subdivided into N elementary resistors, each of length l/N and cross-sectional area A/N (see Figure 36.6). Thus

$$R = \varrho \frac{l}{A} = \varrho \frac{(l/N)}{(A/N)}$$

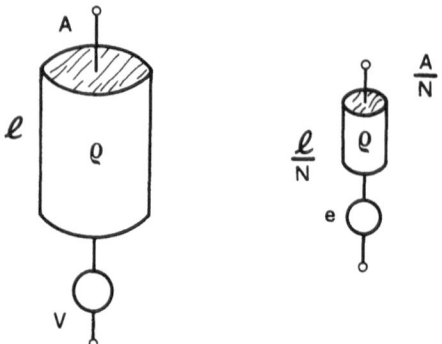

Figure 36.6. Two resistors made of the same material but scaled to different volumes. $V(t)$ is the noise source associated with the larger resistor and $e(t)$ is the noise source associated with a resistor reduced in length and cross-sectional area by a factor N.

where ϱ is the resistivity of the volume conductor and each elementary resistor. By the previous reasoning, the open-circuit voltage noise S_V from the volume conductor is related to the voltage noise S_e from one of the elementary resistors of which it is composed by the relationship

$$S_V(f) = S_e(f)$$

N may be regarded as a large number, reducing the elementary noise source to molecular dimensions. Measurement of the macroscopic parameter S_V gives information about the underlying microscopic behavior of the molecular subunits. This is the primary aim of noise analysis in biological membranes. It must be stressed that in this analysis $V(t) \neq e(t)$; the macroscopic behavior does not instantaneously mimic the microscopic but their average properties are the same.

d. Noise Sources Distributed in a Membrane

Consider the two-dimensional circuit of repeating units r and e (Figure 36.7). There is no loss in generality in writing this circuit as the one-dimensional array since the addition of any new element is always in parallel with all the others.

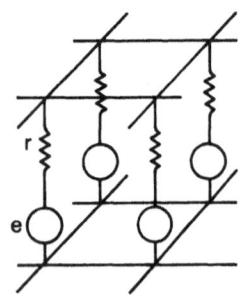

Figure 36.7. A parallel array of noise sources in a two-dimensional membrane.

For two elements, the open-circuit voltage of Figure 36.7 is

$$V = \frac{e_1 r_2 + e_1 r_2}{r_1 + r_2}$$

For three elements,

$$V = \frac{e_1 r_2 r_3 + e_2 r_1 r_3 + e_3 r_1 r_2}{r_1 r_2 + r_2 r_3 + r_3 r_1}$$

and so on. Let each unit noise source have the same average properties; the open-circuit voltage noise for two elements is

$$S_V = \frac{S_e r^2 + S_e r^2}{(2r)^2} = \frac{1}{2} S_e$$

For three elements

$$S_V = \tfrac{1}{3} S_e$$

and so on. Thus for N units having the same average properties and distributed in parallel in a membrane, the open-circuit voltage noise is

$$S_V = \frac{1}{N} S_e$$

Consider each noise source confined to an area a. If the total membrane area is A then

$$N = \frac{A}{a}$$

and

$$S_V = \frac{a}{A} S_e$$

S_e and a are properties only of the elementary noise sources. (They may be regarded either as distributed or discrete.) Voltage spectral density is inversely proportional to membrane area.

PROBLEM. Show that the closed-circuit current noise from the membrane in Figure 36.7 is

$$S_I(f) = N S_i(f)$$

where $S_i(f) = S_e / r^2$ is the spectral density of an elementary current noise source. Evidently

$$S_I(f) = \frac{A}{a} S_i(f)$$

Current spectral density is directly proportional to membrane area.

PROBLEM. Assume each noise source within a membrane to be instantaneously correlated with every other source:

$$e(t) = e(t) = e(t) = \cdots$$

Show that in this case $S_V = S_e$ (or $S_I = S_i$), independent of membrane area. For correlated noise sources, membranes of different area would generate the same noise amplitude. For instanstaneously identical noise sources, the addition of parallel elements changes the noise amplitude and the source resistance in a way which exactly cancels the effect of adding new elements. For independent noise sources, the noise adds in the square while the resistance changes linearly. In the former case, the sources are added and then squared; in the latter case, the voltage sources are squared and then added.

PROBLEM. Let each noise source in the membrane be a parallel rc circuit with voltage source $e(t)$ in the r branch. Show that in this case, as in the pure resistive case, N random, independent rc noise sources in parallel give

$$S_V = \frac{1}{N} S_e$$

Also show that

$$S_I = NS_e\left(\frac{1 + \omega^2\tau^2}{r^2}\right)$$

where $\tau = rc$. If C_m and R_m are the specific membrane resistances (Section 27) show that the density of noise sources is given by

$$\frac{N}{A} = \frac{r}{R_m} = \frac{C_m}{c}$$

For a density of $100/\mu^2$, show that a biological membrane (Section 27) would have

$$r \simeq 10^{13}\,\Omega, \qquad c \simeq 10^{-16}\,\text{F}$$

Here, r is regarded as uniform over the elementary area a of each noise source. If the actual noise source is heterogeneous, with a conductance pore located at the center of area a, the elementary resistance will be larger than calculated above.

37. Power Spectral Density

In sections 35 and 36 we discussed the voltage spectral density (voltage noise) $S_V(f)$, and the current spectral density (current noise) $S_I(f)$, from a variety of circuits. Here we introduce the power spectral density $S_W(f)$.

Consider a branched network with an arbitrary arrangement of noise sources and impedances having but two points of access. The network may always be reduced to an equivalent circuit between those same two points, consisting of a noiseless impedance Z in series with a noise source $e(t)$. The open circuit voltage is

$$\bar{V}(f) = \bar{e}(f)$$

and the voltage noise is

$$S_V(f) = S_e(f) \ominus \text{V}^2/\text{Hz}$$

The closed-circuit current is

$$\bar{I}(f) = \frac{\bar{e}(f)}{Z}$$

and the current noise is

$$S_I(f) = \frac{S_e(f)}{|Z|^2} \ominus \text{A}^2/\text{Hz}$$

The power spectral density may be defined as

$$S_W(f) = \frac{2}{T} |\bar{I}\bar{V}| = \frac{2}{T} \frac{|\bar{e}|^2}{|Z|}$$

in the limit of large T (see Section 34). By previous arguments,

$$S_W(f) = \frac{S_V(f)}{|Z|} = S_I(f)|Z| \ominus \frac{\text{W}}{\text{Hz}}$$

The term "power spectrum" is sometimes loosely used for the voltage or current spectral density. This is because the voltage and current spectral densities are squared quantities directly proportional to power, whereas the frequency composition of signals (Section 32) is linear in either voltage or current.

PROBLEM. Show that the power spectral density from a parallel RC circuit with noise source $e(t)$ in the R branch is given by

$$S_W(f) = \frac{S_e(f)}{R(1 + \omega^2 \tau^2)^{1/2}}$$

where $\tau = RC$. Compare S_W with S_V and S_I for the same circuit.

The power spectral density defined above is of theoretical interest only. In practice, either the current or voltage is measured.

4

Noise Analysis

38. *Filtering*

The measurement of spectral density is closely connected to the concept of filtering. Here we introduce some simple circuits used for this purpose.

The transfer function of a circuit is defined as the ratio of its output to its input. For example, consider the circuit in Figure 38.1. From Figure 38.1

$$\frac{\bar{V}}{\bar{E}} = \frac{1}{1 + i\omega\tau_2} \equiv Y_2$$

Y is the transfer function in the frequency domain. As usual, $\tau_2 = R_2C_2$. Once Y is known, the output may be found for any input; thus

$$\bar{V} = Y_2\bar{E}$$

The bars in the formulas above refer to the minus-i Fourier transform, e.g.,

$$\bar{V}(f) = \int_{-\infty}^{\infty} V(t)e^{-i\omega t}\, dt$$

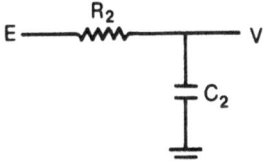

Figure 38.1. Low-pass *RC* filter.

115

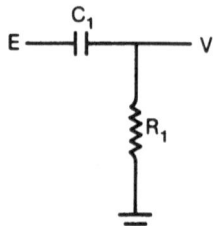

Figure 38.2. High-pass *RC* filter.

Let the symbol $\rightarrowtail$ stand for the plus-*i* Fourier transform; then

$$V(t) = \int_{-\infty}^{\infty} \bar{V}(f)e^{i\omega t}\, df$$

may be written

$$V(t) = \overrightarrow{\bar{V}(f)} = \overrightarrow{Y\bar{E}}$$

To obtain the output in the time domain, take the minus-*i* transform of the input, multiply it by the transfer function, and take the plus-*i* transform of the product. The crossed-bar operation may be thought of as canceling the bar operation. Several examples illustrating this procedure are given in Section 26.

The circuit in Figure 38.1 is called a low-pass network since high frequencies are shorted to ground to a greater degree than low frequencies ($Y_2 = 1$ for $\omega = 0$ and $Y_2 = 0$ for $\omega \to \infty$). The circuit in Figure 38.2 is called a high-pass network. From Figure 38.2

$$\frac{\bar{E} - \bar{V}}{1/i\omega C_1} = \frac{\bar{V}}{R_1}$$

Let $\tau_1 = R_1 C_1$; then

$$Y_1 = \frac{i\omega\tau_1}{1 + i\omega\tau_1}$$

The output for arbitrary input is found by the same rules given above for the low-pass network.

a. Buffered RC Bandpass Filter

Consider a bandpass filter composed of a buffered low-pass and a high-pass section (Figure 38.3). An operational amplifier which acts as a perfect buffer is introduced between the two sections. The two stages are nonloading and their individual transfer functions may be multiplied. The transfer function for the buffered low-pass filter is

$$Y = Y_2 Y_1 = \frac{1}{1 + i\omega\tau_2} \frac{i\omega\tau_1}{1 + i\omega\tau_1}$$

The position of the two stages is unimportant.

Let $E(t)$ represent a source of random noise. The effect of the buffered RC filter is to attenuate the spectral density of the noise at low and high frequencies. Consider the special case

$$\tau_2 = \tau_1 = \tau$$

This filter is called a L filter and the transfer function is

$$L = \frac{i\omega\tau}{(1 + i\omega\tau)^2}$$

Consider a noise source $E(t)$ at the input of the L filter. The output in the frequency domain is

$$\bar{V}(f) = L\bar{E}(f)$$

The spectral density of the output is therefore (Section 35)

$$S_V(f) = \frac{2}{T} | L\bar{E} |^2 = | L |^2 \left(\frac{2}{T} | \bar{E} |^2 \right)$$

or

$$S_V(f) = | L |^2 S_E(f)$$

in the limit of large T. The spectral density at the output is the spectral density at the input times the modulus square of the filter transfer function. This is true regardless of the particular form of the filter as long as the transfer function represents the complete circuit. In the special case considered above,

$$| L |^2 = \frac{\omega^2\tau^2}{(1 + \omega^2\tau^2)^2}$$

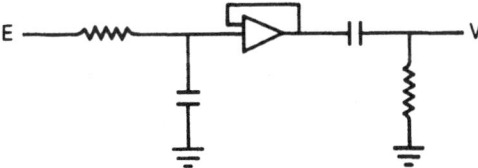

Figure 38.3. Buffered RC band-pass filter.

PROBLEM.   Plot $|L|^2$ versus f on log–log paper for $\tau = 1$ msec, $10 < f < 10^4$ Hz. Show analytically that $|L|^2$ has a maximum value when

$$\omega = \frac{1}{\tau} \quad \text{or} \quad f = \frac{1}{2\pi RC}$$

and that at this frequency $|L|^2 = 1/4$.

b. Transfer Functions and Impedance

In general

$$S_V(f) = |Y|^2 S_E(f)$$

where E and V represent the input and output spectra. $|Y|^2$ may be measured by injecting noise of a known spectral shape and observing the spectrum of the output noise. When the input spectrum is white (constant) then

$$S_E(f) = A \ominus \text{V}^2/\text{Hz}$$

and

$$|Y|^2 = \frac{1}{A} S_V(f)$$

In this case, the shape of the output spectrum is identical to the shape of the filter transfer function modulus square.

The transfer function of a circuit is related to the circuit's impedance. The network in Figure 38.4 has open-circuit voltage $\bar{V}$; the closed-circuit current (Section 35) is

$$\bar{I} = \frac{\bar{E}Z_2}{Z_1 Z_2} = \frac{\bar{E}}{Z_1}$$

I is the current that would flow in an external branch formed by connecting

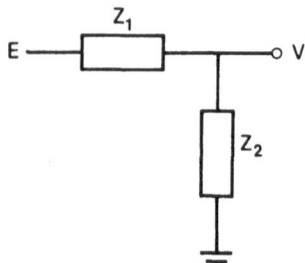

Figure 38.4. General input/output circuit to illustrate the relationship between the circuit impedance Z (with respect to V) and the circuit transfer function.

to ground the point at which V would be measured. The circuit impedance is

$$Z = \bar{V}/\bar{I}$$

and since

$$Y = \bar{V}/\bar{E}$$

then

$$Z = Y\bar{E}/\bar{I} = YZ_1$$

The impedance of the circuit is the transfer function times the impedance of the input branch. For example, the low-pass circuit in Figure 38.1 has impedance

$$Z = R_2 Y_2 = \frac{R_2}{1 + i\omega\tau_2}$$

PROBLEM. Relate the impedance (with respect to V) of the high-pass circuit in Figure 38.2 to its transfer function.

PROBLEM. Derive the impedance (with respect to V) of the buffered high-pass, low-pass filter in Figure 38.3.

Since the transfer function and the circuit impedance are proportional, the network impedance may also be measured by comparing input and output noise spectra; it is the modulus square that is obtained in this way. For example, let $S_E(f)$ be the input to a low-pass RC filter. The output is

$$S_V = |Y|^2 S_E = |Z|^2 S_E / R^2$$

If $S_E = A$ has the units of V²/Hz is a white noise source, then

$$|Z|^2 = \left(\frac{R^2}{A}\right) S_V \ominus \Omega^2$$

Multiplying the output spectral density by an appropriate factor gives the system impedance. $S_E = A$ is only a convenience. In principle, any known input may be used.

It might be thought that information about the system impedance must be known before measuring the impedance by comparing input and output noise, e.g., R in the example above. This is circumvented by applying the noise through a known impedance. For example, apply E to Z through R_0 (Figure 38.5): Z is unknown. R_0 is known and is considered

Figure 38.5. Evaluating an unknown impedance Z by injecting a white noise source E through a known resistor R_0 and measuring the output noise V.

part of the input circuit. The impedance that is measured, however, is between the point at which V is measured and ground. This includes R_0 in parallel with Z, or

$$\frac{R_0 Z}{R_0 + Z}$$

Since R_0 is known, Z can be evaluated. For $R_0 \gg Z$,

$$\frac{R_0 Z}{R_0 + Z} \simeq Z$$

This condition is frequency dependent.

c. Unbuffered RC Bandpass

Consider the circuit in Figure 38.6. The two stages load one another and the transfer function is not the product of the separate transfer functions.

To calculate the transfer function, replace the first stage (that which precedes the solid dot in Figure 38.6) by its equivalent (Figure 38.7). In Figure 38.7

$$Z' = \frac{R_1}{1 + i\omega\tau_1}$$

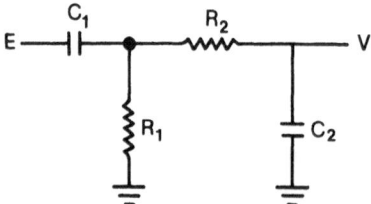

Figure 38.6. Unbuffered RC bandpass filter.

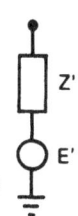

Figure 38.7. Equivalent circuit of the first stage of the unbuffered bandpass filter in Figure 38.6.

and

$$\bar{E}' = \frac{\bar{E}}{1 + i\omega\tau_1}$$

This equivalent is in series with R_2. From the cross-product rule for open-circuit voltage (Section 31),

$$\bar{V} = \frac{\bar{E}'(1/i\omega C_2)}{Z' + R_2 + (1/i\omega C_2)}$$

Substituting for $\bar{E}'$ and solving for $\bar{V}/\bar{E}$ gives the transfer function for the unbuffered RC bandpass filter;

$$\frac{\bar{V}}{\bar{E}} = Y = \frac{i\omega\tau_1}{(1 + i\omega\tau_2)(1 + i\omega\tau_1) + i\omega R_1 C_2} \qquad \text{(unbuffered)}$$

Recall the transfer function for the buffered filter is given by

$$Y = \frac{i\omega\tau_1}{(1 + i\omega\tau_2)(1 + i\omega\tau_1)} \qquad \text{(buffered)}$$

The two results differ by a cross term in the denominator.

PROBLEM. Show that the buffered RC transfer function modulus square may also be written in the form

$$|Y|^2 = \frac{f_2^2 f^2}{(f^2 + f_1^2)(f^2 + f_2^2)}$$

where $f_1 = 1/2\pi\tau_1$ and $f_2 = 1/2\pi\tau_2$. f_1 is called the lower cutoff frequency and f_2 the upper cutoff frequency. Show that the analogous expression for the unbuffered case is

$$|Y|^2 = \frac{f_2^2 f^2}{(f^2 + f_1^2)(f^2 + f_2^2) + (2\pi f)^2 R_1^2 C_2^2 f_1^2 f_2^2}$$

Plot these two expressions versus f on log–log paper for $0.1 < f < 10$ Hz. Let $f_1 = 10$ Hz and $f_2 = 1000$ Hz.

d. L Filter Compared for Buffered and Unbuffered Case

Let $f_1 = f_2$ in the above expressions. Then

$$| L |^2 = \frac{f_2^2 f^2}{f^4 + (2f_2^2) f^2 + f_2^4} \qquad \text{(buffered)}$$

$$| L |^2 = \frac{f_2^2 f^2}{f^4 + (7f_2^2) f^2 + f_2^4} \qquad \text{(unbuffered)}$$

PROBLEM. Plot $| L |^2$ versus f on log–log paper for the two cases above. Let $f_2 = 100$ Hz; plot between $f = 1$ and 10,000 Hz. In both cases, $f = f_2$ maximizes $| L |^2$; note that for $f = f_2$, buffered $| L |^2 = 1/4$, while unbuffered $| L |^2 = 1/9$. To compare shapes, scale the amplitudes so that each is of equal magnitude at its maximum value.

e. The Q Filter

Consider the circuit in Figure 38.8. L and C in parallel have impedance $i\omega L/(1 - \omega^2 LC)$. By the cross-product rule, the open-circuit voltage is given by

$$\bar{V} = \frac{\bar{E} i\omega L/(1 - \omega^2 LC)}{i\omega L/(1 - \omega^2 LC) + R}$$

Therefore

$$Q = \frac{\bar{V}}{\bar{E}} = \frac{i\omega L}{R(1 - \omega^2 LC) + i\omega L}$$

and

$$| Q |^2 = \frac{\omega^2 L^2}{R^2(1 - \omega^2 LC)^2 + \omega^2 L^2}$$

Let

$$\omega_0^2 = 1/LC$$

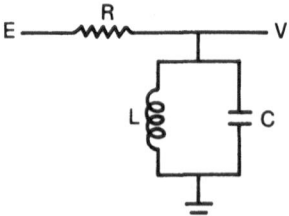

Figure 38.8. The *Q* filter.

Then the transfer function modulus square for the Q filter may be written

$$|Q|^2 = \frac{1}{1 + \mathcal{Q}^2(\omega_0/\omega - \omega/\omega_0)^2}$$

where

$$\mathcal{Q} = \omega_0 RC = \frac{R}{(L/C)^{1/2}}$$

PROBLEM. Show that $|Y|^2$ for a Q filter has a maximum value when $\omega = \omega_0$. Let

$$\omega_0 = 2\pi f_0$$

and show that an equivalent form is

$$|Q|^2 = \frac{1}{1 + \mathcal{Q}^2(f_0/f - f/f_0)^2}$$

where

$$f_0 = \frac{\mathcal{Q}^2}{2\pi RC} = \frac{(L/C)^{1/2}}{2\pi C}$$

Plot $|Q|^2$ versus f on log–log paper; let $\mathcal{Q} = 10$ and $f_0 = 10$ Hz; plot between $f = 1$ and 100 Hz.

PROBLEM. Show that the transfer function modulus square of the buffered L filter may be written

$$|L|^2 = \frac{1}{(f_0/f + f/f_0)^2} = \frac{f^2 f_0^2}{(f^2 + f_0^2)^2}$$

where $f_0 = 1/2\pi\tau$. Compare $|L|^2$ with $|Q|^2$ by plotting the two against frequency; scale such that their maximum values are equal. Let $\mathcal{Q} = 10$, $f_0 = 10$ Hz and plot between $f = 1$ and 100 hertz.

Filters with high $\mathcal{Q}$ are called narrow-band filters. Although f_0 is proportional to $\mathcal{Q}^2$, f_0 and $\mathcal{Q}$ may be adjusted independently. For example, let L and C in the network be set to give a particular value of f_0; R may be adjusted to give any value of $\mathcal{Q}$. By changing L, f_0 may be shifted to a new position of the frequency axis; R may be adjusted either to keep $\mathcal{Q}$ constant or to keep the area under $|Q|^2$ constant. These are called constant Q filters or constant bandwidth filters.

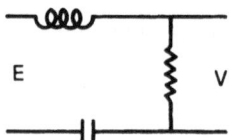

Figure 38.9. Alternative network for the Q filter.

The buffered L filter is equivalent to the Q filter (except for a constant factor) when $\mathcal{Q} = 1/2$. Thus

$$\frac{1}{1 + (\tfrac{1}{2})^2(f_0/f - f/f_0)^2} = \frac{4f^2f_0^2}{4f^2f_0^2 + (f_0^2 - f^2)^2} = \frac{4f^2f_0^2}{f^4 + 2f_0^2f^2 + f_0^2}$$

The shapes of the two filters are identical when normalized to their maximum value. The L filter is a broadband filter (low $\mathcal{Q}$) and has constant $\mathcal{Q}$ for different f_0.

PROBLEM.   Consider the circuit shown in Figure 38.9, where E is the input and V the output. Show that the transfer function modulus square is given by

$$\left|\frac{\bar{V}}{\bar{E}}\right|^2 = \frac{1}{1 + \mathcal{Q}^2(f_0/f - f/f_0)^2}$$

where $\mathcal{Q} = \omega_0(L/R)$ and $\omega_0^2 = 1/LC$. The input side of the capacitor may be tied to ground; V is measured differentially across R.

39. Measurement of Spectral Density

Spectral density $S(f)$ may be estimated by passing the noise signal through a bank of filters and measuring the rms value at each output.

Consider a narrow-band Q filter. Schematically the process of filtering may be viewed as selecting a range of frequencies from an arbitrary input.

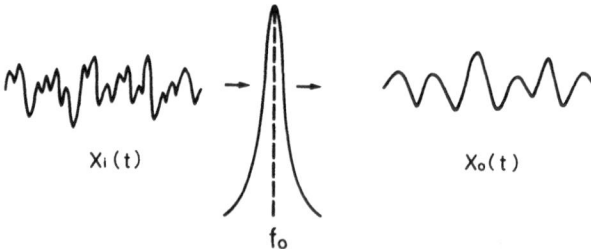

$x_i(t)$ $x_o(t)$

f_0

Figure 39.1. Noise signal $x_i(t)$ filtered by a narrow-band Q filter centered at f_0. $x_o(t)$ is the output.

The selection depends on the center frequency f_0 of the filter and the filter bandwidth Δf. Let x be the input and x_0 be the output of the filter. (See Figure 39.1). The zero crossings of the output $x_0(f)$ occur at an average rate of about f_0 (the mean interval between the zero crossings is about $1/f_0$).

The output in the frequency domain is

$$\bar{x}_0(f) = Q\bar{x}(f)$$

and therefore

$$S_0(f) = |Q|^2 S(f)$$

$S(f)$ is the spectral density we wish to measure.

Define the spectral density of the input signal at the frequency $f = f_0$ as

$$S(f_0) = \frac{\int_0^\infty S_0(f)\,df}{\int_0^\infty |Q|^2\,df} = \frac{\int_0^\infty |Q|^2 S(f)\,df}{\int_0^\infty |Q|^2\,df}$$

This is reasonable since as the Q filter becomes narrow

$$|Q|^2 \rightarrow \delta(f - f_0)$$

and the preceding formula becomes the identity $S(f_0) = S(f_0)$.

The integral in the denominator of $S(f_0)$ is called the bandwidth of the filter and is the area under $|Q|^2$:

$$\Delta f = \int_0^\infty |Q|^2\,df$$

PROBLEM.   Show that when

$$|Q|^2 = 1 \Big/ \left[1 + \mathcal{Q}^2\left(\frac{f_0}{f} - \frac{f}{f_0}\right)^2\right]$$

then

$$\int_0^\infty |Q|^2\,df = \frac{\pi}{2}\frac{f_0}{\mathcal{Q}}$$

The bandwidth of a Q filter is related to its center frequency f_0 and sharpness factor $\mathcal{Q}$ by

$$\Delta f = \frac{\pi}{2}\frac{f_0}{\mathcal{Q}}$$

$\mathcal{Q} = $ constant implies that Δf increases linearly with f_0.

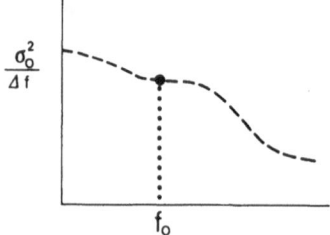

Figure 39.2. Point spectral density obtained by narrow-band filtering. σ_0^2 is the variance of the output and Δf is the bandwidth of the filter centered at f_0.

The integral in the numerator of $S(f_0)$ is the variance of the output signal:

$$\sigma_0{}^2 = \int_0^\infty S_0(f)\, df$$

The spectral density of $x(t)$ at $f = f_0$ may now be written

$$S(f_0) = \frac{\sigma_0{}^2}{\Delta f}$$

Since σ_0 is the rms value of the output signal, the practical formula becomes

$$S(f_0) = \frac{2\mathcal{Q}}{\pi f_0}\,(\text{rms})_0{}^2$$

Consider a noise signal passed through a series of Q filters with various center frequencies, e.g., $f_0 = 2$, 5, 10, and 20 Hz. The rms value is measured at each output, squared, multiplied by the factor $(2\,\mathcal{Q}/\pi)$, and divided by the filter center frequency. This value is plotted against the center frequency (see Figure 39.2). The series of points so obtained pass through $S(f)$.

PROBLEM.   A voltage noise source is measured with an amplifier having 1000 gain. The noise is passed through seven Q filters, having $\mathcal{Q} = 10$ and center frequencies $f_0 = 1$, 2, 5, 10, 20, 50, and 100 Hz. The rms values measured at the filter outputs are 3.3, 3.3, 3.0, 2.7, 2.4, 2.1, and 1.8 V. Plot the voltage spectral density of the noise source V²/Hz vs. Hz on log–log paper.

40. *Effect of Filter Bandwidth*

In the previous section the procedure for estimating spectral density was based on filtering the noise and measuring the rms value at the filter output. Here the effect of filter bandwidth on spectral estimates is calculated for several important cases.

a. White Noise

Suppose that the noise source has spectral density

$$S(f) = A$$

If the noise is a voltage source, A has the units of V²/Hz. Consider the noise to be analyzed by a filter bank with center frequencies infinitely dense of the frequency axis. The discrete variable f_0 becomes the continuous variable q; each filter has the transfer function

$$|Q|^2 = \frac{1}{1 + \mathcal{Q}^2(q/f - f/q)^2}$$

$\mathcal{Q}$ is finite and adjacent filter bandwidths overlap.

The measured spectral density is

$$S(q) = \frac{\int_0^\infty |Q|^2 S(f)\, df}{\int_0^\infty |Q|^2\, df} = A$$

since $S(f) = A$ comes out of the integral. $S(q)$ versus q is constant; therefore a white noise source is not distorted by filter bandwidth.

b. 1/f Noise

Let a noise source have spectral density

$$S(f) = B/f$$

Note that a voltage noise source has B in units of V². To calculate $S(q)$ it is required to integrate $\int_0^\infty (B/f)|Q|^2\, df$:

$$\int_0^\infty \frac{(B/f)\, df}{1 + \mathcal{Q}^2(q/f - f/q)^2} = B \int_0^\infty \frac{q^2 f\, df}{\mathcal{Q}^2 f^4 + (1 - 2\mathcal{Q}^2)q^2 f^2 + \mathcal{Q}^2 q^4}$$

$$\equiv B \int_0^\infty \frac{q^2 f\, df}{af^4 + bf^2 + c}$$

Let $u = f^2$ and $du = 2f\, df$. Then

$$\int_0^\infty \frac{f\, df}{af^4 + bf^2 + c} = \frac{1}{2} \int_0^\infty \frac{du}{au^2 + bu + c}$$

There are two conditions for evaluating the integral in u: For $4ac > b^2$

$$\int_0^\infty \frac{du}{au^2 + bu + c} = \frac{2}{(4ac - b^2)^{1/2}} \tan^{-1} \frac{2au + b}{(4ac - b^2)^{1/2}} \Big|_0^\infty$$

For $4ac < b^2$

$$\int_0^\infty \frac{du}{au^2 + bu + c} = \frac{1}{(b^2 - 4ac)^{1/2}} \ln \left[\frac{2au + b - (b^2 - 4ac)^{1/2}}{2au + b + (b^2 - 4ac)^{1/2}} \right] \Big|_0^\infty$$

In both formulas, $a = \mathcal{Q}^2$, $b = q^2(1 - 2\mathcal{Q}^2)$, and $c = q^4\mathcal{Q}^2$, therefore

$$4ac = 4q^4\mathcal{Q}^4$$

and

$$b^2 = 4q^4\mathcal{Q}^4 + q^4(1 - 4\mathcal{Q}^2)$$

The two conditions of integration become

$$4ac > b^2, \qquad (1 - 4\mathcal{Q}^2) < 0, \qquad \mathcal{Q} > \tfrac{1}{2}$$
$$4ac < b^2, \qquad (1 - 4\mathcal{Q}^2) > 0, \qquad \mathcal{Q} < \tfrac{1}{2}$$

For narrow-band filters, $\mathcal{Q} > 1/2$. All real positive values of $\mathcal{Q}$ are possible. Evaluating all $\mathcal{Q} > 1/2$ cases first:

$$\int_0^\infty \frac{du}{au^2 + bu + c} = \frac{2}{(4ac - b^2)^{1/2}} \left[\frac{\pi}{2} - \tan^{-1} \frac{b}{(4ac - b^2)^{1/2}} \right]$$

Substituting for a, b, and c gives the final results for $\mathcal{Q} > 1/2$.

$$\int_0^\infty \frac{B}{f} |Q|^2 \, df = \frac{B}{(4\mathcal{Q}^2 - 1)^{1/2}} \left[\frac{\pi}{2} - \tan^{-1} \frac{1 - 2\mathcal{Q}^2}{(4\mathcal{Q}^2 - 1)^{1/2}} \right]$$

In a similar way, it may be shown that for $\mathcal{Q} < 1/2$,

$$\int_0^\infty \frac{B}{f} |Q|^2 \, df = \frac{B/2}{(1 - 4\mathcal{Q}^2)^{1/2}} \ln \left[\frac{(1 - 2\mathcal{Q}^2) + (1 - 4\mathcal{Q}^2)^{1/2}}{(1 - 2\mathcal{Q}^2) - (1 - 4\mathcal{Q}^2)^{1/2}} \right]$$

Recall that $\int_0^\infty |Q|^2 \, df = (\pi q)/(2\mathcal{Q})$.

The measured spectral densities of $1/f$ noise analyzed with Q filters of finite bandwidth are as follows: For $Q > 1/2$

$$S(q) = \frac{B}{q} \frac{\mathcal{Q}}{(4\mathcal{Q}^2 - 1)^{1/2}} \left[1 + \frac{2}{\pi} \tan^{-1} \frac{2\mathcal{Q}^2 - 1}{(4\mathcal{Q}^2 - 1)^{1/2}} \right]$$

For $Q < 1/2$

$$S(q) = \frac{B}{q} \frac{\mathcal{Q}/\pi}{(1-4\mathcal{Q}^2)^{1/2}} \ln\left[\frac{(1-2\mathcal{Q}^2)+(1-4\mathcal{Q}^2)^{1/2}}{(1-2\mathcal{Q}^2)-(1-4\mathcal{Q}^2)^{1/2}}\right]$$

In both cases, the effect of finite bandwidth is to alter the amplitude, but not the shape, of $1/f$ noise. The correct form of the measured noise, namely, B/q, is undistorted by finite bandwidth although its magnitude is multiplied by a constant factor which depends on $\mathcal{Q}$.

PROBLEM. Show that the two cases treated above converge when $\mathcal{Q} = 1/2$; when $\mathcal{Q} = 1/2$, $S(q) = 2B/q$.

PROBLEM. Investigate the factor $S(q)/(B/q)$ for $1/2 < \mathcal{Q} < \infty$.

c. Lorentzian Noise

Let a noise source have spectral density

$$S(f) = \frac{K\theta}{1 + \omega^2\theta^2}$$

K has units of V² if the noise is from a voltage source. Note that $\omega^2\theta^2 = (f/f_\theta)^2$; f_θ is the cutoff frequency associated with time constant θ.

It is required to integrate

$$\int_0^\infty \frac{K\theta}{1+\omega^2\tau^2}\,|Q|^2\,df = K\theta q\int_0^\infty \frac{a^2}{a^2+u^2}\,\frac{u^2\,du}{u^4\mathcal{Q}^2+(1-2\mathcal{Q}^2)u^2+\mathcal{Q}^2}$$

$$= K\theta q\left(\frac{a^2}{\mathcal{Q}^2}\right)\int_0^\infty \frac{u^2\,du}{u^6+pu^4+ru^2+a^2}$$

where

$$u = f/q, \qquad\qquad a = f_\theta/q$$
$$p = a^2 + (1/\mathcal{Q}^2 - 2), \qquad r = 1 + a^2(1/\mathcal{Q}^2 - 2)$$

The integral is now in standard form (BH $\#$ 20-11), namely,

$$\int_0^\infty \frac{u^2\,du}{u^6+pu^4+ru^2+a^2} = \frac{\pi}{x_L^2(x_L - p) - 2a}$$

where x_L is the largest root of the equation

$$(x^2 - p)^2 - 8ax - 4r = 0$$

The spectral density is obtained upon division by the filter bandwidth $(\pi q)/(2\mathscr{Q})$. Finally, the effect of finite bandwidth on a Lorentzian spectrum measured with a continuous set of Q filters is

$$S(q) = \frac{2K\theta}{\mathscr{Q}} \frac{a^2}{x_L(x_L{}^2 - p) - 2a}$$

The solution, though formally correct, is awkward. The quantities a, x_L, and p all depend on q, and x_L is the solution of a fourth-order equation.

PROBLEM. Consider the value of the measured Lorentzian spectrum $S(q)$ at the point $q = f_\theta$. Show that when $\mathscr{Q} = 1/2$,

$$S(f_\theta) = K/8$$

PROBLEM. Show that when $\mathscr{Q} = 1/2$, the equation for determining x_L becomes

$$(x^2 - a^2)^2 + 4(x^2 + a^2) - 8ax = 0$$

which has the largest root $x_L = a$. Substitute $x_L = a$ into $S(q)$ for measured Lorentzian spectrum and compare with the integral spectral density for Lorentzian noise (Section 49).

41. The Convolution Theorem

In previous sections, the spectral density was used to describe the average properties of a random noise signal. This description is based in the frequency domain and the spectral density is measured by filtering. An alternative description of random noise, based in the time domain, is called the correlation function. The spectral density and the correlation function are related through the convolution theorem.

The convolution of two functions $f(x)$ and $g(x)$ is defined as the integral

$$\int_{-\infty}^{\infty} f(x)g(u - x)\, dx = h(u)$$

In words, the convolution of $f(x)$ with $g(x)$ states the following: Consider the mirror image of $g(x)$, i.e., $g(-x)$. Multiply $g(-x)$ by $f(x)$ and evaluate the area under the product. Shift $g(-x)$ by u and repeat the operation. Area plotted against u is the convolution of $f(x)$ and $g(x)$.

An alternate form of notation for convolution is obtained by changing the variable of integration to a dummy variable u, thus

$$\int_{-\infty}^{\infty} f(u)g(x-u)\,du = h(x)$$

or

$$h(x) = f(x) * g(x)$$

which reads "$f(x)$ convolved with $g(x)$." This choice of variable leaves the result of convolution, $h(x)$, a function of the original variabile x. An excellent introduction to convolution is given in Bracewell (1965), Chap. 3.

PROBLEM. Show that the convolution operation is:

(a) commutative: $f * q = g * f$

(b) associative: $e * (f * g) = (e * f) * g$

(c) distributive: $(e + f) * g = e * g + e * f$

PROBLEM. Consider the truncated exponential functions given, for $t > 0$, by

$$f(t) = e^{-t/\tau_1} \quad \text{and} \quad g(t) = e^{-t/\tau_2}$$

Both $f(t)$ and $g(t) = 0$ when $t < 0$. Plot $f(t)$ and $g(-t)$. Show that their convolution is

$$h(t) = \tau_1 \tau_2 \frac{e^{-t/\tau_1} - e^{-t/\tau_2}}{\tau_1 - \tau_2}$$

Show that when $\tau_1 = \tau_2$, $f * g = te^{-t/\tau}$. Plot this function for $\tau = 1$ msec and $0 < t < 10$ msec.

The units of the convolution integral differ from the units of the original functions. In the above problem, for example, if g and f have the units of V, then $g * f$ has the units of V² sec.

The convolution theorem states the following: If $e_1(t)$ has the Fourier transform $\bar{e}_1(f)$, and $e_2(t)$ the Fourier transform $\bar{e}_2(f)$, then

$$\overline{e_1(t) * e_2(t)} = \bar{e}_1(f)\bar{e}_2(f)$$

The Fourier transform of the convolution is the product of the Fourier transforms.

The theorem may be demonstrated by expanding the above expression:

$$\int_{-\infty}^{\infty}\left[\int_{-\infty}^{\infty}e_1(u)e_2(t-u)\,du\right]e^{-i\omega t}\,dt = \int_{-\infty}^{\infty}e_1(u)\int_{-\infty}^{\infty}e_2(t-u)e^{-i\omega t}\,dt\,du$$

The shift theorem states

$$\int_{-\infty}^{\infty}e_2(t-u)e^{-i\omega t}\,dt = \int_{-\infty}^{\infty}e_2(t-u)e^{-i\omega(t-u)}e^{-i\omega u}\,d(t-u)$$

$$= \bar{e}_2(f)e^{-i\omega u}$$

where u is regarded as constant over t integration. Using the shift theorem,

$$\int_{-\infty}^{\infty}e_1(u)\int_{-\infty}^{\infty}e_2(t-u)e^{-i\omega t}\,dt\,du = \int_{-\infty}^{\infty}e_1(u)\bar{e}_2(f)e^{-i\omega u}\,du$$

$$= \bar{e}_1(f)\bar{e}_2(f)$$

PROBLEM. Demonstrate the convolution theorem for the case

$$e_1(t) = e_2(t) = e^{-t/\tau}, \qquad t > 0$$

SOLUTION. From a previous result $e_1 * e_2 = te^{-t/\tau}$. It is required to obtain

$$\int_{-\infty}^{\infty}te^{-t/\tau}e^{-i\omega t}\,dt \quad \text{and} \quad \int_{-\infty}^{\infty}e^{-t/\tau}e^{-i\omega t}\,dt$$

From CF # I-438 (note that $p \equiv i\omega$)

$$\int_{-\infty}^{\infty}e^{-\beta g}e^{-i\omega g}\,dg = \frac{1}{i\omega + \beta}, \qquad 0 < g$$

In the same table (# 442) one finds

$$\int_{-\infty}^{\infty}\pm ge^{\mp\beta g}e^{-i\omega g}\,dg = \frac{1}{(i\omega \pm \beta)^2}, \qquad 0 < \pm g$$

Applying these formulas to our case gives

$$\int_{-\infty}^{\infty}te^{-t/\tau}e^{-i\omega t}\,dt = \frac{1}{(i\omega + 1/\tau)^2} = \frac{\tau^2}{(1 + i\omega\tau)^2}$$

and

$$\left(\int_{-\infty}^{\infty}e^{-t/\tau}e^{-i\omega t}\,dt\right)\left(\int_{-\infty}^{\infty}e^{-t/\tau}e^{-i\omega t}\,dt\right) = \left(\frac{\tau}{1 + i\omega\tau}\right)\left(\frac{\tau}{1 + i\omega\tau}\right)$$

The last two expressions are equivalent.

Note that in this case the Fourier transform is a complex number. Its real and imaginary parts are

$$\frac{1 - \omega^2\tau^2}{(1 - \omega^2\tau^2)^2 + 4\omega^2\tau^2} \quad \text{and} \quad \frac{-2i\omega\tau}{(1 - \omega^2\tau^2)^2 + 4\omega^2\tau^2}$$

42. The Correlation Theorem

An operation closely related to convolution is correlation. The correlation of two functions $f(x)$ and $g(x)$ is defined as

$$\int_{-\infty}^{\infty} f(u)g(u + u) \, du = c(x)$$

In this case, $f(u)$ is multiplied, not by the mirror image of $g(u)$, but by $g(u)$ itself; as before, the area of the product is found as a function of their relative displacement x.

PROBLEM. Show that the correlation operation is not commutative, i.e.,

$$\int_{-\infty}^{\infty} f(u)g(x + u) \, du \neq \int_{-\infty}^{\infty} f(x + u)g(u) \, du$$

Show that the convolution operation is commutative:

$$\int_{-\infty}^{\infty} f(u)g(x - u) \, du = \int_{-\infty}^{\infty} f(x - u)g(u) \, du$$

PROBLEM. Show that the correlation theorem (analogous to the convolution theorem) is as follows: If $e_1(t)$ and $e_2(t)$ have Fourier transforms $\bar{e}_1(f)$ and $\bar{e}_2(f)$, then

$$\bar{c}(f) = \overset{\frown}{e_1}(f)\bar{e}_2(f)$$

where the crossed bar stands for the plus-i Fourier transform and the uncrossed bar for the minus-i transform. In this notation, the convolution theorem reads

$$\bar{h}(f) = \bar{e}_1(f)\bar{e}_2(f)$$

PROBLEM. Consider the truncated exponential functions

$$f(t) = e^{-t/\tau_1} \quad \text{and} \quad g(t) = e^{-t/\tau_2}$$

Both $f(t)$ and $g(t) = 0$ for $t < 0$. Show that the convolution of these functions is

$$c(t) = \tau_1\tau_2 \frac{e^{t/\tau_1} - e^{-t/\tau_2}}{\tau_1 + \tau_2}$$

PROBLEM. Demonstrate the convolution theorem for

$$f(t) = g(t) = e^{-t/\tau}$$

PROBLEM. Show that

$$c(x) = \int_{-\infty}^{\infty} f(u)f(x+u)\,du \le \int_{-\infty}^{\infty} f^2(u)\,du$$

The correlation of a function with itself is maximum at $c(0)$ since the area under the self product is maximum for no relative shift. [A proof is given in Bracewell (1965), p. 48.]

PROBLEM. Show that

$$\int_{-\infty}^{\infty} f(u)f(x+u)\,du = \int_{-\infty}^{\infty} f(u)f(-x+u)\,du$$

i.e., the correlation of a function with itself is even.

The convolution and correlation operations are cross operations which generally deal with two different functions. The self-, or autoconvolution, is given by

$$h(x) = \int_{-\infty}^{\infty} f(u)f(x-u)\,du$$

The autocorrelation is given by

$$c(x) = \int_{-\infty}^{\infty} f(u)f(x+u)\,du$$

By definition, the plus-i Fourier transformation of $e(t)$ is

$$\overset{+}{e}(f) = \int_{-\infty}^{\infty} e(t)e^{i\omega t}\,dt = \bar{e}^*(f)$$

when $e(t)$ is real. Therefore

$$\overset{+}{e}(f)\bar{e}(f) = \bar{e}^*(f)\bar{e}(f) = |\,\bar{e}(f)\,|^2$$

In this case, the correlation theorem becomes

$$\bar{c}(f) = |\bar{e}(f)|^2$$

Taking the plus-*i* transformation of both sides of this equation gives

$$c(t) = \int_{-\infty}^{\infty} |e(f)|^2 e^{i\omega t}\, df$$

PROBLEM. Show that when $t = 0$, the above form of the correlation theorem reduces to Rayleigh's theorem (Section 34).

PROBLEM. Show that an alternative form of the correlation theorem is

$$c(t) = 2 \int_{0}^{\infty} |\bar{e}(f)|^2 \cos \omega t\, df$$

when $|\bar{e}|^2$ is an even function.

In Section 34 the spectral density of a signal $e(t)$ extending over time period T was defined by

$$S(f) = \frac{2}{T} |\bar{e}(f)|^2$$

Recall that this formula applies for $f > 0$. Inserting this definition into the result of the last problem gives

$$c(t) = T \int_{0}^{\infty} S(f) \cos \omega t\, df \qquad (42.1)$$

The self-correlation of a function is equal to the cosine transformation of the spectral density of the function multiplied by its duration. In this expression, t extends over all time, $-\infty < t < \infty$, whereas f extends only over positive values, $0 < f < \infty$.

Consider an arbitrary function $e(t)$, which is zero everywhere except in the range $0 < t < T$ (Figure 42.1). Note that $e(t)$ is represented in the

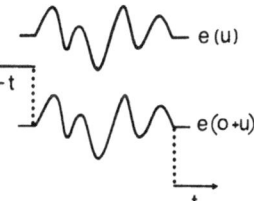

Figure 42.1. A schematic representation of $e(t)$ in the dummy variable space u with t as the shift variable.

dummy variable space u, and t is the shift variable. For self-convolution,

$$c(t) = c(-t)$$

The practical implication of this result is that one obtains the same self-correlation by shifting to the right as to the left. To obtain the self-correlation over all t, only the half-operation is done (say for $t > 0$). The result is then reflected about the origin to obtain $c(t)$ for all t. This fact usually results in the self-correlation being written only for positive values of t.

It is evident from the above example that the self-product of a function of finite duration vanishes when the shift exceeds the duration (Figure 42.1). Therefore

$$c(t) = \int_{-\infty}^{\infty} e(u)e(t+u)\, du = \int_{0}^{T} e(u)\, d(t+u)\, du, \qquad t > 0$$

Combining this expression with Equation (42.1), one obtains for $t > 0$ and $f > 0$,

$$\frac{1}{T} \int_{0}^{T} e(u)e(t+u)\, du = \int_{0}^{\infty} S(f) \cos \omega t\, df$$

The expression on the left-hand side is called the autocorrelation function $C(t)$. Thus

$$C(t) = \frac{1}{T} \int_{0}^{T} e(u)e(t+u)\, du$$

PROBLEM.   Show that the autocorrelation function is also given by

$$C(t) = \int_{-\infty}^{\infty} SS(f) e^{i\omega t}\, df$$

where $SS(f)$ is the double-sided spectral density defined for $-\infty < f < \infty$.

If $e(t)$ has the units of V then $C(t)$ is in V². The maximum value of the autocorrelation occurs for zero shift; $C(0)$ is identical to the variance:

$$C(0) = \frac{1}{T} \int_{0}^{T} e^2(u)\, du = \frac{1}{T} \int_{0}^{T} e^2(t)\, dt$$

The choice of u as a dummy variable is, of course, arbitrary. It is common to use τ for the shift variable and retain t for the function being correlated, thus

$$C(\tau) = \frac{1}{T} \int_0^T e(t)e(\tau + t)\, dt$$

In this notation, for $\tau > 0$ and $f > 0$,

$$C(\tau) = \int_0^\infty S(f) \cos \omega\tau \, df$$

PROBLEM. Show that for $\tau > 0$ and $f > 0$, the inversion formula for Fourier transformation

$$\overleftarrow{x}(f) = x(\tau)$$

leads to the expression

$$S(f) = 4 \int_0^\infty C(\tau) \cos \omega\tau \, d\tau$$

43. Measurement of Correlation Functions

The correlation function of any signal may be obtained by the shift and multiply operation implied by the definition

$$C(\tau) = \frac{1}{T} \int_0^T e(t)e(\tau + t)\, dt$$

For example, consider the function

$$e(t) = e_0 \sin \omega_0 t$$

for $0 < t < T$. As in our previous examples (Section 33), ω_0 is the frequency of the sine wave and is related to the period by

$$\omega_0 = 2\pi f_0 = \frac{2\pi}{T}$$

The correlation function for the sine wave of finite duration is therefore

$$C(\tau) = \frac{e_0^2}{T} \int_0^T \sin \omega_0 t \sin \omega_0(\tau + t)\, dt$$

From the identity $\sin(A + B) = \sin A \cos B + \cos A \sin B$, the expression for the correlation function becomes

$$C(\tau)\frac{e_0^2}{T}\left(\cos \omega_0\tau \int_0^T \sin^2 \omega_0 t \, dt + \sin \omega_0\tau \int_0^T \sin \omega_0 t \cos \omega_0 t \, dt\right)$$

The integrals are of standard form. Let $u = \omega_0 t$, then

$$\int_0^T \sin^2 \omega_0 t \, dt = \frac{1}{\omega_0}\int_0^T \sin^2 u \, du = \frac{1}{\omega_0}\left(\frac{u}{2} - \frac{1}{4}\sin 2u\right)_0^T$$

and

$$\int_0^T \sin \omega_0 t \cos \omega_0 t \, dt = \frac{1}{\omega_0}\int_0^T \sin u \cos u \, du = \frac{1}{\omega_0}\left(\frac{1}{2}\sin^2 u\right)_0^T$$

Since $\omega_0 = 2\pi/T$, the second integral vanishes. Finally

$$C(\tau) = \frac{e_0^2}{\omega_0 T}\cos \omega_0\tau\left[\left(\frac{\omega_0 t}{2} - \frac{1}{4}\sin 2\omega_0 t\right)_0^T\right]$$

The quantity in brackets is equal to π. Therefore, the correlation function of a sine wave that is zero everywhere except for $0 \leq t \leq T$ is a cosine wave

$$C(\tau) = \frac{e_0^2}{2}\cos \omega_0\tau$$

zero everywhere except $0 \leq \tau \leq T$.

PROBLEM. Consider a sine wave $e_0 \sin \omega_0 t$ defined for $-T \leq t \leq T$. Derive the correlation function for this function.

PROBLEM. Calculate the spectral density of $e(t) = e_0 \sin \omega_0 t$ ($0 \leq t \leq T$) from the correlation function. Compare with a direct calculate from $e(t)$.

SOLUTION. The spectral density is given by

$$S(f) = 4\int_0^\infty C(\tau)\cos \omega\tau \, d\tau = 2e_0^2\int_0^T \cos \omega_0\tau \cos \omega\tau \, d\tau$$

By direct calculation,

$$\bar{e}(f) = \int_{-\infty}^\infty e_0 \sin \omega_0 t e^{-i\omega t} \, dt = e_0\int_0^T \sin \omega_0 t(\cos \omega t - i \sin \omega t) \, dt$$

and

$$S(f) = \frac{2}{T} \mid \bar{e}(t) \mid^2 \qquad (0 < f < \infty)$$

(see Section 34). The problem is completed by calculating the integrals and substituting $\bar{e}(t)$ into the last expression.

PROBLEM. For the problem above, show that $S(f_0) = B^2 T$ is an immediate result, since in this case

$$S(f_0) = 2B^2 \int_0^T \cos^2 \omega_0 \tau \, d\tau$$

and

$$\bar{e}(f_0) = -ie_0 \int_0^T \sin^2 \omega_0 t \, dt$$

PROBLEM. Consider the correlation function for $e(t) = e_0 \sin \omega_0 t$ for arbitrary length signal $0 < t < T_1$, where $T_1 > T$ is considered as a variable. Investigate the case $T_1 \rightarrow \infty$.

The examples above illustrate that a finite signal in real time (t) has a finite correlation function in correlation time (τ); the spectral density, however, extends to all frequencies. The remarkable feature is that two such apparently diverse operations (shifting in the time domain versus filtering in the frequency domain) provide the same description of an arbitrary signal. The two descriptions, $C(\tau)$ and $S(f)$, are regarded as the same since one may be transformed into the other. The basic reason behind the similarity of shifting and filtering operations is found in the shift theorem (Section 41); the shift theorem states that for arbitrary $e(t)$,

$$\int_{-\infty}^{\infty} e(t - \tau)e^{-i\omega t} \, dt = \bar{e}(f)e^{-i\omega\tau}$$

where $\bar{e}(f)$ is the Fourier transform of $e(t)$. The operation on the left is performed on a time-shifted version of $e(t)$. On the right, multiplication of $\bar{e}(f)$ by $e^{-i\omega\tau}$ corresponds to the filtering operation.

An Approximate Method for Obtaining Correlation Functions. The correlation function of an arbitrary signal may be approximated by the following method. Consider the signal to be given by the set of values

$$e_0, e_1, e_2, e_3, \dots, e_n$$

where each e_i represents the value of $e(t)$ at t_i. [Do not confuse the use of e_0 here—to represent the value of the signal at $t = 0$—with its previous use as the maximum value of a signal. There is no restriction that the first point in $e(t)$ be the largest value the signal may attain.] The values e_i are all equally spaced.

As an example, consider a signal $e(t)$ represented by four equally spaced numbers e_0, e_1, e_2, and e_3. The correlation function

$$C(\tau) = \frac{1}{T} \int_0^T e(t)e(\tau + t) \, dt$$

is approximated by forming the sum of the self-product after shifting, thus

Δt

$\overgroup{}$

$e_0 e_1 e_2 e_3$

$e_0 e_1 e_2 e_3 \qquad\qquad e_0 e_0 + e_1 e_1 + e_2 e_2 + e_3 e_3$

$\quad e_0 e_1 e_2 e_3 \qquad\qquad\quad e_1 e_0 + e_2 e_1 + e_3 e_2$

$\undergroup{}$
$\Delta \tau \; e_0 e_1 e_2 e_3 \qquad\qquad\qquad e_2 e_0 + e_3 e_1$

$\quad\quad e_0 e_1 e_2 e_3 \qquad\qquad\qquad\qquad e_3 e_0$

In each case, the product is made between the shifted set of numbers and the original, unshifted set. The four points of the correlation function are

$$C_0 = (e_0 e_0 + e_1 e_1 + e_2 e_2 + e_3 e_3) \frac{\Delta t}{T}$$

$$C_1 = (e_1 e_0 + e_2 e_1 + e_3 e_2) \frac{\Delta t}{T}$$

$$C_2 = (e_2 e_0 + e_3 e_1) \frac{\Delta t}{T}$$

$$C_3 = (e_3 e_0) \frac{\Delta t}{T}$$

Δt is the spacing between the values e_0, e_1, etc. $\Delta \tau$ is the shift. In this example, the correlation points C_0, C_1, etc., are separated by spacing $\Delta \tau = \Delta t$.

Evidently, as the spacing between points in real time (Δt) becomes smaller, the approximation to the integral becomes better. The correlation shift ($\Delta \tau$) is not required to be equal to Δt; however, it makes little sense for $\Delta \tau$ to be less than Δt, and information is lost if it is greater.

EXAMPLE.   Consider the function $e(t) = B \cos \omega_0 t$ between $0 < t < T$ represented by the points

$t = 0$	$T/4$	$T/2$	$3T/4$	T
$e(t) = B$	0	$-B$	0	B

In this case, $\Delta t / T = 1/4$. The correlation function is represented by the points

$\tau = 0$	$T/4$	$T/2$	$3T/4$	T
$C(\tau) = \frac{3}{4} B^2$	0	$-\frac{1}{2} B^2$	0	$\frac{1}{4} B^2$

PROBLEM.   Repeat the above example for $\Delta t = T/8$ and $T/16$. Compare the point correlation functions for these two cases with the theoretical correlation function $(B^2/2) \cos \omega_0 t$.

PROBLEM.   Show that the convolution integral is approximated by forming the serial product for the set of numbers e_0, e_1, e_2, e_3

$$e_0 e_0$$
$$e_0 e_1 + e_1 e_0$$
$$e_0 e_2 + e_1 e_1 + e_2 e_0$$
$$e_0 e_3 + e_1 e_2 + e_2 e_1 + e_3 e_0$$

(See Bracewell, 1965, p. 30.)

44. Correlation Functions: Examples

The correlation function was derived from the convolution theorem in Sections 41 and 42. The correlation function is an alternative to the spectral density for describing the average properties of a random noise signal. The two functions are related by the equations

$$C(\tau) = \int_0^\infty S(f) \cos \omega \tau \, df \tag{44.1}$$

$$S(f) = 4 \int_0^\infty C(\tau) \cos \omega \tau \, d\tau \tag{44.2}$$

These are the measured quantities. It is implied that $0 < \tau < \infty$ and

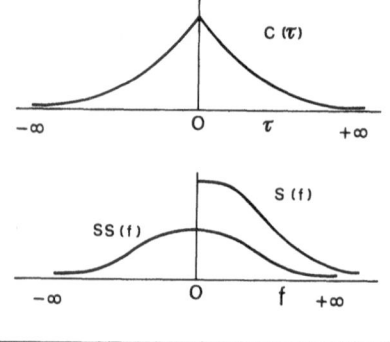

Figure 44.1. Diagram to illustrate the symmetry of the correlation function (C) and the relationship between the double-sided spectral density (SS) and the single-sided spectral density (S). Measured functions are $C(\tau)$, $\tau \geq 0$ and $S(f)$, $f \geq 0$.

$0 < f < \infty$; thus $C(\tau)$ and $S(f)$ are single-sided functions. Note that the double-sided correlation function is simply $C(\tau)$ reflected about the vertical axis; however, the double-sided spectral density is one-half the single-sided spectral density. This is illustrated in Figure 44.1.

The variable τ is used in the correlation function to distinguish the function from one dependent on real time. Thus, the noise signal $e(t)$ has correlation function $C(\tau)$. It is evident that $C(\tau)$ represents average (not instantaneous) properties of the signal. For example, if $\tau = 0$

$$C(\tau) = \frac{1}{T} \int_0^T e(t)e(\tau + t)\, dt$$

becomes

$$C(0) = \frac{1}{T} \int_0^T e^2(t)\, dt = \sigma^2$$

which is the variance of the noise. [Note that the correspondence between $C(0)$ and σ^2 also follows from the first formula of this section, since $\cos \omega\tau = 1$ when $\tau = 0$.]

Each point in $C(\tau)$ is a variance, just as each point in $S(f)$ is a variance (divided by a narrow bandwidth at that frequency). The difference is that each point in $C(\tau)$ contains information from the entire signal $e(t)$; points in the correlation function are highly correlated with one another. However, each point in $C(\tau)$ contains new information as a result of the shift and multiply operation. The spectral density derives from a filter and multiply operation. Ideally, each point in the spectral density is uncorrelated with every other point and contains independent information about $e(t)$.

Although $C(\tau)$ and $S(f)$ result from rather different operations on $e(t)$, the two contain the same average information, and one description may always be derived from the other. Some examples will serve to illustrate this point.

a. *White Noise*

Consider the spectral density (e.g., A has the units of V^2/Hz)

$$S(f) = A$$

Then

$$C(\tau) = \int_0^\infty A \cos \omega\tau \, df$$

Recall the definition of the delta function (Section 25)

$$\delta(\tau) = \int_{-\infty}^\infty e^{-i\omega\tau} \, df$$

$$= \int_{-\infty}^\infty (\cos \omega\tau - i \sin \omega\tau) \, df$$

Since the sine is odd and the cosine is even, the delta function may also be written (for $0 < f < \infty$),

$$\delta(\tau) = 2 \int_0^\infty \cos \omega\tau \, df$$

The correlation function corresponding to white noise is a delta function at the origin

$$C(\tau) = \tfrac{1}{2}A \, \delta(\tau)$$

PROBLEM. Show that the above expression implies $S(f) = A$ from Equation (44.2).

b. *1/f Noise*

Consider the spectral density (e.g., B has the units of V^2)

$$S(f) = B/f$$

Then

$$C(\tau) = \int_0^\infty \frac{B}{f} \cos \omega\tau \, df$$

Let $u = \omega\tau$. Since $\omega = 2\pi f$, the correlation function for $1/f$ noise may also be written

$$C(\tau) = B \int_0^\infty \frac{\cos u}{u} \, du$$

TABLE 44.1. A Short Table of the Cosine Integral (Defined in the Text) as a Function of the Limit of Integration x

$-\text{Ci}(x)$	x
∞	0
4.0280	0.01
3.3349	0.02
2.4191	0.05
1.7279	0.10
1.0422	0.20
0.1779	0.50
-0.3374	1.00

The integral on the right-hand side is related to the cosine integral $\text{Ci}(x)$, where x is the lower limit of integration. $\text{Ci}(x)$ is defined as

$$\text{Ci}(x) = -\int_x^\infty \frac{\cos u}{u}\, du$$

$\text{Ci}(x)$ is a tabulated function; a complete table is given in Jahnke and Emde (1945), Table I. A short version is given in Table 44.1. $\text{Ci}(x)$ is a decaying function of x. The first zero crossing occurs for

$$0.61 < x < 0.62$$

The function continues in a damped oscillation about zero, with a second zero crossing for $x \simeq 3.4$, a third for $x \simeq 6$, and so on.

Consider $1/f$ noise with a lower cutoff frequency f_0; thus

$$S(f) = B/f, \qquad f_0 < f < \infty$$

The correlation function is

$$C(\tau) = -B\,\text{Ci}(2\pi f_0 \tau)$$

PROBLEM. Let $B = 10^{-8}$ V². Let the lower cutoff frequency for $1/f$ noise be $f_0 = 1$ Hz. Plot the correlation function for

$$\frac{0.01}{2\pi} < \tau < \frac{1.0}{2\pi}\ \text{sec}$$

Repeat for $f_0 = 0.5$ and 2 Hz.

PROBLEM.   Show that $C(\tau) = -B\,\mathrm{Ci}(2\pi f_0\tau)$ implies $S(f) = B/f$ from Equation (44.2).

c. Lorentzian Noise

Consider the spectral density (e.g., K has the units of V^2)

$$S(f) = \frac{K\theta}{1 + \omega^2\tau^2}$$

The correlation function is

$$C(\tau) = \int_0^\infty \frac{K\theta}{1 + \theta^2\omega^2}\cos\omega\tau\,df$$

The right-hand side of this equation is nearly in the standard form (BH # 160-5):

$$\int_0^\infty \frac{\cos px\,dx}{q^2 + x^2} = \frac{\pi}{2q}e^{-pq}$$

PROBLEM.   Use the above integral to show that the correlation function of Lorentzian noise is given by

$$C(\tau) = \tfrac{1}{4}Ke^{-\tau/\theta}$$

where $S(0) = K\theta$ and θ is related to the cutoff frequency f_θ by

$$f_\theta = \frac{1}{2\pi\theta}$$

PROBLEM.   Show by direct integration that the variance of Lorentzian noise is

$$\sigma^2 = \int_0^\infty S(f)\,df = \tfrac{1}{4}K$$

PROBLEM.   Show that $C(\tau) = \tfrac{1}{4}Ke^{-\tau/\theta}$ implies $S(f) = K\theta/(1 + \omega^2\theta^2)$ from Equation (44.2).

45. *Integral Spectra*

In Section 40 the effect of filter bandwidth on narrow-band estimates of spectral densities was discussed for three important cases. If the narrow-band filter is the Q filter, the spectral density estimate is

$$S(q) = \frac{\int_0^\infty S(f)\,|\,Q\,|^2\,df}{\int_0^\infty |\,Q\,|^2\,df}$$

The denominator is the constant factor $\pi q/2\mathcal{Q}$. The center frequency of the analyzing filter is considered a continuous variable q. For $1/f$ noise

$$S(f) = \frac{B}{f}, \qquad S(q) = \frac{B}{q}\,F(\mathcal{Q})$$

where

$$F(\mathcal{Q}) = \frac{\mathcal{Q}}{\pi ir}\ln\!\left(\frac{s-ir}{s+ir}\right), \qquad \mathcal{Q} < \frac{1}{2}$$

$$= \frac{2}{\pi} = 0.6366\cdots, \qquad \mathcal{Q} = \frac{1}{2}$$

$$= \frac{\mathcal{Q}}{r}\left(1 + \frac{2}{\pi}\tan^{-1}\frac{s}{r}\right), \qquad \mathcal{Q} > \frac{1}{2}$$

and $r = (4\mathcal{Q}^2 - 1)^{1/2}$, $s = 2\mathcal{Q}^2 - 1$

Figure 45.1 is a plot of $F(\mathcal{Q})$ against $\mathcal{Q}$. The bottom curve is for

$$0 \le \mathcal{Q} \le 4$$

The top curve is for

$$4 \le \mathcal{Q} \le 32$$

The value $\mathcal{Q} = 1/2$ is shown as a dashed line. The spectral shape of $1/f$ noise is unchanged by this operation. For

$$\mathcal{Q} > 8, \qquad F(\mathcal{Q}) > 0.98$$

which implies less than 2% error in the amplitude of the measured $1/f$ noise for $\mathcal{Q} > 8$.

To introduce the integral spectrum, recall that the operation implied by

$$\int_0^\infty S(f)\,|\,Q\,|^2\,df$$

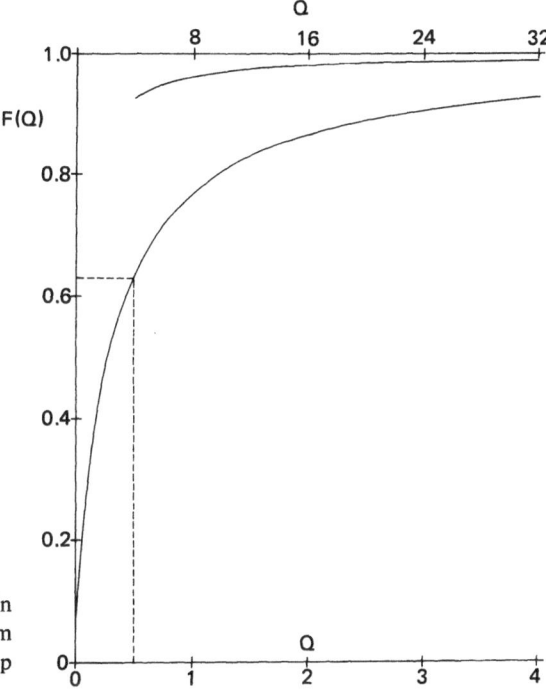

Figure 45.1. $F(\mathcal{Q})$, defined in the text, for $0 \leq \mathcal{Q} \leq 4$ (bottom curve) and $4 \leq \mathcal{Q} \leq 32$ (top curve).

stands for filtering, multiplying, and averaging in the time domain. In order to see this, consider the outcome of a single filtering operation. For an arbitrary signal $e(t)$ between $0 < t < T$, the Fourier transform is $\bar{e}(f)$; $\infty < f < \infty$. The output from any one filter is

$$\bar{e}_q = \bar{e}(f)Q$$

where Q is the transfer function of the filter. Taking the square modulus, and integrating over all frequencies, gives

$$\int_{-\infty}^{\infty} |\bar{e}(f)Q|^2 \, df$$

Confining the integration to positive frequencies (real filters) and dividing by the signal duration T gives

$$\frac{2}{T} \int_0^{\infty} |\bar{e}(f)Q|^2 \, df = \int_0^{\infty} S(f) |Q|^2 \, df$$

since

$$|\bar{e}Q|^2 = |\bar{e}|^2 |Q|^2$$

By Rayleigh's theorem (Section 34)

$$\int_{-\infty}^{\infty} | \bar{e}(f)Q |^2 \, df = \int_{0}^{T} e_q^2(t) \, dt$$

where $e_q(t)$ is the filter output, $0 < t < T$. Therefore

$$\int_{0}^{\infty} S(f) \, | \, Q \, |^2 \, df = \frac{1}{T} \int_{0}^{T} e_q^2(t) \, dt$$

and

$$S(q) = \frac{(1/T) \int_{0}^{T} e_q^2(t) \, dt}{\int_{0}^{\infty} | \, Q \, |^2 \, df}$$

The numerator is the variance at the output of a Q filter with center frequency q.

It should be evident that the last step—division by the filter bandwidth—is superfluous; all of the analysis occurs through filtering, multiplying, and averaging the signal. Therefore, one may define an integral spectrum by the numerator only; thus

$$I(q) = \int_{0}^{\infty} S(f) \, | \, Q \, |^2 \, df$$

For example,

$$S(f) = \frac{B}{f} \rightarrow I(q) = \frac{\pi B}{2Q} F(\mathcal{Q})$$

$1/f$ noise appears flat as an integral spectrum. Conversely, if one obtains the same variance at each filter output (i.e., for every value of q) the noise must be $1/f$. Furthermore, this result is independent of $\mathcal{Q}$. For example, if

$$\mathcal{Q} = 1/2, \qquad F(\mathcal{Q}) = 2/\pi$$

and

$$S(f) = B/f \rightarrow I(q) = 2B$$

If the noise source is a voltage, then

$$S(f) \ominus V^2/\text{Hz}$$

but

$$I(q) \ominus V^2$$

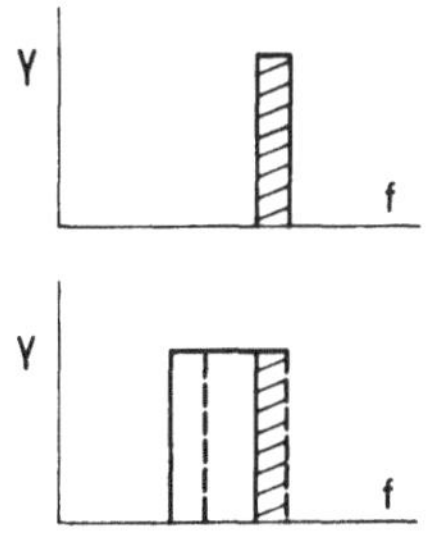

Figure 45.2. Ideal rectangular filters. In the upper diagram a single narrow-band filter is illustrated. In the lower diagram, two broadband filters are shown; the second filter (dashed lines) is moved to the right by an amount equal to the bandpass of the narrow-band filter in the top diagram.

The integral spectrum is a variance plotted against frequency; the spectral density is a variance divided by filter bandwidth plotted against frequency.

In the above example, the Q-filter analysis of $1/f$ noise was independent of $\mathcal{Q}$, hence of filter bandwidth. The only effect of filter bandwidth is to change the factor of B, but not the dependence of $I(q)$ on q. $\mathcal{Q} = 1/2$ implies a broadband filter (Section 38); since q is a continuous variable, the filters must overlap. A simple example will illustrate that no information is lost due to this overlap.

For purposes of discussion only, consider filters that are perfect rectangles in the frequency domain. There are two ways to achieve narrow-band filtering: one may construct a set of narrow filters, or a set of broadband filters which overlap everywhere except in a narrow range. This is illustrated in Figure 45.2. Y is the transfer function of these idealized filters. The shaded areas are identical in the two cases, and the narrow-band filter (Figure 45.2, top) and broadband filter (Figure 45.2, bottom) are formally equivalent. We shall prove this analytically below.

The analyzing filter may be of any form. A convenient choice is the L filter (Section 38) since it is easily constructed and has a relatively simple transfer function. Recall that an L filter is a low-pass RC filter in series with a high-pass CR filter; the transfer function (if these two stages are nonloading) has square modulus

$$|L|^2 = \frac{\omega^2\tau^2}{(1 + \omega^2\tau^2)^2} = \frac{f^2 f_0^2}{(f^2 + f_0^2)^2}$$

where $\tau = RC$ and $f_0 = 1/2\pi RC$ is the center frequency of the filter. Consider f_0 as a continuous variable l, i.e., the filters are infinitely close to one another. Then

$$|L|^2 = \frac{f^2 l^2}{(f^2 + l^2)^2}$$

and

$$I(l) = \int_0^\infty S(f)\,|L|^2\,df$$

By a previous argument, $I(l)$ results from a filtering, multiplying, and averaging operation. For a random signal $e(t)$, $0 \leq t \leq T$,

$$I(l) = \frac{1}{T} \int_0^T e_l^2(t)\, dt$$

where $e_l(t)$ is the output from the filter with center frequency l. When $e(t)$ is in V, $I(l)$ is in V².

Figure 45.3 shows a set of seven narrow-band (top) and seven broad-band (bottom) filters. The vertical scales are $|Q|^2$ and $|L|^2$, respectively. The frequency axis is logarithmic. Note that the maximum value of the Q filter, square modulus, is unity, and that

$$|L|^2_{\max} = \tfrac{1}{4} |Q|^2_{\max}$$

Recall that

$$\int_0^\infty |Q|^2\, df = \frac{\pi}{2} \frac{f_0}{Q}$$

and

$$\int_0^\infty |L|^2\, df = \frac{\pi}{4} f_0$$

The center frequencies of the filters in Figure 45.3 are

$$f_0 = 10,\ 20,\ 50,\ 100,\ 200,\ 500,\ 1000\ \text{Hz}$$

In the limit of filters infinitely close to one another,

$$f_0 \to q, \qquad Q \text{ filter}$$

$$f_0 \to l, \qquad L \text{ filter}$$

The filters in each set appear to have identical shapes on a log-frequency scale; their actual width increases proportionally with frequency.

The use of broadband filters with large overlap does not result in a loss of information. $I(l)$ contains the same information as $S(f)$. Broadband filters lose information about the spectral density of a noise source only when they do not overlap. Figure 45.4 illustrates this point. The shaded area in Figure 45.4 is larger than in Figure 45.2. Information about spectral density may be lost since the variance from each filter in Figure 45.4 contains new information from a broader range than in Figure 45.2.

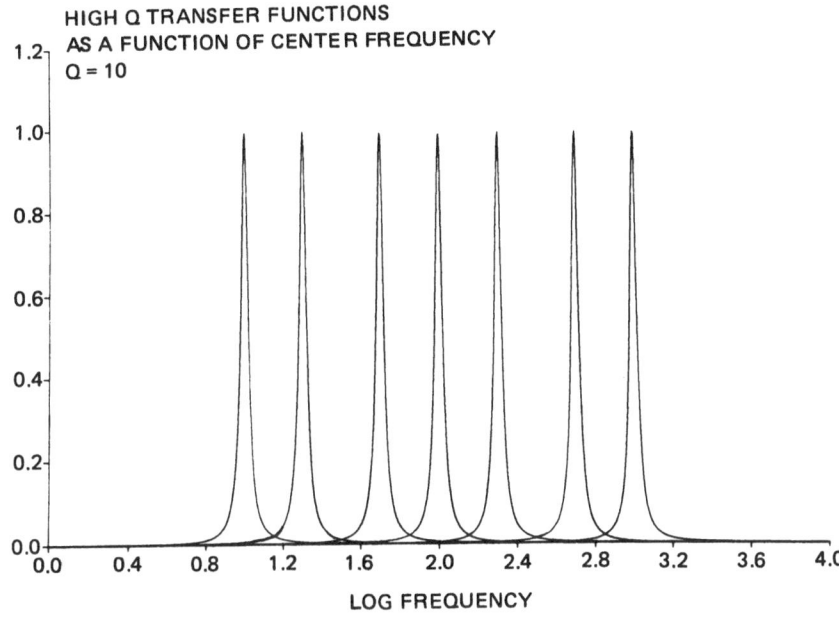

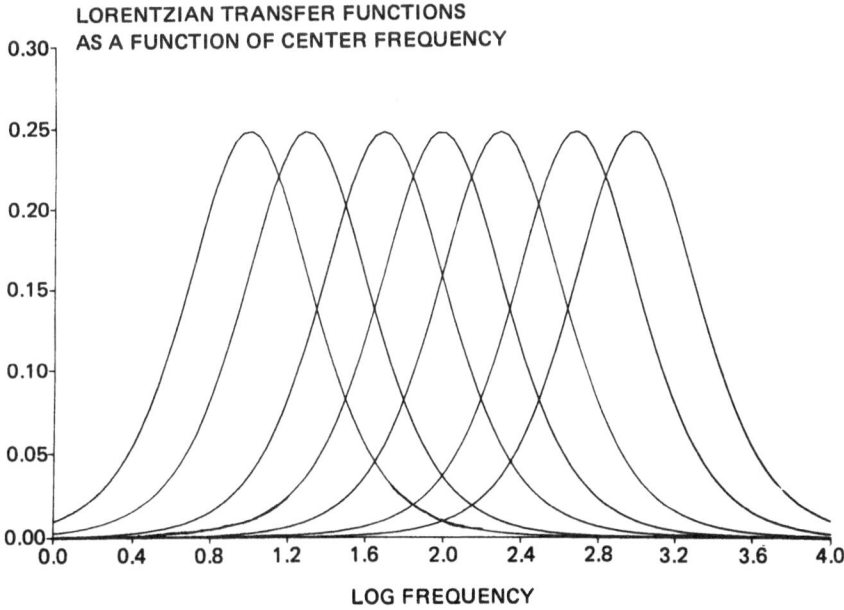

Figure 45.3. A set of narrow-band Q filters (top) compared with L filters (bottom) at the same center frequencies. In each case, the modulus square of the filter transfer function is plotted against log f. The center frequencies are 10, 20, 50, 100, 200, 500, and 1000 Hz. (Courtesy B. A. Sokol.)

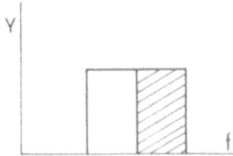

Figure 45.4. Nonoverlapping broadband filters.

Another integral transformation, the correlation function, also contains information equivalent to the spectral density. Compare the formulas

$$C(\tau) = \int_0^\infty S(f) \cos(2\pi\tau f)\, df$$

and

$$I(l) = \int_0^\infty S(f) \left(\frac{fl}{f^2 + l^2} \right)^2 df$$

In each case, the spectral density is multiplied by a specified kernel and the integral is taken over all frequencies. The first equation above may be solved for $S(f)$ and so may the second. Before deriving the inversion formula for $I(l)$, we consider the relationship between $I(l)$ and $C(\tau)$.

46. Relationship between the Integral Spectrum and the Correlation Function

Since

$$S(f) = 4 \int_0^\infty C(\tau) \cos \omega\tau\, df$$

then by definition

$$I(l) = 4 \int_0^\infty \int_0^\infty C(\tau) \cos \omega\tau \left(\frac{fl}{f^2 + l^2} \right)^2 df\, d\tau$$

$$= 4 \int_0^\infty C(\tau) \int_0^\infty \cos \omega\tau \left(\frac{fl}{f^2 + l^2} \right)^2 df\, d\tau$$

It is required to integrate

$$\int_0^\infty \cos(2\pi\tau f) \left(\frac{f/l}{1 + f^2/l^2} \right)^2 df = l \int_0^\infty \frac{r^2 \cos(2\pi\tau l r)\, dr}{(1 + r^2)^2}$$

where $r = f/l$ and $dr = df/l$. From BH $\#$ 170-8

$$\int_0^\infty \frac{x^2 \cos px}{(q^2 + x^2)^2} \, dx = \frac{\pi}{4q} (1 - pq)e^{-pq}$$

therefore

$$I(l) = 4 \int_0^\infty C(\tau) \left[\frac{\pi l}{4} (1 - 2\pi\tau l)e^{-2\pi\tau l} \right] d\tau$$

$$= \pi l \int_0^\infty (1 - 2\pi\tau l)C(\tau)e^{-2\pi\tau l} \, d\tau$$

Let $s = 2\pi l$. Then

$$I(s) = \frac{1}{2} s \int_0^\infty (1 - s\tau)C(\tau)e^{-s\tau} \, d\tau$$

The Laplace transformation of an arbitrary function $f(t)$ is defined as

$$\tilde{f}(s) = \int_0^\infty f(t)e^{-st} \, d\tau$$

By integration by parts

$$\frac{d}{ds} \tilde{f}(s) = -\int_0^\infty tf(t)e^{-st} \, dt$$

Therefore

$$I(s) = \frac{s}{2} \left[\tilde{C}(s) + s \frac{d}{ds} \tilde{C}(s) \right]$$

where $\tilde{C}(s)$ is the Laplace transform of $C(\tau)$.

A good introduction to Laplace transformation and its relationship to the Fourier transformation is given by S. Goldman (1949) and by Churchill (1972).

EXAMPLE. Consider a spectral density of the form

$$S(f) = \frac{K\theta}{1 + \omega^2\theta^2}$$

Then

$$C(\tau) = \int_0^\infty S(f) \cos \omega\tau \, df = \frac{K}{4} e^{-\tau/\theta}$$

To calculate the integral spectrum, it is required to evaluate the Laplace transform of an exponential.

By direct integration

$$\int_0^\infty e^{-\tau/\theta} e^{-s\tau}\, d\tau = \int_0^\infty e^{-(1/\theta+s)\tau}\, d\tau = \frac{1}{1/\theta + s}$$

Therefore

$$C(\tau) = \frac{K}{4} e^{-\tau/\theta} \rightarrow \tilde{C}(s) = \frac{K}{4} \frac{\theta}{1 + \theta s}$$

and

$$\frac{d}{ds}\, \tilde{C}(s) = \frac{K\theta}{4} \left[\frac{-\theta}{(1 + \theta s)^2} \right]$$

Finally, the integral spectrum of the Lorentzian spectral density is

$$I(s) = \frac{K\theta}{4} \frac{s}{2} \left[\frac{1}{1 + \theta s} - \frac{s\theta}{(1 + \theta s)^2} \right] = \frac{K\theta}{4} \frac{s}{2} \frac{1}{(1 + \theta s)^2}$$

This equation may be rewritten as a function of l, and the cutoff frequency of the Lorentzian f_θ by the definitions

$$s = 2\pi l \quad \text{and} \quad \theta = \frac{1}{2\pi f_\theta}$$

Substituting these expressions into $I(s)$ above gives

$$I(l) = \frac{K\theta\pi}{4} \frac{l}{(1 + l/f_\theta)^2}$$

This is the integral spectrum corresponding to $C(\tau) = \frac{1}{4} K e^{-\tau/\theta}$.

If $e(t)$ is a voltage, $S(f)$ is a power spectrum, $C(\tau)$ is a correlation function, and $I(l)$ is an integral spectrum, then

$$S(f) \ominus V^2/Hz \quad \text{vs. Hz}$$

$$C(\tau) \ominus V^2 \quad \text{vs. sec}$$

$$I(l) \ominus V^2 \quad \text{vs. Hz}$$

The integral spectrum has the vertical axis of the correlation function and the horizontal axis of the power spectrum. In some ways, the integral spectrum may be thought of as an intermediate between the other two functions.

47. Examples of Integral Spectra

a. White Noise

Consider the spectral density (e.g., $A \ominus V^2/Hz$)

$$S(f) = A$$

Then

$$I(l) = \int_0^\infty A\left(\frac{fl}{f^2 + l^2}\right)^2 df = Al \int_0^\infty \left(\frac{r}{1 + r^2}\right)^2 dr$$

where $r = f/l$. From BH # 170-8, (with $p = 0$ and $q = 1$)

$$\int_0^\infty \left(\frac{r}{1 + r^2}\right)^2 dr = \frac{\pi}{4}$$

Therefore, the integral spectrum corresponding to white noise is

$$I(l) = \frac{A\pi}{4} l$$

b. 1/f Noise

Consider the spectral density (e.g., $B \ominus V^2$)

$$S(f) = B/f$$

Then

$$I(l) = \int_0^\infty \frac{B}{f} \left(\frac{fl}{f^2 + l^2}\right)^2 df = B \int_0^\infty \frac{r}{(1 + r^2)^2} dr$$

From GR # 3.241-4 in the Appendix (with $u = 2$, $p = 1$, $q = 1$, $n = 1$, and $a = 2$),

$$\int_0^\infty \frac{r\, dr}{(1 + r^2)^2} = \frac{1}{2}$$

Therefore the integral spectrum corresponding to $1/f$ noise is

$$I(l) = B/2$$

c. 1/f^α Noise

Consider a spectral density of the form

$$S(f) = B/f^\alpha$$

where

$$-1 < \alpha < 3$$

The integral spectrum is

$$I(l) = \int_0^\infty \frac{B}{f^\alpha} \left(\frac{fl}{f^2 + l^2} \right)^2 df = Bl^{1-\alpha} \int_0^\infty \frac{r^{2-\alpha} \, dr}{(1 + r^2)^2}$$

where $r = f/l$. From GR # 3.241-4,

$$\int_0^\infty \frac{x^{u-1} \, dx}{(p + qx^a)^{n+1}} = \frac{1}{ap^{n+1}} \left(\frac{p}{q} \right)^{u/a} \frac{\Gamma(u/a)\Gamma(1 + n - u/a)}{\Gamma(1 + n)}$$

for

$$0 < u/a < n + 1$$

The symbol $\Gamma(\ )$ stands for the gamma function. A good introduction to the gamma function is given by Goldman (1966). If m is positive, then

$$\Gamma(m) = \int_0^\infty x^{m-1} e^{-x} \, dx$$

For integral values of m,

$$\Gamma(m) = (m - 1)!$$

For half-integral values (i.e., m integral)

$$\Gamma\left(\frac{m + 1}{2} \right) = \frac{1 \cdot 3 \cdot 5 \cdots (m - 1)}{2} \pi^{1/2}$$

To use the general expression GR # 3.241-4, set

$$p = q = 1, \qquad n = 1$$
$$a = 2, \qquad u = 3 - \alpha$$

Then

$$0 < \frac{u}{a} < n + 1 \rightarrow 0 < \frac{3 - \alpha}{2} < n + 1$$

or

$$-1 < \alpha < 3$$

which was the original condition on α. In this range, therefore,

$$\int_0^\infty \frac{r^{2-\alpha}\,dr}{(1+r^2)^2} = \frac{1}{2}\frac{\Gamma[(3-\alpha)/2]\Gamma[2-(3-\alpha)/2]}{\Gamma(2)}$$

Note that $\Gamma(2)=1$.

Using this result, the integral spectrum for B/f^α noise is

$$I(l) = \frac{B}{2}\Gamma\left(\frac{1-\alpha}{2}\right)\Gamma\left(\frac{3-\alpha}{2}\right)l^{1-\alpha}$$

$$= \frac{B}{2}b(\alpha)l^{1-\alpha}, \qquad -1 < \alpha < 3$$

The factor $b(\alpha)$ is a number whose value depends only on α. For example, when $\alpha = 0$

$$b(\alpha) = \Gamma\left(\frac{1}{2}\right)\Gamma\left(\frac{3}{2}\right) = \frac{\pi}{2}$$

and

$$I(l) = \frac{B\pi}{4}l$$

in agreement with our previous result for white noise. When $\alpha = 1$,

$$b(\alpha) = \Gamma(0)\Gamma(1) = 1$$

and

$$I(l) = \frac{B}{2}$$

in agreement with our previous result for $1/f$ noise.

The units of B will depend on α. When $\alpha = 0$, B must have the units V^2/Hz; when $\alpha = 1$, B has the units of V^2.

PROBLEM. Find the integral spectrum corresponding to $\alpha = -1, -\frac{1}{2}, \frac{1}{2}, \frac{3}{2}$, and 2. What are the units of B in each case?

d. Lorentzian Noise

Consider the spectral density

$$S(f) = \frac{K\theta}{1+\omega^2\theta^2}$$

(e.g., K has the units of V²); θ is related to the cutoff frequency by

$$f_\theta = \frac{1}{2\pi\theta}$$

The integral spectrum is

$$I(l) = \int_0^\infty \frac{K\theta}{1 + \omega^2\theta^2} \left(\frac{fl}{f^2 + l^2}\right)^2 df$$

$$= \frac{K}{2\pi} u \int_0^\infty \frac{1}{u^2 + r^2} \left(\frac{r}{1 + r^2}\right)^2 dr$$

where $u = f_\theta/l$. From **BH # 20-11**,

$$\int_0^\infty \frac{1}{u^2 + r^2} \left(\frac{r}{1 + r^2}\right)^2 dr = \frac{\pi}{4} \frac{1}{(1 + u)^2}$$

Substituting this integral into the expression for $I(l)$ gives

$$I(l) = \frac{Ku}{2\pi} \frac{\pi}{4} \frac{1}{(1 + u)^2} = \frac{K}{8} \frac{u}{(1 + u)^2}$$

Replacing u by f_θ/l, the integral spectrum for Lorentzian noise is

$$I(l) = \frac{K}{8} \frac{1/u}{(1 + 1/u)^2} = \frac{K}{8} \frac{l/f_\theta}{(1 + l/f_\theta)^2}$$

This result agrees with the example given in Section 46, where the integral spectrum of Lorentzian noise was obtained through the Laplace transformation of the correlation function.

48. Inversion of the Integral Spectrum

The general relationship between the integral spectrum and the spectral density is given by the integral equation

$$I(l) = \int_0^\infty S(f)K(l, f) df$$

where

$$K(l, f) = \frac{(f/l)^2}{[1 + (f/l)^2]^2}$$

It is required to find $S(f)$ given $I(l)$.

One approach is to transform the integral equation into a convolution integral (Section 41). (I am indebted to Mike Schlessinger for pointing out the advantage of this approach.)

Consider the transformation

$$f = e^{x/2}$$

$$l = e^{y/2}$$

Then the kernel of the integral equation, $K(l, f)$, becomes

$$K(l, f) = \frac{e^{x-y}}{(1 + e^{x-y})^2} = K(x - y)$$

The kernel is a symmetric function, i.e.,

$$K(x - y) = K(y - x)$$

The change in variable of integration also changes the limits of integration; as

$$f \to 0, \qquad x \to -\infty$$

$$f \to \infty, \qquad x \to \infty$$

and

$$df = \tfrac{1}{2} e^{x/2}\, dx$$

The expression for the integral spectrum becomes

$$I(y) = \frac{1}{2} \int_{-\infty}^{\infty} e^{x/2} S(x) K(y - x)\, dx$$

which is the standard form of the convolution integral.

The convolution theorem (Section 41) states that if $f(x)$ and $g(x)$ are convolved, i.e.,

$$\int_{-\infty}^{\infty} f(x) g(y - x)\, dx = h(y) = f(y) * g(y)$$

then

$$\bar{f}(z) \bar{g}(z) = \bar{h}(z)$$

where

$$\bar{f}(z) = \int_{-\infty}^{\infty} f(y) e^{-i2\pi yz}\, dy$$

etc., are the Fourier transforms of $f(y)$, $g(y)$, and $h(y)$. Note that the functions f and g, originally in x space, were transformed to y space to apply the theorem (see Section 41).

Solving for $\bar{f}(z)$,

$$\bar{f}(z) = \frac{\bar{h}(z)}{\bar{g}(z)}$$

The original function $f(x)$ may be retrieved from the plus-i transformation

$$f(y) = \int_{-\infty}^{\infty} \bar{f}(z) e^{i2\pi yz} \, dz$$

Since $f(y)$ is simply $f(x)$ written with y replacing x, one may also write

$$f(x) = \int_{-\infty}^{\infty} \bar{f}(z) e^{i2\pi xz} \, dz$$

Apply this result to the inversion of the integral spectrum written in y space:

$$\bar{I}(z) = \tfrac{1}{2} \overline{e^{z/2} S(z)} \, \overline{K(z)}$$

and

$$S(x) = 2e^{-x/2} \int_{-\infty}^{\infty} \frac{\bar{I}(z)}{\bar{K}(z)} e^{i2\pi xz} \, dz$$

Replacing $e^{x/2}$ by f gives the required result.

In the particular transformation we are considering (the L filter)

$$K(x) = \frac{e^x}{(1 + e^x)^2}$$

In y space, replace x by y everywhere; the Fourier transform is

$$K(z) = \int_{-\infty}^{\infty} \frac{e^y}{(1 + e^y)^2} e^{-i2\pi yz} \, dy$$

Replace the exponential by its Euler expansion; since $K(y) = K(-y)$ is an even function,

$$\bar{K}(z) = 2 \int_{0}^{\infty} \frac{e^y}{(1 + e^y)^2} \cos(2\pi yz) \, dy$$

From BH # 264-16,

$$\int_{0}^{\infty} \frac{\cos py}{(e^{qy} + 1)^2} e^{qy} \, dy = \frac{1}{q^2} \frac{p\pi}{e^{p\pi/q} - e^{-p\pi/q}}$$

Let $p = 2\pi z$ and $q = 1$. Then

$$\bar{K}(z) = \frac{4\pi^2 z}{e^{2\pi^2 z} - e^{-2\pi^2 z}} = \frac{2\pi^2 z}{\sinh(2\pi^2 z)}$$

Substituting $\bar{K}(z)$ into the above expression for $S(x)$ gives

$$S(x) = 2e^{-x/2} \int_{-\infty}^{\infty} \frac{\sinh(2\pi^2 x)}{(2\pi^2 x)} \bar{I}(z) e^{i2\pi xz} \, dz$$

This is the required result. In summary, to find the spectral density $S(f)$ from the integral spectrum $I(l)$:

(a) Change $I(l)$ to $I(y)$ by the transformation $l = e^{y/2}$.

(b) Calculate $\bar{I}(z)$ from $I(y)$.

(c) Multiply $\bar{I}(z)$ by $\sinh(2\pi^2 z)/(2\pi^2 z)$.

(d) Calculate the plus-i Fourier transform of this product.

(e) Multiply by $2e^{-x/2}$ to obtain $S(x)$.

(f) Change $S(x)$ to $S(f)$ by the transform $f = e^{x/2}$.

PROBLEM. In Section 46, the integral spectrum $I(l)$ was related to the correlation function $C(\tau)$ through

$$I(s) = \frac{1}{2} s \left[\tilde{C}(s) + s \frac{d}{ds} \tilde{C}(s) \right]$$

where $s = 2\pi l$ and

$$\tilde{C}(s) = \int_0^{\infty} C(\tau) e^{-s\tau} \, d\tau$$

Solve the differential equation for $\tilde{C}(s)$ in terms of $I(s)$. A table of inverse Laplace transformations may be used to obtain $C(\tau)$ from $\tilde{C}(s)$.

49. The Integral Spectral Density

The integral spectral density is obtained by dividing the spectral density by the filter bandwidth, in analogy with the spectral density (Section 39). For the L filter, the integral spectral density is

$$\frac{\int_0^{\infty} S(f) |L|^2 \, df}{\int_0^{\infty} |L|^2 \, df} = \frac{I(l)}{\pi l/4}$$

As an example, consider Lorentzian noise. The integral spectrum corresponding to

$$S(f) = \frac{K\theta}{1 + (f/f_\theta)^2}$$

is

$$I(l) = \frac{K\theta\pi}{4} \frac{l}{(1 + l/f_\theta)^2}$$

The integral spectral density of the Lorentzian is therefore

$$\frac{4}{\pi l} I(l) = \frac{K\theta}{(1 + l/f_\theta)^2}$$

This expression bears an interesting resemblance to $S(f)$.

The spectral density and the integral spectral density for Lorentzian noise are compared in Figure 49.1.

The upper curve in Figure 49.1 is $S(f)$; the lower curve is $4I(l)/\pi l$. For each curve,

$$K\theta = 1000, \qquad f_\theta = (1000)^{1/2}$$

f_θ is indicated by the arrow. Note that the spectral density and the integral spectral density have the same units, e.g., V^2/Hz. The two functions approach each other at low and high frequencies, but diverge near the cutoff frequency f_θ.

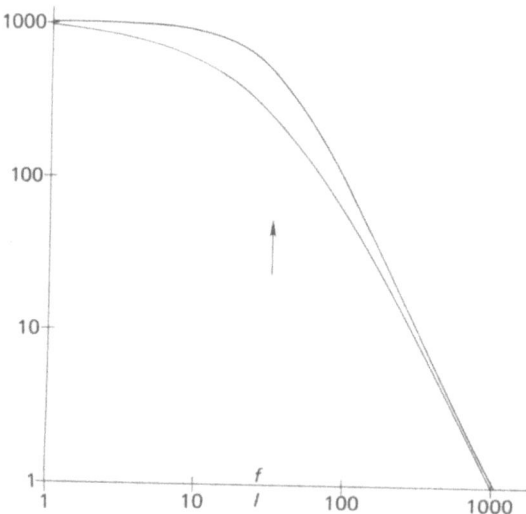

Figure 49.1. Spectral density (upper curve) and integral spectral density (lower curve) of Lorentzian noise. The arrow indicates the cutoff frequency of the Lorentzian.

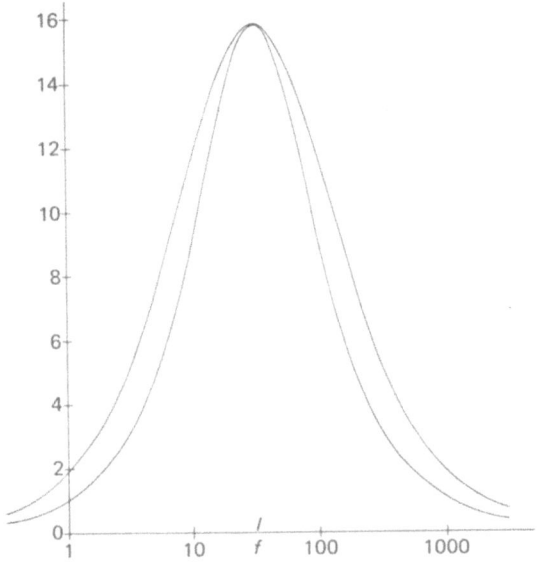

Figure 49.2. The integral spectrum (upper curve) and the function $fS(f)$ (lower curve) for a Lorentzian.

PROBLEM. Calculate the integral spectral density for $S(f) = A$ and $S(f) = B/f$.

Consider the function $fS(f)$. For Lorentzian noise

$$fS(f) = \frac{K\theta f}{1 + (f/f_\theta)^2}$$

This function is compared with

$$I(l) = \frac{\pi}{4} \frac{K\theta l}{(1 + l/f_\theta)^2}$$

in Figure 49.2. The two functions are made equal at f_θ to emphasize their difference in sharpness. The functions that are actually plotted are

$$\frac{2l}{(1 + l/f_\theta)^2} \quad \text{(upper)} \quad \propto I(l)$$

and

$$\frac{f}{1 + (f/f_\theta)^2} \quad \text{(lower)} \quad \propto fS(f)$$

with $f_\theta = (1000)^{1/2}$. Both functions have their maximum value at the cutoff frequency. $fS(f)$ and $I(l)$ have the same units, e.g., volts.

Figure 49.2 is a convenient way to plot $I(l)$ for Lorentzian noise; the amplitude is plotted linearly and the frequency logarithmically. Apart from a multiplicative constant in the amplitude, the only effect of the infinitely narrow filters used to obtain $fS(f)$ is the relative sharpness shown in Figure 49.2. Each point in $fS(f)$ is statistically independent; the points in $I(l)$, like those in $C(\tau)$, are highly correlated. The frequency resolution of the spectral density is greater than the frequency resolution of the integral spectrum. On the other hand, the amplitude resolution of the integral spectrum is greater than that of the spectral density (Section 45). Both contain the same information.

50. Multiple Lorentzians

The theory of membrane noise involves the sum of several Lorentzians (Section 74). Here we consider two Lorentzians and how the spectrum of their sum depends on their relative amplitude and cutoff frequency.

A Lorentzian spectral density may be written in the form

$$S(f) = \frac{K\theta}{1 + \omega^2\theta^2} = \frac{cf_\theta}{f_\theta^2 + f^2}$$

where

$$c = \frac{K}{2\pi} \quad \text{and} \quad f_\theta = \frac{1}{2\pi\theta}$$

The "amplitude" of a Lorentzian is often taken as its value as f approaches zero. In this case, $S(0)$ depends on f_θ, thus

$$S(0) = \frac{c}{f_\theta} = K\theta$$

On the other hand, the variance of the noise is

$$\sigma^2 = \int_0^\infty S(f)\, df = c \int_0^\infty \frac{dr}{1 + r^2}$$

where $r = f/f_\theta$. The variance depends only on c:

$$\sigma^2 = \frac{\pi}{2}\, c = \frac{K}{4}$$

Therefore, the "amplitude" and variance of Lorentzian noise are related by

$$\sigma^2 = \frac{S(0)}{4\theta}$$

A Lorentzian integral spectrum may be written in the form

$$I(l) = \frac{\pi}{4}\frac{cf_\theta l}{(f_\theta + l)^2}$$

When $l = f_\theta$, $I(l)$ is maximum. Thus,

$$I(f_\theta) = \frac{\pi}{16}c = \frac{1}{8}\sigma^2$$

Therefore, two Lorentzian integral spectra of equal "amplitude" have equal variance regardless of the cutoff frequency.

Consider the sum of two Lorentzians, i.e.,

$$S(f) = \frac{c'f_\theta'}{(f_\theta')^2 + f^2} + \frac{c''f_\theta''}{(f_\theta'')^2 + f^2}$$

The integral spectrum corresponding to this sum is

$$I(l) = \frac{\pi}{4}\left[\frac{c'f_\theta'l}{(f_\theta' + l)^2} + \frac{c''f_\theta''l}{(f_\theta'' + l)^2}\right]$$

In order to investigate the shape of $I(l)$ as a function of l, consider the equation

$$\frac{dI(l)}{dl} = 0$$

The roots of this equation are the inflection points of $I(l)$, e.g., the values of l for which $I(l)$ goes through maxima or minima. Below are given values of l that satisfy this equation for a few special cases.

Let

$$p = \frac{c''}{c'} \quad \text{and} \quad q = \frac{f_\theta''}{f_\theta'}$$

Then the condition $dI/dl = 0$, for $I(l)$ the sum of two Lorentzian integral spectra, gives the quartic equation

$$(l - f_\theta)(l + qf_\theta)^3 + pq(l - qf_\theta)(l + f_\theta)^3 = 0$$

Note that the ratio of the amplitudes of the two Lorentzian spectral densities is

$$\frac{S''(0)}{S'(0)} = \frac{p}{q}$$

The ratio of the amplitudes of the two Lorentzian integral spectra at their cutoff frequencies is

$$\frac{I''(f_\theta'')}{I'(f_\theta')} = p$$

This is also the ratio of the variance of the two noise components.

a. Equal Cutoff Frequencies (q = 1), Arbitrary Amplitude

In this case, the quartic equation becomes

$$(l - f_\theta)(l + f_\theta)^3(1 + p) = 0$$

The only real positive root is

$$l = f_\theta$$

which corresponds to the maximum value of $I(l)$. The sum of two Lorentzians with equal cutoff frequencies has a maximum value at that frequency, independent of their relative amplitude.

b. Different Cutoff Frequencies (q = 3), Same Variance (p = 1)

In this example,

$$f_\theta'' = 3f_\theta' \quad \text{and} \quad S''(0) = \tfrac{1}{3}S'(0)$$

For purposes of calculation, let $f_\theta' = 1$ Hz. Substituting $p = 1$ and $q = 3$ into the quartic equation gives

$$(l - 1)(l + 3)^3 + 3(l - 3)(l + 1)^3 = 0$$

or

$$l^4 + 2l^3 - 6l - 9 = 0$$

By inspection, a root of this equation is

$$l = 3^{1/2} = 1.732$$

This result could have been predicted from geometric considerations: $p = 1$ implies that the two components have the same amplitude at their respective cutoff frequencies; thus

$$I''(f_\theta'') = I'(f_\theta')$$

Since I'' and I' are symmetric on a log l plot, their sum will give a maximum (or minimum) halfway between f_θ'' and f_θ' on a logarithmic scale. For

$$f_\theta' = 1 \text{ Hz}, \qquad f_\theta'' = 3 \text{ Hz}$$

the midpoint on a log l scale occurs when

$$l = (1 \times 3)^{1/2}$$

or the geometric mean of the two frequencies.

PROBLEM. Show that the integral spectrum of a Lorentzian is symmetric on a log l scale. Hint: Let $u = \log l$ and $u_\theta = \log f_\theta$. Consider the two cases

$$u = u_\theta \pm v$$

PROBLEM. Consider case (b) above. Let $c = 1 \text{ V}^2$ and $f_\theta = 1 \text{ Hz}$. Plot the two Lorentzian integral spectra, and their sum, on linear–log paper in the range $0.1 < l < 30 \text{ Hz}$.

PROBLEM. Repeat the above problem for $p = 1$, $q = 5$ (expand the frequency axis accordingly).

c. Resolution of Two Lorentzians

In the previous examples, the two Lorentzians were sufficiently similar that their sum resulted in a function $I(l)$ with only one peak. Here we investigate the conditions for which the Lorentzians are resolved, i.e., two peaks appear in the sum. This corresponds to two real positive roots of the quartic equation derived from the condition $dI/dl = 0$.

Rearranging the quartic into factors of l gives

$$al^4 + bl^3 + cl^2 + dl + e = 0$$

where

$$a = (1 + pq)$$
$$b = f_0[(3q - 1) + pq(3 - q)]$$
$$c = 3qf_0^2(q - 1)(1 - p)$$
$$d = qf_0^3[p(1 - 3q) + q(q - 3)]$$
$$e = -q^2f_0^4(q + p)$$

Of the four roots of this equation, one is always real and negative and one is always real and positive. The former has no physical significance; the latter is the value of l for which $I(l)$ is maximum. Depending on the values of q and p, the other two roots are either complex [$I(l)$ has only one peak] or real [$I(l)$ has two peaks separated by a relative minimum].

Table 50.1 gives the smallest value of q (at a set value of p) for which two peaks are resolved. The quartic equation was solved numerically using Newton's method for determining the zeros of a polynomial (courtesy of B. A. Sokol).

The value of p is the ratio of the variance of the second to the first Lorentzian; q is the ratio of their cutoff frequencies. The real positive roots are the frequencies at which the sum $I(l)$ has its first peak, its relative minimum, and its second peak.

For example, let the first Lorentzian have twice the variance of the second:

$$c' = 2c'' \qquad (p = 0.5)$$

TABLE 50.1. *A Summary of the Relative Shapes Two Lorentzian Functions Must Have, When Analyzed as Integral Spectra, in Order to Appear as Separate Peaks*

$p = c''/c'$	$q = f_0''/f_0'$	Real positive roots (Hz)		
0.1	265	10.0	119	144
0.25	100	10.2	457	512
0.5	45	10.9	21	210
1	14	33.3	37.4	42
2	45	21.4	21.8	412
4	100	19.5	21.9	981
8	210	18.6	22.2	2090
10	265	18.5	22.3	2640

If $f_\theta' = 10$ Hz, then the value of the second cutoff frequency at which the two Lorentzians are just resolved is

$$f_\theta'' = 450 \text{ Hz}$$

Values of f_θ'' below 450 Hz result in a sum with only one peak; values of f_θ'' above 450 Hz give two peaks with increasing clarity. The ratio of amplitudes of the spectral density in this case is

$$S'(0) = 90S''(0)$$

Although the actual frequencies are 10 and 450 Hz, the apparent frequencies are 10.9 and 210 Hz.

Evidently two Lorentzians must be significantly different in order to be resolved by the criterion used here. In order to detect underlying components in a spectrum composed of several Lorentzians, curve-fitting procedures must be used. This is because Lorentzians are broad, highly dispersed spectra, not well characterized by a single frequency.

PROBLEM. Plot the individual Lorentzian integral spectra, and their sum, for the cases $p = 0.25$ and 4. Use the conditions given in the above table. For each curve, calculate the ratio $S'(0)/S''(0)$.

PROBLEM. An interesting exercise is to consider the resolution of two Lorentzians using the function

$$fS(f) = \frac{c'f_\theta'f}{(f_\theta')^2 + f^2} + \frac{c''f_\theta''f}{(f_\theta'')^2 + f^2}$$

This leads to a different quartic equation and a separate set of solutions. One might expect slightly improved resolution using $fS(f)$ over $I(l)$, based on Figure 49.2. The advantages of spectral densities, however, lie mainly with spectra containing sharp peaks, not with smooth spectra such as the Lorentzian.

PROBLEM. Consider the spectral density of the sum of two sine waves arbitrarily close in frequency and amplitude; thus

$$S(f) \propto \delta(f - f') + \delta(f - f'')$$

Show that the corresponding integral spectrum is distinct from one describing a single sine wave.

PROBLEM. Derive the inversion formula for the integral spectral density.

SOLUTION. For brevity, define the integral spectral density as

$$ISD(l) = \frac{4}{\pi l} I(l)$$

By definition (Section 48)

$$ISD(l) = \frac{4}{\pi l} \int_0^\infty S(f)K(l, f) \, df$$

Let $f = e^{x/2}$ and $l = e^{y/2}$; then

$$ISD(y) = \frac{2}{\pi} \int_{-\infty}^\infty S(x)e^{-(y-x)/2}K(y - x) \, dx$$

Following the previous development of the inversion formula for the integral spectrum,

$$\overline{ISD}(z) = \frac{2}{\pi} \bar{S}(z)\overline{e^{-z/2}K(z)}$$

Solving for the spectral density gives

$$S(x) = \frac{\pi}{2} \int_{-\infty}^\infty \frac{\overline{ISD}(z)}{\overline{e^{-z/2}K(z)}} e^{i2\pi xz} \, dz$$

This is the general solution; substituting $x = 2 \ln f$ gives the final result.

The particular transformation filter we have been considering (the L filter) has

$$K(x) = \frac{e^x}{(1 + e^x)^2}$$

Therefore

$$\overline{e^{-z/2}K(z)} = \int_{-\infty}^\infty \frac{e^{-y/2}e^y}{(1 + e^y)^2} e^{-i2\pi yz} \, dy$$

The integrand may be rearranged to give

$$\overline{e^{-z/2}K(z)} = \frac{1}{4} \int_{-\infty}^\infty e^{-y/z} \operatorname{sech}^2(y/2)e^{-i2\pi yz} \, dy$$

From CF # I-614.1,

$$\int_{-\infty}^\infty \frac{\pi}{4\gamma^2} e^{-\lambda g} \operatorname{sech}^2\left(\frac{\pi g}{2\gamma}\right) e^{-i2\pi fg} \, dg = \frac{i2\pi f + \lambda}{\sin[\gamma(i2\pi f + \lambda)]}$$

Let $\lambda = 1/2$, $\gamma = \pi$, $g = y$, and $f = z$; then

$$\overline{e^{-z/2}K(z)} = \frac{\pi(i2\pi z + 1/2)}{\sin[\pi(i2\pi z + 1/2)]}$$

Finally

$$S(x) = \frac{\pi}{2} \int_{-\infty}^{\infty} \frac{\sin[\pi(i2\pi z + 1/2)]}{\pi(i2\pi z + 1/2)} \overline{ISD}(z)e^{i2\pi xz}\, dz$$

To find $S(f)$ calculate $\overline{ISD}(z)$ from $I(l)$ and perform the indicated transformation.

PROBLEM. Derive the relationship between the integral spectral density and the correlation function.

51. Examples of the Inversion Formula

In Section 48 a general expression for the calculation of the spectral density from the integral spectrum was given. These two representations of random noise may be thought of as transform pairs; other transformation pairs are the real and the imaginary parts of a filter transfer function (the Hilbert transformation), the spectral density and the correlation function (the Fourier transformation), etc. Here we consider some specific examples that illustrate spectral density and integral spectrum pairs.

a. White Noise

The integral spectrum of white noise is (Section 47)

$$I(l) = \frac{A\pi}{4}\, l$$

where A, for example, has the units of V^2/Hz. We shall transform the integral spectral density of white noise (Section 49):

$$\frac{4}{\pi l} I(l) = A$$

From Section 49 the spectral density is related to the integral spectral density by

$$S(x) = \frac{\pi}{2} \int_{-\infty}^{\infty} \frac{\sin[\pi(i2\pi z + 1/2)]}{\pi(i2\pi z + 1/2)} \overline{ISD}(z)e^{i2\pi xz}\, dz$$

where $x = 2 \ln f$. For white noise

$$\overline{ISD}(z) = \int_{-\infty}^{\infty} A e^{-i2\pi yz} \, dy = A\delta(z)$$

Substituting this result into the previous expression gives

$$S(x) = \frac{\pi}{2} A\left(\frac{\sin \pi/2}{\pi/2}\right) = A$$

Since the result is not a function of x,

$$S(f) = A$$

which is the required result; an integral spectrum with $I(l) = A\pi/4l$ transforms to a spectral density $S(f) = A$.

b. 1/f Noise

The integral spectrum for B/f noise is (Section 47)

$$I(l) = B/2$$

where B, for example, has the units of V^2. From Section 48 the spectral density is related to the integral spectrum by

$$S(x) = 2e^{-x/2} \int_{-\infty}^{\infty} \frac{\sinh(2\pi^2 x)}{(2\pi^2 x)} \bar{I}(z)e^{i2\pi xz} \, dz$$

where $x = 2 \ln f$. For B/f noise

$$\bar{I}(z) = \int_{-\infty}^{\infty} \left(\frac{B}{2}\right) e^{-i2\pi yz} \, dy = \frac{B}{2} \delta(z)$$

Substituting this expression into the last gives

$$S(x) = 2e^{-x/2}\left(\frac{B}{2}\right) = Be^{-x/2}$$

since $(\sinh u/u)$ goes to one as u goes to zero. Substituting $f = e^{x/2}$, results in

$$S(f) = B/f$$

as required.

In the previous two examples, the rather complicated calculations associated with the integral transformations have been circumvented by arranging the problem so that a delta function appears in the integrand. In the following example, this is not possible.

c. Lorentzian Noise

The integral spectrum corresponding to the spectral density $K/(1+\omega^2\theta^2)$, where K and θ are constant and $\omega = 2\pi f$, is given by (Section 47)

$$I(l) = \frac{K}{8}\frac{l/f_\theta}{(1+l/f_\theta)^2}$$

Making the substitutions

$$l = e^{y/2}$$

$$f_\theta = e^{y_\theta/2}$$

gives

$$I(l) = \frac{K}{8}\frac{e^{(y-y_\theta)/2}}{(1+e^{(y-y_\theta)/2})^2}$$

It is required to find $\bar{I}(z)$, where

$$\bar{I}(z) = \frac{K}{8}\int_{-\infty}^{\infty}\frac{e^{(y-y_\theta)/2}}{(1+e^{(y-y_\theta)/2})^2}e^{-i2\pi yz}\,dy$$

In section 48 we showed that

$$\int_{-\infty}^{\infty}\frac{e^y}{1+e^y}e^{-i2\pi yz}\,dy = \frac{2\pi^2 z}{\sinh(2\pi^2 z)}$$

Applying the shift theorem (CF #I-207) and the scaling theorem (CF #I-205), the equation above may be applied to the evaluation of $\bar{I}(z)$; thus

$$I(z) = \frac{K}{8}(2e^{-i2\pi y_\theta z})\frac{2\pi^2(2z)}{\sinh[2\pi^2(2z)]}$$

or

$$\bar{I}(z) = \frac{K}{4}e^{-i2\pi y_\theta z}\frac{4\pi^2 z}{\sinh(4\pi^2 z)}$$

Substituting this expression into the inversion formula derived in Section

48 gives

$$S(x) = 2e^{-x/2} \int_{-\infty}^{\infty} \frac{\sinh(2\pi^2 z)}{(2\pi^2 z)} \left[\frac{K}{4} e^{-i2\pi y_\theta z} \frac{(4\pi^2 z)}{\sinh(4\pi^2 z)} \right] e^{i2\pi x z} \, dz$$

$$= e^{-x/2} K \int_{-\infty}^{\infty} \frac{\sinh(2\pi^2 z)}{\sinh(4\pi^2 z)} e^{i2\pi(x-y_\theta)z} \, dz$$

From the expansions

$$\sinh u = \tfrac{1}{2}(e^u - e^{-u})$$

$$\cosh u = \tfrac{1}{2}(e^u + e^{-u})$$

it may be shown that

$$\frac{\sinh(2\pi^2 z)}{\sinh(4\pi^2 z)} = \frac{1}{2\cosh(2\pi^2 z)}$$

Substituting this identity into the last expression for $S(x)$ gives

$$S(x) = e^{-x/2} K \int_{-\infty}^{\infty} \frac{1}{2\cosh(2\pi^2 z)} e^{i2\pi(x-y_\theta)z} \, dz$$

$$= e^{-x/2} K \int_{0}^{\infty} \frac{1}{\cosh(2\pi^2 z)} \cos 2\pi(x - y_\theta)z \, dz$$

This follows since $\cosh(2\pi^2 z)$ is an even function of z.
    From BH # 264-14,

$$\int_{0}^{\infty} \frac{\cos pz}{\cosh qz} \, dz = \frac{\pi}{2q} \frac{1}{\cosh(p\pi/2q)}$$

Let $p = 2\pi(x - y_\theta)$ and $q = 2\pi^2$; then

$$S(x) = e^{-x/2} K \left(\frac{1}{4\pi} \right) \frac{1}{\cosh[(x - y_\theta)/2]}$$

$$= \frac{K}{2\pi} \frac{e^{-x/2}}{e^{(x-y_\theta)/2} + e^{-(x-y_\theta)/2}}$$

Substituting $x = 2 \ln f$ and $y_\theta = 2 \ln f_\theta$ gives

$$S(f) = \frac{K}{2\pi} \frac{1/f}{(f/f_\theta) + (f_\theta/f)}$$

$$= \frac{K}{2\pi} \frac{1/f_\theta}{1 + (f/f_\theta)^2}$$

Since $\omega = 2\pi f$ and $\theta = 2\pi f_\theta$, the above equation reduces to

$$S(f) = \frac{K\theta}{1 + \omega^2\theta^2}$$

as required.

PROBLEM. Use the inversion formula for the integral spectral density, derived in Section 49, to invert the ISD for a Lorentzian

$$ISD(l) = \frac{K}{2\pi} \frac{1/f_\theta}{(1 + l/f_\theta)^2}$$

52. Correlation Functions of Filtered Noise

In Section 38 a simple broadband filter, consisting of a low-pass RC network coupled to a high-pass CR network, was introduced. If the two filter stages are separated by a buffer amplifier, the transfer function is

$$Y = \frac{1}{1 + i\omega\tau_2} \frac{i\omega\tau_1}{1 + i\omega\tau_1}$$

where $\tau_2 = R_2C_2$ is the time constant of low-pass stage, with cutoff frequency

$$f_2 = \frac{1}{2\pi R_2C_2}$$

and $\tau_1 = C_1R_1$ is the time constant of the high-pass stage, with cutoff frequency

$$f_1 = \frac{1}{2\pi R_1C_1}$$

Using the above definitions, the transfer function for the buffered high-pass/low-pass RC filter may be written

$$Y = \frac{if/f_1}{(1 + if/f_1)(1 + if/f_2)}$$

Calculating the square modulus of Y gives

$$|Y|^2 = \frac{f_2^2 f^2}{(f^2 + f_1^2)(f^2 + f_2^2)}$$

PROBLEM. Plot $|Y|^2$ as a function of f on log–log paper; let $f_2 = 10 f_2$ and $f_1 = 1$ Hz. Consider the limiting cases $f_1 = 0$ and $f_2 \to \infty$.

The square modulus of Y may be thought of as a unitless transfer function for noise spectra; if the input spectral density to such a filter is

$$S(f) \qquad \text{(input)}$$

the output spectral density is

$$S(f) \, |Y|^2 \qquad \text{(output)}$$

The correlation function of noise with spectral density $S(f)$ is given by (Section 43)

$$C(\tau) = \int_0^\infty S(f) \cos \omega\tau \, df$$

Therefore, the correlation function of noise with spectral density $S(f)$ filtered by Y is

$$C(\tau) = \int_0^\infty S(f) \, |Y|^2 \cos \omega\tau \, df$$

In order to demonstrate the usefulness of this relationship, we calculate the effect of a broadband RC filter on the correlation function. (See DeFelice and Sokol, 1976a,b and Kolb *et al.*, 1976.)

a. White Noise

For $S(f) = A$,

$$C(\tau) = \int_0^\infty A \, \frac{f_2^2 f^2}{(f^2 + f_1^2)(f^2 + f_2^2)} \cos \omega\tau \, df$$

The appropriate cosine transform is given by BH #175-2. [This same transform in BMP #1.2-18 contains a typographical error.] From Bierens de Haan (1939),

$$\int_0^\infty \frac{x^2 \cos px \, dx}{(q^2 + x^2)(r^2 + x^2)} = \frac{\pi}{2(q^2 - r^2)} (q e^{-pq} - r e^{-pr})$$

Substituting $p = 2\pi\tau$, $x = f$, $q = f_1$ and $r = f_2$ gives the correlation function as

$$C(\tau) = A \, \frac{\pi}{2} f_2^2 \, \frac{f_2 e^{-2\pi\tau f_2} - f_1 e^{-2\pi\tau f_1}}{f_2^2 - f_1^2}$$

When $f_1 = f_2$, the broadband *RC* filter becomes the *L* filter discussed in Section 38. In this case

$$C(\tau) = A \frac{\pi}{4} f_2 (1 - 2\pi\tau f_2) e^{-2\pi \tau f_2}$$

PROBLEM. Show that the variance of the filtered white noise in the two cases considered above ($f_1 \neq f_2$ and $f_1 = f_2$) is

$$\frac{A\pi f_2^2}{2(f_1 + f_2)} \qquad (f_1 \neq f_2)$$

and

$$\frac{A\pi f_2}{4} \qquad (f_1 = f_2)$$

PROBLEM. Plot the two correlation functions derived above for $f_1 \neq f_2$ and $f_1 = f_2$. Show that the first zero crossing occurs when

$$\tau_0 = \frac{\ln f_2/f_1}{f_2 - f_1} \qquad (f_1 \neq f_2)$$

and

$$\tau_0 = \frac{1}{2\pi f_2} \qquad (f_1 = f_2)$$

Do any other zero crossings occur?

PROBLEM. Consider the limiting case of the correlation function for $f_1 \neq f_2$ as $f_1 \to 0$ and $f_2 \to \infty$.

b. $1/f^\alpha$ Noise

For the spectral density B/f^α it is required to calculate

$$C(\tau) = \int_0^\infty \frac{B}{f^\alpha} \frac{f_2^2 f^2}{(f^2 + f_1^2)(f^2 + f_2^2)} \cos \omega\tau \, df$$

To carry out this calculation reduce the integrand to a sum of simpler fractions by using the Heaviside expansion theorem. [This theorem is explained clearly in Goldman (1966), pp. 68–76.]

The Heaviside Expansion Theorem. The fraction

$$f(u) = \frac{a(u)}{b(u)}$$

where $a(u)$ and $b(u)$ are polynomials in u, may be written as

$$f(u) = \frac{K_1}{u - u_1} + \frac{K_2}{u - u_2} + \cdots$$

where $u_1, u_2, \ldots$, are the roots of the equation

$$b(u) = 0$$

and $K_1, K_2, \ldots$, are given by

$$K_1 = (u - u_1)\,\frac{a(u)}{b(u)}\Big|_{u=u_1}$$

etc. This form of the expansion assumes no multiple roots of $b(u) = 0$; $u_1, u_2, \ldots$, are distinct. Also, $b(u)$ must be a higher degree polynomial than $a(u)$.

Rearranging the expression for the correlation function gives

$$C(\tau) = Bf_2{}^2 \int_0^\infty \frac{f^{2-\alpha}}{(f^2 + f_1{}^2)(f^2 + f_2{}^2)}\cos \omega\tau\,df$$

The fraction in the integrand is expanded using the Heaviside theorem; thus

$$\frac{f^{2-\alpha}}{(f^2 + f_1{}^2)(f^2 + f_2{}^2)} = \frac{K_1}{f - if_1} + \frac{K_2}{f + if_1} + \frac{K_3}{f - if_2} + \frac{K_4}{f + if_2}$$

The K's are given, for $(2 - \alpha) < 4$, by the expressions

$$K_1 = (f - if_1)\,\frac{f^{2-\alpha}}{(f^2 + f_1{}^2)(f^2 + f_2{}^2)}\Big|_{f=if_1} = \frac{f_1{}^{1-\alpha}(i^{1-\alpha})}{2(f_2{}^2 - f_1{}^2)}$$

$$K_2 = K_1{}^* = \frac{f_1{}^{1-\alpha}}{2(f_2{}^2 - f_1{}^2)i^{1-\alpha}}$$

$$K_3 = (f - if_2)\,\frac{f^{2-\alpha}}{(f^2 + f_1{}^2)(f^2 + f_2{}^2)}\Big|_{f=if_2} = \frac{f_2{}^{1-\alpha}(i^{1-\alpha})}{2(f_1{}^2 - f_2{}^2)}$$

$$K_4 = K_3{}^* = \frac{f_2{}^{1-\alpha}}{2(f_1{}^2 - f_2{}^2)i^{1-\alpha}}$$

Combining the first two terms gives

$$\frac{K_1}{f - if_1} + \frac{K_2}{f + if_1} = \frac{f_1{}^{1-\alpha}}{2(f_2{}^2 - f_1{}^2)}\left(\frac{i^{1-\alpha}}{f - if_1} + \frac{i^{\alpha-1}}{f + if_1}\right)$$

The second two terms give

$$\frac{K_3}{f - if_2} + \frac{K_4}{f + if_2} = \frac{f_2^{1-\alpha}}{2(f_1^2 - f_2^2)} \frac{i^{1-\alpha}}{f - if_2} + \frac{i^{\alpha-1}}{f + if_2}$$

Let $1 - \alpha = m$, then

$$\frac{i^m}{f - if_1} + \frac{i^{-m}}{f + if_1} = \frac{f(i^m + i^{-m}) + f_1(i^m - i^{-m})}{f^2 + f_1^2}$$

Since $i^m = \exp(im\pi/2)$, and $\exp(i\theta) = \cos\theta + i\sin\theta$, it follows that

$$i^m + i^{-m} = 2\cos\frac{m\pi}{2}$$

and

$$i^m - i^{-m} = 2\sin\frac{m\pi}{2}$$

Substitution of these identities into the original expansion gives the following expression:

$$\frac{f^{2-\alpha}}{(f^2 + f_1^2)(f^2 + f_2^2)}$$
$$= \frac{1}{f_2^2 - f_1^2} \left\{ \frac{f_1^{1-\alpha}}{f^2 + f_1^2} \left[f\cos\frac{(1-\alpha)\pi}{2} - f_1\sin\frac{(1-\alpha)\pi}{2} \right] \right.$$
$$\left. - \frac{f_2^{1-\alpha}}{f^2 + f_2^2} \left[f\cos\frac{(1-\alpha)\pi}{2} - f_2\sin\frac{(1-\alpha)\pi}{2} \right] \right\}$$

The cosine transformations of $1/(f^2 + f_1^2)$ and $f/(f^2 + f_1^2)$ are given in Bierens de Haan (1939), 160-(5) and (6), respectively. The final expression for broadband-limited B/f^α noise ($\alpha > -2$) is

$$C(\tau) = \frac{Bf_2^2}{2(f_2^2 - f_1^2)} \left\{ \sin\frac{\alpha\pi}{2} \left[f_2^{1-\alpha}E(2\pi\tau f_2) - f_1^{1-\alpha}E(2\pi\tau f_1) \right] \right.$$
$$\left. + \pi\cos\frac{\alpha\pi}{2} \left[f_2^{1-\alpha}e^{-2\pi\tau f_2} - f_1^{1-\alpha}e^{-2\pi\tau f_1} \right] \right\}$$

where we have defined

$$E(x) = e^{-x}\overline{\mathrm{Ei}}(x) + e^x\mathrm{Ei}(-x)$$

$\overline{\mathrm{Ei}}(x)$ and $\mathrm{Ei}(-x)$ are the tabulated exponential integral functions, defined in Jahnke and Emde (1945). A thorough treatment can be found in Lowan (1940). Some properties of these special functions are given in Section 53.

PROBLEM. Show that when $\alpha = 0$, the above expression for $C(\tau)$ reduces to the white noise case given in Section 52.a. Show that when $\alpha = 1$, $C(\tau)$ reduces to the expression

$$C(\tau) = \frac{Bf_2^2}{2(f_2^2 - f_1^2)} \left[E(2\pi\tau f_2) - E(2\pi\tau f_1) \right]$$

The units of B depend on α.

For $\alpha = 1$ and $f_1 = f_2$, one can proceed more directly since the integral to be evaluated is

$$\int_0^\infty B \frac{f \cos \omega\tau \, df}{(f^2 + f_2^2)^2}$$

Integration by parts leads to

$$\int_0^\infty \frac{f \cos(2\pi\tau f) \, df}{(f^2 + f_2^2)^2} = -\frac{1}{2} \frac{\cos 2\pi\tau f}{f^2 + f_2^2} \bigg|_0^\infty - \pi\tau \int_0^\infty \frac{\sin(2\pi\tau f) \, df}{f^2 + f_2^2}$$

The sine transformation of $1/(f^2 + f_2^2)$ is given in BH # 160-(3). For $\alpha = 1$ and $f_1 = f_2$, the expression for L-filtered $1/f$-noise correlation functions is

$$C(\tau) = \frac{B}{2} [1 - \pi\tau f_2 \bar{E}(2\pi\tau f_2)]$$

where $\bar{E}(x)$ is

$$\bar{E}(x) = e^{-x}\bar{\mathrm{Ei}}(x) - e^x\mathrm{Ei}(-x)$$

For $\alpha = 1$ and $\tau = 0$, use the approximations

$$\bar{\mathrm{Ei}}(x) \simeq \mathrm{Ei}(-x) \simeq \ln \gamma x$$

where

$$\gamma = 1.781072 \cdots$$

and $\ln \gamma$ is Euler's constant. For $\alpha = 1$ and $\tau = 0$, the variance for broadband-filtered B/f noise is

$$C(0) = B \frac{f_2^2}{f_2^2 - f_1^2} \ln \frac{f_2}{f_1}$$

which, for $f_1 = f_2$, reduces to

$$C(0) = B/2$$

This last expression is the variance of L-filtered $1/f$ noise. Recall that $B/2$ is also the amplitude of the integral spectrum of $1/f$ noise, which is constant for all values of $f_1 = f_2 = l$.

PROBLEM.   Show that for $\alpha = 2$ the broadband-filtered correlation function is

$$C(\tau) = \frac{B\pi}{2} \frac{f_2/f_1}{f_2{}^2 - f_1{}^2} (f_2 e^{-2\pi \tau f_1} - f_1 e^{-2\pi \tau f_2})$$

For $\alpha = 2$ and $f_1 = f_2$

$$C(2\pi\tau) = \frac{B\pi}{2} \frac{1}{f_2} (1 + 2\pi\tau f_2) e^{-2\pi \tau f_2}$$

For $\alpha = 2$, $\tau = 0$, the variance of B/f^2 noise is

$$C(0) = \frac{B\pi}{2} \frac{f_2/f_1}{f_2 + f_1}$$

which, for $f_1 = f_2$, reduces to

$$C(0) = \frac{B\pi}{4} \frac{1}{f_2}$$

PROBLEM.   Investigate the broadband-filtered ($f_1 \neq f_2$) and L-filtered ($f_1 = f_2$) correlation functions for

$$\alpha = \tfrac{1}{2} \quad \text{and} \quad \alpha = \tfrac{1}{2}$$

Derive expressions for the correlation functions and the variances expected for these categories of noise. The units of B depend on α.

For general α the correlation function is given by ($\alpha > -2$, $f_1 = f_2$)

$$C(\tau) = \frac{B}{4} f_2 \Bigg\{ \sin \frac{\alpha\pi}{2} \left[f_2{}^{1-\alpha} \left(\frac{2}{f_2} - 2\pi\tau \bar{E}(2\pi\tau f_2) \right) + (1 - \alpha) f_2{}^{-\alpha} E(2\pi\tau f_2) \right]$$

$$+ \cos \frac{\alpha\pi}{2} [(1 - \alpha) f_2{}^{-\alpha} - 2\pi\tau f_2{}^{1-\alpha}] e^{-2\pi \tau f_2} \Bigg\}$$

For general α, $C(0) \to \infty$ since $E(u)$ goes to infinity at $u = 0$ (see Section

53). Only the cases $\alpha = 0$, 1, and 3, of those we have considered, have finite mean square values when band-limited by L filters.

c. Lorentzian Noise

For $S(f) = K\theta/(1 + \omega^2\theta^2)$, the expression to be evaluated is

$$C(\tau) = \int_0^\infty \frac{K\theta}{1 + (f/f_\theta)^2} \frac{f_2^2 f^2}{(f^2 + f_1^2)(f^2 + f_2^2)} \cos 2\pi\tau f \, df$$

where $\theta = 1/2\pi f_\theta$. The fraction

$$\frac{f^2}{(f^2 + f_\theta^2)(f^2 + f_1^2)(f^2 + f_2^2)}$$

may be expanded as in Section 52.b:

$$\frac{f^2}{(f^2 + f_\theta^2)(f^2 + f_1^2)(f^2 + f_2^2)}$$
$$= \frac{K_1}{f - if_\theta} + \frac{K_2}{f + if_\theta} + \frac{K_3}{f - if_1} + \frac{K_4}{f + if_1} + \frac{K_5}{f - if_2} + \frac{K_6}{f + if_2}$$
$$= \frac{(K_1 + K_2)f + if_\theta(K_1 - K_2)}{f^2 + f_\theta^2} + \frac{(K_3 + K_4)f + if_1(K_3 - K_4)}{f^2 + f_1^2}$$
$$+ \frac{(K_5 + K_6)f + if_2(K_5 - K_6)}{f^2 + f_2^2}$$

Using the Heaviside expansion theorem,

$$K_1 = (f - if_\theta) \frac{f^2}{(f^2 + f_\theta^2)(f^2 + f_1^2)(f^2 + f_2^2)}\bigg|_{f=if_\theta}$$
$$= \frac{-f_\theta^2}{2if_\theta(f_1^2 - f_\theta^2)(f_2^2 - f_\theta^2)}$$

Since $K_2 = K_1{}^*$, and K_1 is pure imaginary, $K_1 + K_2 = 0$, but

$$K_1 - K_2 = \frac{-f_\theta^2}{if_\theta(f_1^2 - f_\theta^2)(f_2^2 - f_\theta^2)}$$

By analogy,

$$K_3 - K_4 = \frac{-f_1^2}{if_1(f_\theta^2 - f_1^2)(f_2^2 - f_1^2)}$$

and

$$K_5 - K_6 = \frac{-f_2^2}{if_2(f_\theta^2 - f_2^2)(f_1^2 - f_2^2)}$$

Since $K_3 + K_4$ and $K_5 + K_6$ are also zero, the full expansion becomes equal to

$$-\frac{f_\theta^2}{(f^2 + f_\theta^2)(f_1^2 - f_\theta^2)(f_2^2 - f_\theta^2)} - \frac{f_1^2}{(f^2 + f_1^2)(f_\theta^2 - f_1^2)(f_2^2 - f_1^2)}$$

$$-\frac{f_2^2}{(f^2 + f_2^2)(f_\theta^2 - f_2^2)(f_1^2 - f_\theta^2)}$$

The cosine transformations of $1/(f^2 + f_\theta^2)$ etc., are given in BH # 160-(5).

The final expression for the broadband-filtered, Lorentzian noise correlation function is

$$C(\tau) = \frac{K\theta\pi}{2} f_\theta^2 f_2^2 \left[\frac{-f_\theta e^{-2\pi \tau f_\theta}}{(f_\theta^2 - f_1^2)(f_\theta^2 - f_2^2)} - \frac{f_1 e^{-2\pi \tau f_1}}{(f_1^2 - f_\theta^2)(f_1^2 - f_2^2)} \right.$$
$$\left. - \frac{f_2 e^{-2\pi \tau f_2}}{(f_2^2 - f_1^2)(f_2^2 - f_\theta^2)} \right]$$

Note that $K\theta\pi/2 = K/4f_\theta$.

Case 1. $f_2 \to \infty$.   In this case, the above expression for $C(\tau)$ reduces to

$$C(\tau) = \frac{K}{4} f_\theta \left(\frac{f_1 e^{-2\pi \tau f_1} - f_\theta e^{-2\pi \tau f_\theta}}{f_1^2 - f_\theta^2} \right)$$

which, for $\tau = 0$, becomes

$$C(0) = \frac{K}{4} \left(\frac{f_\theta}{f_1 + f_\theta} \right)$$

Case 2. $f_1 \to 0$.   In this case, $C(\tau)$ reduces to

$$C(\tau) = \frac{K}{4} f_2 \left(\frac{f_\theta e^{-2\pi \tau f_2} - f_2 e^{-2\pi \tau f_\theta}}{f_\theta^2 - f_2^2} \right)$$

which, for $\tau = 0$, becomes

$$C(0) = \frac{K}{4} \frac{f_2}{f_\theta + f_2}$$

Case 3. $f_1 \to 0$, $f_2 \to \infty$.   In this case, $C(\tau)$ reduces to

$$C(\tau) = \frac{K}{4} e^{-2\pi \tau f_\theta}$$

which, for $\tau = 0$, becomes

$$C(0) = \frac{K}{4}$$

Case 4. $f_1 = f_2$. In this case,

$$C(\tau) = \frac{K}{4} f_0 f_2^2 \left[\frac{f_2 e^{-2\pi\tau f_2} - f_\theta e^{-2\pi\tau f_\theta}}{(f_2^2 - f_\theta^2)^2} - \frac{(1 - 2\pi\tau f_2)e^{-2\pi\tau f_2}}{2f_2(f_2^2 - f_\theta^2)} \right]$$

which, for $\tau = 0$, reduces to

$$C(0) = \frac{K}{8} \frac{f_\theta f_2}{(f_\theta + f_2)^2}$$

Case 5. $f_1 = f_2 = f_\theta$. In this case,

$$C(\tau) = \frac{K}{32} e^{-2\pi\tau f_2}[1 + 2\pi\tau f_2 - (2\pi\tau f_2)^2]$$

which, for $\tau = 0$, becomes

$$C(0) = \frac{K}{32}$$

53. The Exponential Integral

In the previous section we derived expressions for correlation functions of filtered noise. For noise of the type

$$S(f) = B/f^\alpha \qquad (\alpha > -2)$$

filtered by broadband *RC* filters, the correlation functions were obtained in terms of exponential integrals. Here we derive properties of the exponential integral.

There is a difference in notation in the literature regarding these functions. Three standard works are compared; these are Jahnke and Emde (1945), A. Lowand (1940), and Geller and Ng (1969).

Jahnke and Emde (JE). The following definitions are given:

$$-\mathrm{Ei}(-x) = -\int_{-\infty}^{-x} \frac{e^t}{t} dt = \int_x^\infty \frac{e^{-t}}{t} dt \qquad (53.1)$$

$$\overline{\mathrm{Ei}}(x) = \mathrm{li}\, e^x = \int_0^{e^x} \frac{dt}{\ln t}$$

Let $\ln t = u$; then

$$\overline{\mathrm{Ei}}(x) = \int_{-\infty}^x \frac{e^u}{u} du \qquad (53.2)$$

Then $u = x$ implies $t = e^x$ and $u = -\infty$ implies $t = 0$.

A. Lowand (AL). In this paper, the definitions are

$$-\mathrm{Ei}(-x) = \int_x^\infty \frac{e^{-t}}{t}\, dt \qquad (53.3)$$

which is identical to Equation (53.1), and

$$\mathrm{Ei}(x) = \int_{-\infty}^x \frac{e^t}{t}\, dt \qquad (53.4)$$

which is identical to Equation (53.2).

Geller and Ng (GN). Finally, GN define

$$E_1(x) = \int_x^\infty \frac{e^{-t}}{t}\, dt, \qquad x < 0 \qquad (53.5)$$

which is the negative of Equations (53.1) and (53.3), and

$$\mathrm{Ei}(x) = -\int_{-x}^\infty \frac{e^{-t}}{t}\, dt$$

$$= \int_{-\infty}^x \frac{e^t}{t}\, dt, \qquad x > 0 \qquad (53.6)$$

The three sets of notation used for the exponential integral are summarized below. All three sets now describe the same function.

JE	AL	GN
$\mathrm{Ei}(-x)$	$\mathrm{Ei}(-x)$	$-E_1(x)$
$\overline{\mathrm{Ei}}(x)$	$\mathrm{Ei}(x)$	$\mathrm{Ei}(x)$

It seems preferable to write the two expressions with the same integrand. We define the functions (Figure 53.1)

$$\mathrm{Ei}(-x) = \int_{-\infty}^{-x} \frac{e^t}{t}\, dt \qquad (53.7)$$

$$\mathrm{Ei}(x) = \int_{-\infty}^x \frac{e^t}{t}\, dt \qquad (53.8)$$

as the exponential integral. Figure 53.1 shows e^t/t vs. t for negative and

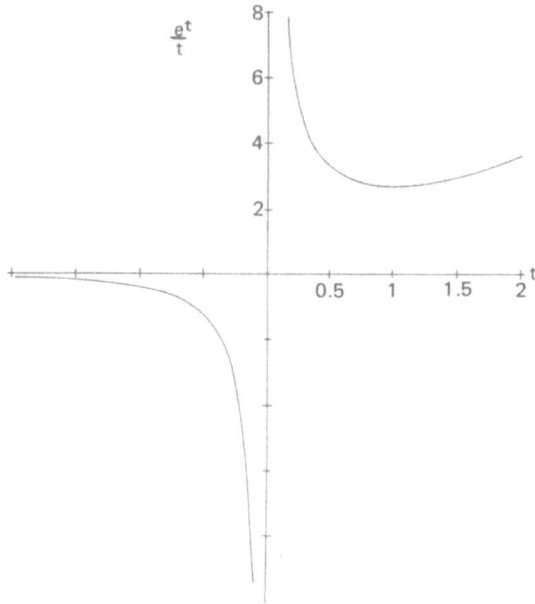

Figure 53.1. The function e^t/t versus t for negative and positive values of t.

positive values of the argument. From such a graph, the meaning of Equations (53.7) and (53.8) becomes clear. $\text{Ei}(-x)$ is the area under e^t/t between $t = -\infty$ and some value $t = -x$ restricted to the left-hand side of the graph. Thus, $\text{Ei}(-x)$ is always negative. $\text{Ei}(x)$ is the area under e^t/t between $t = -\infty$ and some value $t = x$ restricted to the right-hand side of the graph. $\text{Ei}(x)$ may be negative, zero, or positive.

From Geller and Ng (1969), # 3.2-1, (using our notation) the condition

$$\text{Ei}(-x) + \text{Ei}(x) = 0$$

exists when

$$x = 0.372507\cdots$$

This is the value of x at which the area on the right-hand side of Figure 53.1 is equal to the area on the left-hand side.

In Section 52 we found it useful to define the following combinations of the exponential integral and the exponential function:

$$E(x) = e^{-x}\text{Ei}(x) + e^{x}\text{Ei}(-x) \tag{53.9}$$

$$\bar{E}(x) = e^{-x}\text{Ei}(x) - e^{x}\text{Ei}(-x) \tag{53.10}$$

PROBLEM. Show that

$$\frac{d\,\mathrm{Ei}(-x)}{dx} = \frac{e^{-x}}{x} \tag{53.11}$$

and

$$\frac{d\,\mathrm{Ei}(x)}{dx} = \frac{e^{x}}{x} \tag{53.12}$$

Equations (53.11) and (53.12) lead directly to

$$\frac{dE(x)}{dx} = \frac{2}{x} - \bar{E}(x) \tag{53.13}$$

and

$$\frac{d\bar{E}(x)}{dx} = -\,E(x) \tag{53.14}$$

From Equation (53.9), E is an even function of x:

$$E(-x) = e^{x}\mathrm{Ei}(-x) + e^{-x}\mathrm{Ei}(x)$$

$$= E(x)$$

From Equation (53.10), $\bar{E}$ is odd:

$$\bar{E}(-x) = e^{x}\mathrm{Ei}(-x) - e^{-x}\mathrm{Ei}(x)$$

$$= -\bar{E}(x)$$

SOLUTION. The derivatives of $\mathrm{Ei}(x)$, $\mathrm{Ei}(-x)$, $\bar{E}(x)$, and $E(x)$. By definition

$$\mathrm{Ei}(x) = \int_{-\infty}^{x} \frac{e^{u}}{u}\, du$$

From the definition of the derivative

$$\frac{d}{dx}\mathrm{Ei}(x) = \lim_{h \to 0} \frac{\mathrm{Ei}(x+h) - \mathrm{Ei}(x)}{h}$$

$$= \lim_{h \to 0} \frac{\int_{x}^{x+h} (e^{u}/u)\, du}{h}$$

Doing the indicated integration

$$\int \frac{e^u}{u}\, du = \log u + \frac{u}{1!} + \frac{u^2}{2 \cdot 2!} + \frac{u^3}{3 \cdot 3!} + \cdots$$

and therefore

$$\frac{d}{dx}\, \mathrm{Ei}(x)$$

$$= \lim_{h=0} \frac{\left[\log(x+h) + \dfrac{(x+h)}{1!} + \dfrac{(x+h)^2}{2 \cdot 2!} + \cdots\right] - \left[\log x + \dfrac{x}{1!} + \dfrac{x^2}{2 \cdot 2!} + \cdots\right]}{h}$$

By Taylor's expansion

$$\log(x + h) = \log(x) + h\left(\frac{1}{x+h}\right) + \cdots$$

Thus

$$\frac{d}{dx}\, \mathrm{Ei}(x) = \lim_{h=0} \frac{1}{h} \left\{ \left[\log x + \frac{h}{(h+x)} + \cdots + (x+h) + \frac{x^2 + 2xh + h^2}{2 \cdot 2!} + \cdots \right] \right.$$

$$\left. - \left[\log x + x + \frac{x^2}{2 \cdot 2!} + \cdots \right] \right\}$$

$$= \lim_{h=0} \frac{1}{h} \left(\frac{h}{h+x} + \cdots + h + \frac{2xh + h^2}{2 \cdot 2!} + \cdots \right)$$

Therefore

$$\frac{d}{dx}\, \mathrm{Ei}(x) = \lim_{h=0} \left(\frac{1}{x+h} + \cdots + 1 + \frac{x}{2!} + \cdots \right)$$

$$= \frac{1}{x} + 1 + \frac{x}{2!} + \cdots$$

By definition

$$e^x = 1 + x + \frac{x^2}{2!} + \cdots$$

and therefore

$$\frac{d}{dx}\, \mathrm{Ei}(x) = \frac{e^x}{x}$$

By a similiar argument, it can be shown that

$$\frac{d}{dx}\,\text{Ei}(-x) = \frac{e^{-x}}{x}$$

By definition

$$E(x) = e^{-x}\text{Ei}(x) + e^{x}\text{Ei}(-x)$$

$$\bar{E}(x) = e^{-x}\text{Ei}(x) - e^{x}\text{Ei}(-x)$$

Therefore

$$\frac{dE(x)}{dx} = e^{-x}\left(\frac{e^x}{x}\right) - \text{Ei}(x)e^{-x} + e^{x}\left(\frac{e^{-x}}{x}\right) + \text{Ei}(-x)e^{x}$$

$$= \frac{2}{x} - \bar{E}(x)$$

and

$$\frac{d\bar{E}(x)}{dx} = e^{-x}\left(\frac{e^x}{x}\right) - \bar{\text{E}}\text{i}(x)e^{-x} - e^{x}\left(\frac{e^{-x}}{x}\right) - \text{Ei}(-x)e^{x}$$

$$= -E(x)$$

PROBLEM.   Show that differential equations having $E(x)$ and $\bar{E}(x)$ as solutions are

$$\frac{d^2 E}{dx^2} - E = -\frac{2}{x^2}$$

$$\frac{d^2 \bar{E}}{dx^2} - \bar{E} = -\frac{2}{x}$$

Graphs of $E(x)$ and $\bar{E}(x)$ are shown in Figures 53.2 and 53.3. Note that $E(x)$ diverges at the origin. $\bar{E}(x)$ is zero at the origin; however, the slope of $\bar{E}(x)$ at $x = 0$ is infinite.

Values of $E(x)$ and $\bar{E}(x)$ are given in Table 53.1 for $0 \le x \le 50$. These values have been derived using the numerical methods given for $\text{Ei}(x)$ and $\text{Ei}(-x)$ by Lowand (1940).

In Lowand's method of computation, infinite series representations are used to calculate $\text{Ei}(x)$ and $\text{Ei}(-x)$; the series is terminated to give a selected accuracy. The series representations are

$$\text{Ei}(x) = e^{x}\left(\frac{1}{x} + \frac{1}{x^2} + \frac{2!}{x^3} + \frac{3!}{x^4} + \cdots\right)$$

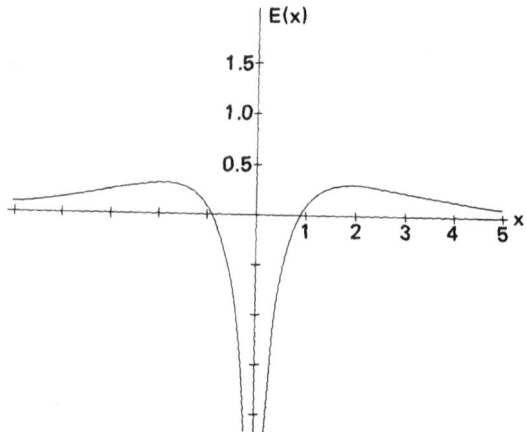

Figure 53.2. The function $E(x) = e^{-x}\text{Ei}(x) + e^{x}\text{Ei}(-x)$ defined in Section 52. As $x \to 0$, $E(x) \to \infty$.

and

$$-\text{Ei}(-x) = e^{x}\left(\frac{1}{x} - \frac{1}{x^2} + \frac{2!}{x^3} - \frac{3!}{x^4} + \cdots\right)$$

Ei(x) and Ei$(-x)$ are then substituted into Equations (53.9) and (53.10) to derive Table 53.1 (courtesy of Barry A. Sokol).

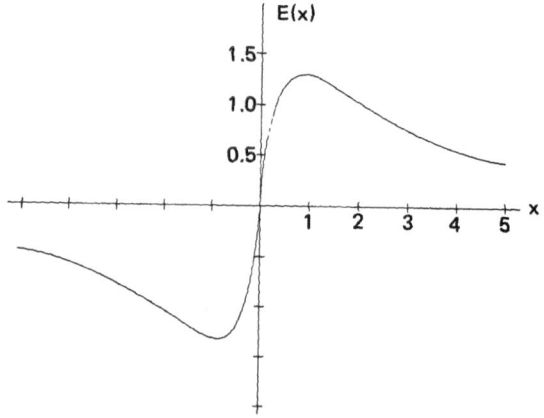

Figure 53.3. The function $\bar{E}(x) = e^{-x}\text{Ei}(x) - e^{x}\text{Ei}(-x)$ defined in Section 52. As $x \to 0$, $d\bar{E}(x)/dx \to \infty$.

TABLE 53.1. Values of the Functions E(x) and Ē(x),ᵃ Defined in the Text, for x between 0 and 1 and for x between 1 and 50 ᵇ

x	$E(x)$	$\bar{E}(x)$
0.0000	∞	0.00000 + 00
0.0100	−0.80565 + 01	0.10056 + 00
0.0200	−0.66715 + 01	0.17341 + 00
0.0300	−0.58627 + 01	0.23580 + 00
0.0400	−0.52899 + 01	0.29143 + 00
0.0500	−0.48468 + 01	0.34203 + 00
0.0600	−0.44858 + 01	0.38864 + 00
0.0700	−0.41816 + 01	0.43193 + 00
0.0800	−0.39191 + 01	0.47241 + 00
0.0900	−0.36885 + 01	0.51042 + 00
0.1000	−0.34830 + 01	0.54626 + 00
0.1100	−0.32980 + 01	0.58015 + 00
0.1200	−0.31300 + 01	0.61228 + 00
0.1300	−0.29762 + 01	0.64280 + 00
0.1400	−0.28345 + 01	0.67184 + 00
0.1500	−0.27034 + 01	0.69952 + 00
0.1600	−0.25815 + 01	0.72594 + 00
0.1700	−0.24676 + 01	0.75118 + 00
0.1800	−0.23609 + 01	0.77532 + 00
0.1900	−0.22606 + 01	0.79842 + 00
0.2000	−0.21661 + 01	0.82055 + 00
0.2100	−0.20769 + 01	0.84176 + 00
0.2200	−0.19924 + 01	0.86210 + 00
0.2300	−0.19122 + 01	0.88162 + 00
0.2400	−0.18360 + 01	0.90036 + 00
0.2500	−0.17634 + 01	0.91835 + 00
0.2600	−0.16942 + 01	0.93564 + 00
0.2700	−0.16282 + 01	0.95225 + 00
0.2800	−0.15651 + 01	0.96821 + 00
0.2900	−0.15047 + 01	0.98356 + 00
0.3000	−0.14468 + 01	0.99831 + 00
0.3100	−0.13912 + 01	0.10125 + 01
0.3200	−0.13379 + 01	0.10261 + 01
0.3300	−0.12867 + 01	0.10393 + 01
0.3400	−0.12375 + 01	0.10519 + 01
0.3500	−0.11901 + 01	0.10640 + 01
0.3600	−0.11444 + 01	0.10757 + 01

continued overleaf

TABLE 53.1 (*continued*)

x	$E(x)$	$\bar{E}(x)$
0.3700	$-0.11004 + 01$	$0.10869 + 01$
0.3800	$-0.10580 + 01$	$0.10977 + 01$
0.3900	$-0.10171 + 01$	$0.11081 + 01$
0.4000	$-0.97760 + 00$	$0.11181 + 01$
0.4100	$-0.93945 + 00$	$0.11276 + 01$
0.4200	$-0.90257 + 00$	$0.11368 + 01$
0.4300	$-0.86692 + 00$	$0.11457 + 01$
0.4400	$-0.83245 + 00$	$0.11542 + 01$
0.4500	$-0.79908 + 00$	$0.11623 + 01$
0.4600	$-0.76679 + 00$	$0.11702 + 01$
0.4700	$-0.73551 + 00$	$0.11777 + 01$
0.4800	$-0.70522 + 00$	$0.11849 + 01$
0.4900	$-0.67587 + 00$	$0.11918 + 01$
0.5000	$-0.64741 + 00$	$0.11984 + 01$
0.5100	$-0.61982 + 00$	$0.12047 + 01$
0.5200	$-0.59306 + 00$	$0.12108 + 01$
0.5300	$-0.56711 + 00$	$0.12166 + 01$
0.5400	$-0.54192 + 00$	$0.12222 + 01$
0.5500	$-0.51747 + 00$	$0.12274 + 01$
0.5600	$-0.49373 + 00$	$0.12325 + 01$
0.5700	$-0.47068 + 00$	$0.12373 + 01$
0.5800	$-0.44829 + 00$	$0.12419 + 01$
0.5900	$-0.42654 + 00$	$0.12463 + 01$
0.6000	$-0.40541 + 00$	$0.12505 + 01$
0.6100	$-0.38488 + 00$	$0.12544 + 01$
0.6200	$-0.36492 + 00$	$0.12582 + 01$
0.6300	$-0.34552 + 00$	$0.12617 + 01$
0.6400	$-0.32666 + 00$	$0.12651 + 01$
0.6500	$-0.30832 + 00$	$0.12682 + 01$
0.6600	$-0.29048 + 00$	$0.12712 + 01$
0.6700	$-0.27313 + 00$	$0.12740 + 01$
0,6800	$-0.25625 + 00$	$0.12767 + 01$
0.6900	$-0.23983 + 00$	$0.12792 + 01$
0.7000	$-0.22386 + 00$	$0.12815 + 01$
0.7100	$-0.20832 + 00$	$0.12837 + 01$
0.7200	$-0.19319 + 00$	$0.12857 + 01$
0.7300	$-0.17847 + 00$	$0.12875 + 01$
0.7400	$-0.16414 + 00$	$0.12892 + 01$

TABLE 53.1 (*continued*)

x	E(x)	Ē(x)
0.7500	−0.15020 + 00	0.12908 + 01
0.7600	−0.13662 + 00	0.12922 + 01
0.7700	−0.12341 + 00	0.12935 + 01
0.7800	−0.11054 + 00	0.12947 + 01
0.7900	−0.98016 − 01	0.12958 + 01
0.8000	−0.85821 − 01	0.12967 + 01
0.8100	−0.73947 − 01	0.12975 + 01
0.8200	−0.62385 − 01	0.12981 + 01
0.8300	−0.51126 − 01	0.12987 + 01
0.8400	−0.40164 − 01	0.12992 + 01
0.8500	−0.29488 − 01	0.12995 + 01
0.8600	−0.19093 − 01	0.12998 + 01
0.8700	−0.89694 − 02	0.12999 + 01
0.8800	0.88860 − 03	0.12999 + 01
0.8900	0.10488 − 01	0.12999 + 01
0.9000	0.19837 − 01	0.12997 + 01
0.9100	0.28940 − 01	0,12995 + 01
0.9200	0.37805 − 01	0.12992 + 01
0.9300	0.46437 − 01	0.12987 + 01
0.9400	0.54843 − 01	0.12982 + 01
0.9500	0.63028 − 01	0.12976 + 01
0.9600	0.70997 − 01	0.12970 + 01
0.9700	0.78757 − 01	0.12962 + 01
0.9800	0.86312 − 01	0.12954 + 01
0.9900	0.93667 − 01	0.12945 + 01
1.0000	0.10083 + 00	0.12935 + 01
1.1000	0.16271 + 00	0.12902 + 01
1.2000	0.20961 + 00	0.12615 + 01
1.3000	0.24466 + 00	0.12387 + 01
1.4000	0.27028 + 00	0.12129 + 01
1.5000	0,28836 + 00	0.11849 + 01
1.6000	0.30041 + 00	0.11554 + 01
1.7000	0.30764 + 00	0.11250 + 01
1.8000	0.31101 + 00	0.10940 + 01
1.9000	0.31130 + 00	0.10629 + 01
2.0000	0.30915 + 00	0.10318 + 01
2.1000	0.30509 + 00	0.10011 + 01

continued overleaf

TABLE 53.1 (*continued*)

x	$E(x)$	$\bar{E}(x)$
2.2000	0.29954 + 00	0.97084 + 00
2.3000	0.29285 + 00	0.94121 + 00
2.4000	0.28530 + 00	0.91230 + 00
2.5000	0.27712 + 00	0.88418 + 00
2.6000	0.26852 + 00	0.85689 + 00
2.7000	0.25964 + 00	0.83048 + 00
2.8000	0.25061 + 00	0.80497 + 00
2.9000	0.24153 + 00	0.78036 + 00
3.0000	0.23249 + 00	0.75666 + 00
3.1000	0.22355 + 00	0.73386 + 00
3.2000	0.21477 + 00	0.71194 + 00
3.3000	0.20618 + 00	0.69090 + 00
3.4000	0.19781 + 00	0.67070 + 00
3.5000	0.18969 + 00	0.65133 + 00
3.6000	0.18183 + 00	0.63275 + 00
3.7000	0.17425 + 00	0.61495 + 00
3.8000	0.16695 + 00	0.59789 + 00
3.9000	0.15994 + 00	0.58155 + 00
4.0000	0.15321 + 00	0.56590 + 00
4.1000	0.14676 + 00	0.55090 + 00
4.2000	0.14059 + 00	0.53654 + 00
4.3000	0.13469 + 00	0.52278 + 00
4.4000	0.12905 + 00	0.50959 + 00
4.5000	0.12367 + 00	0.49696 + 00
4.6000	0.11855 + 00	0.48485 + 00
4.7000	0.11366 + 00	0.47324 + 00
4.8000	0.10900 + 00	0.46211 + 00
4.9000	0.10457 + 00	0.45143 + 00
5.0000	0.10034 + 00	0.44119 + 00
6.0000	0.67880 − 01	0.35841 + 00
7.0000	0.47989 − 01	0.30127 + 00
8.0000	0.35451 − 01	0.26001 + 00
9.0000	0.27222 − 01	0.22895 + 00
10.0000	0.21585 − 01	0.20471 + 00
11.0000	0.17561 − 01	0.18524 + 00
12.0000	0.14587 − 01	0.16924 + 00
13.0000	0.12328 − 01	0.15583 + 00
14.0000	0.10560 − 01	0.14442 + 00

TABLE 53.1 (*continued*)

x	$E(x)$	$\bar{E}(x)$
15.0000	0.91532 − 02	0.13459 + 00
16.0000	0.80134 − 02	0.12603 + 00
17.0000	0.70760 − 02	0.11850 + 00
18.0000	0.62954 − 02	0.11183 + 00
19.0000	0.56381 − 02	0.10587 + 00
20.0000	0.50792 − 02	0.10052 + 00
21.0000	0.46000 − 02	0.95683 − 01
22.0000	0.41858 − 02	0.91295 − 01
23.0000	0.38254 − 02	0.87293 − 01
24.0000	0.35098 − 02	0.83629 − 01
25.0000	0.32318 − 02	0.80261 − 01
26.0000	0.29857 − 02	0.77155 − 01
27.0000	0.27667 − 02	0.74281 − 01
28.0000	0.25711 − 02	0.71614 − 01
29.0000	0.23955 − 02	0.69132 − 01
30.0000	0.22374 − 02	0.66817 − 01
31.0000	0.20944 − 02	0.64652 − 01
32.0000	0.19648 − 02	0.62624 − 01
33.0000	0.18469 − 02	0.60719 − 01
34.0000	0.17392 − 02	0.58926 − 01
35.0000	0.16408 − 02	0.57237 − 01
36.0000	0.15505 − 02	0.55642 − 01
37.0000	0.14674 − 02	0.54134 − 01
38.0000	0.13909 − 02	0.52705 − 01
39.0000	0.13202 − 02	0.51350 − 01
40.0000	0.12547 − 02	0.50063 − 01
41.0000	0.11941 − 02	0.48839 − 01
42.0000	0.11377 − 02	0.47673 − 01
43.0000	0.10852 − 02	0.46562 − 01
44.0000	0.10363 − 02	0.45502 − 01
45.0000	0.99061 − 03	0.44489 − 01
46.0000	0.94789 − 03	0.43520 − 01
47.0000	0.90787 − 03	0.42592 − 01
48.0000	0.87034 − 03	0.41703 − 01
49.0000	0.83509 − 03	0.40850 − 01
50.0000	0.80194 − 03	0.40032 − 01

[a] The last two digits in each $E(x)$ and $\bar{E}(x)$ entry is the power of 10 by which the preceding number is to be multiplied.
[b] Courtesy of B. A. Sokol.

54. The Effect of Finite Time Measurements

In Section 34 the caret notation was introduced to distinguish the true spectral density of a signal and an estimate of the spectral density obtained during a finite time interval. The true spectral density $S(f)$ and the estimate $\hat{S}(f)$ are related by the expression

$$S(f) = \lim_{T \to \infty} \hat{S}(f)$$

$\hat{S}(f)$ differs from $S(f)$ not in a practical way that may be improved by technique, but in a theoretical way due to the statistical nature of the signal. For a noisy signal, $\hat{S}(f)$ represents a particular estimate of $S(f)$ over time T. The next interval has another $\hat{S}(f)$; these tend toward $S(f)$ on average. (The spectral densities of deterministic signals of finite and infinite duration are also distinguished, as shown in Section 32. Here we are concerned with stochastic signals only.)

The concept behind the caret notation may be extended to any function, e.g., the correlation function and the integral spectrum. Define a general property Φ; it is required to quantify the difference between Φ and $\hat{\Phi}$ and to estimate the errors due to measurements made over finite time interval T.

The variance is the average difference between a signal and its mean value (Section 33). Let $x(t)$ be an arbitrary signal. Then

$$\langle x(t) \rangle = \frac{1}{T} \int_0^T x(t)\, dt$$

is its average value. The average value is independent of t but may depend on the particular segment of time T selected out of all possible segments. The square of the difference between a momentary value of $x(t)$ and its average value $\langle x \rangle$ is a measure of the variation between the two; the average value of this quantity is the variance:

$$\sigma^2 = \langle [x(t) - \langle x(t) \rangle]^2 \rangle$$

σ^2 is a particular value of the variance for a specific choice of the time segment T. To be consistent σ^2 should be labeled with a caret to distinguish it from the variance obtained for the average taken over all time:

$$\sigma^2 = \lim_{T \to \infty} \langle [x(t) - \langle x(t) \rangle]^2 \rangle$$

It is cumbersome to include the caret in discussions where it is evident that finite, specific time segments T are being considered.

To retain the same symbol σ^2 for the variation between the particular value and the true value, write

$$\hat{\sigma}^2 = \langle (\hat{\Phi} - \Phi)^2 \rangle$$

where

$$\sigma^2 = \lim_{T \to \infty} \hat{\sigma}^2$$

Note that Φ may itself be a variance, in which case we have defined the variance of the variance. The process may continue endlessly.

Expanding the above expression for $\hat{\sigma}^2$ gives

$$\hat{\sigma}^2 = \langle \hat{\Phi}^2 + \Phi^2 - 2\hat{\Phi}\Phi \rangle$$
$$= \langle \hat{\Phi}^2 \rangle + \langle \Phi^2 \rangle - 2\langle \hat{\Phi}\Phi \rangle$$

Φ cannot depend on time, therefore

$$\langle \Phi^2 \rangle = \Phi^2$$

and

$$-2\langle \hat{\Phi}\Phi \rangle = -2\Phi\langle \hat{\Phi} \rangle$$

Substitution of these expressions into $\hat{\sigma}^2$ gives

$$\hat{\sigma}^2 = \langle \hat{\Phi}^2 \rangle + \Phi^2 - 2\Phi\langle \hat{\Phi} \rangle$$

Consider two cases: In the limit $T \to \infty$, $\hat{\sigma}^2 \to 0$. This follows from the definition of $\hat{\sigma}^2$ and estimates with this property are said to be consistent estimates. Second, consider T sufficiently large (but not infinite) to assume

$$\langle \hat{\Phi} \rangle \simeq \Phi$$

Alternatively, imagine many estimates made over shorter intervals of time and then averaged. In either case, if the average value of the estimate approaches the true value, then

$$\hat{\sigma}^2 \simeq \langle \hat{\Phi}^2 \rangle - \Phi^2$$

The estimate is said to be an unbiased estimate.

Rather than the time average, an average over different samples is also possible. A sample may be one of a set of identical physical objects, or one of a set of identical challenges to the same object. The average estimate over many samples is called the expected value of the estimate. This approach is discussed thoroughly in Bendat and Piersol (1971, p. 64). The equivalence between the time average and the sample average is called the ergodic theorem.

The mean square error of an estimate of any quantity is defined as

$$\varepsilon^2 = \frac{\langle(\hat{\Phi} - \Phi)^2\rangle}{\Phi^2}$$

For unbiased, consistent estimates, the mean square error is approximately given by

$$\varepsilon^2 \simeq \frac{\langle\hat{\Phi}^2\rangle - \Phi^2}{\Phi^2}$$

This form is usually more convenient for calculations. ε approaches zero as T approaches infinity. ε is normalized; the value 1.0 implies 100% error, 0.5 50% error, and so on.

55. *Examples of Error Calculations*

Consider an arbitrary signal $x(t)$. Define the autocorrelation function of $x(t)$ as

$$R(\tau) = \lim_{T\to\infty} \frac{1}{2T} \int_{-T}^{T} x(t)x(t + \tau)\, dt$$

Since the correlation function is symmetric in τ space, consider only the single-sided integral (Section 44)

$$R(\tau) = \lim_{T\to\infty} \frac{1}{T} \int_{0}^{T} x(t)x(t + \tau)\, dt$$

It is understood that τ is confined to positive values. Bendat and Piersol (1971, Chap. 3) define the autocovariance as

$$C(\tau) = R(\tau) - u^2$$

where

$$\hat{u} = \frac{1}{T} \int_{0}^{T} x(t)\, dt$$

is the average value of $x(t)$.

The literature on membrane noise has been careless in distinguishing the autocorrelation and the autocovariance. The difference between the two functions is the dc component of $x(t)$ squared. If the dc component is removed from $x(t)$ before calculating the autocorrelation, the autocorrelation and the autocovariance are equal. Removing the dc component of a signal usually implies filtering; frequencies other than $f = 0$ are attenuated and may affect the final result (Section 52).

In this book and in most of the noise literature, the symbol $C(\tau)$ is used to represent the correlation function. It may or may not include a dc component. The prefix "auto" in autocorrelation is dropped when it is evident that the signal is being compared with itself. Finally, single-sided correlation functions are used most often, although double-sided correlation functions are more convenient in some theoretical discussions.

a. Correlation Function

By definition, the correlation function is

$$C(\tau) = \lim_{T \to \infty} \frac{1}{T} \int_0^T x(t)x(t + \tau)\, dt$$

The mean square error in the correlation function, due to a finite time sample, is defined as

$$\varepsilon^2 \simeq \frac{\langle \hat{C}^2(\tau) \rangle - C^2(\tau)}{C^2(\tau)}$$

It is required to calculate $\langle \hat{C}^2(\tau) \rangle$.

Introduce a subscript to signify the correlation function of the signal x; by definition

$$\langle \hat{C}_x{}^2(\tau) \rangle = \left\langle \left[\frac{1}{T} \int_0^T x(t)x(t + \tau)\, dt \right]\left[\frac{1}{T} \int_0^T x(t)x(t + \tau)\, dt \right] \right\rangle$$

$$= \left\langle \frac{1}{T^2} \int_0^T \int_0^T x(t)x(t + \tau)x(t')x(t' + \tau)\, dt\, dt' \right\rangle$$

The prime notation is introduced to distinguish the two dummy variables t and t' and to write the product as a double integral.

The Case $\tau = 0$. Since $C(0)$ is the variance of $x(t)$, we shall be calculating the mean square error of the variance. Setting $\tau = 0$ in the above expression gives

$$\langle \hat{C}_x{}^2(0) \rangle = \left\langle \frac{1}{T^2} \int_0^T \int_0^T x^2(t)x^2(t')\, dt\, dt' \right\rangle$$

Recall that the angle brackets mean either a time average or an average over different samples. In either case the angle brackets may be taken inside the integral sign since t and t' are dummy variables.

Replacing the angle brackets by the time average gives

$$\langle \hat{C}_x^2(0) \rangle = \frac{1}{\mathscr{E}} \int_0^{\mathscr{E}} \left[\frac{1}{T^2} \int_0^T \int_0^T x^2(t)x^2(t') \, dt \, dt' \right] dt$$

The time base indicated by the square brackets is $\mathscr{E}$; the time base indicated by the caret is T. The equation says to calculate the variance of $x^2(t)$ over T, square this result, and take the average value for many samples, i.e., $\mathscr{E} \gg T$.

Since the order of integration is arbitrary,

$$\langle \hat{C}_x^2(0) \rangle = \frac{1}{T^2} \int_0^T \int_0^T \left[\frac{1}{\mathscr{E}} \int_0^{\mathscr{E}} x^2(t)x^2(t') \, dt \right] dt' \, dt$$

Let $\tau = t' - t$. The integration over t' is performed holding t constant and $dt' = d\tau$ during this integration. Also, $0 < t' < T$ changes to the range $-t < \tau < T - t$. With these substitutions,

$$\langle \hat{C}_x^2(0) \rangle = \frac{1}{T^2} \int_0^T \int_{-t}^{T-t} \left[\frac{1}{\mathscr{E}} \int_0^{\mathscr{E}} x^2(t)x^2(t + \tau) \, dt \right] d\tau \, dt$$

The expression in square brackets is by definition the correlation function of the square of $x(t)$;

$$C_{x^2}(\tau) = \lim_{\mathscr{E} \to \infty} \frac{1}{\mathscr{E}} \int_0^{\mathscr{E}} x^2(t)x^2(t + \tau) \, dt$$

All functions designated by C in this section are single-sided ($\tau > 0$). However, $C_{x^2}(\tau)$ ranges over positive and negative values of τ as a result of the variable transformation. One must either restrict the range of integration such that $\tau > 0$ ($t' > t$) or consider the correlation functions as being double-sided. For the moment, consider all the correlation functions as double-sided and work in the range $-t < \tau < T - t$.

EXERCISE. Show that if $x(t)$ passes through a squaring device, the correlation function of the output, $x^2(t)$, is given by

$$C_{x^2}(\tau) = C_x^2(0) + 2C_x^2(\tau)$$

where C_x is the correlation function of the input $x(t)$. This result is valid for single- or double-sided correlation functions.

SOLUTION. This problem is solved in Rice (1944). The above equation is equivalent to Rice's Equation 3.9-7. The answer is written in terms of the dc and the ac component of the correlation function. $C_x(\tau)$ is the autocovariance defined in Bendat and Piersol (1971).

PROBLEM. Let $x(t)$ be such that $S_x(f) = A$ with the units of V^2/Hz. Calculate the spectral density of $x^2(t)$.

PROBLEM. Let $x(t)$ be such that $S_x(f) = B/f$, where B is in V^2. Calculate the spectral density for $x^2(t)$.

PROBLEM. Let $x(t)$ be such that $S_x(f) = K\theta/(1 + \omega^2\theta^2)$ where K is in V^2. Calculate the spectral density for $x^2(t)$.

Using Rice's result for $C_x^2(\tau)$, we have (double-sided correlation functions)

$$\langle \hat{C}_x^2(0) \rangle = \frac{1}{T^2} \int_0^T \int_{-t}^{T-t} [C_x^2(0) + 2C_x^2(\tau)] \, d\tau \, dt$$

The first integral on the right-hand side is easily evaluated:

$$\frac{1}{T^2} \int_0^T \int_{-t}^{T-t} C^2(0) \, d\tau \, dt = \frac{1}{T^2} C^2(0) \int_0^T \int_0^T dt' \, dt = C^2(0)$$

PROBLEM. The second integral requires some manipulation. It is required to find the value of

$$\frac{2}{T^2} \int_0^T \int_{-t}^{T-t} C^2(\tau) \, d\tau \, dt = \frac{2}{T^2} \int_0^T \int_0^T C^2(t' - t) \, dt' \, dt$$

SOLUTION. Let τ be positive and negative, i.e., $C(\tau)$ is the double-sided function. Consider the plane of integration (t, τ) and the limits of integration $T - \tau$ and $-\tau$. Integration over τ occurs between the lines $\tau = -t$ and $\tau = T - t$ in Figure 55.1. The values along horizontal lines

Figure 55.1. The (t, τ) plane of integration to illustrate the transformation of the double integral over $C^2(t' - t)$ to a single integral over τ.

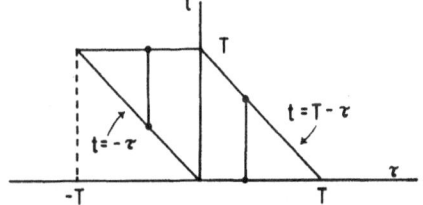

are added over t. The integration may also be over the vertical lines. Two regions of integration, $\tau < 0$ and $\tau > 0$, are considered. For $\tau < 0$, the length of the vertical line is $T - (-\tau) = T + \tau$. For $\tau > 0$, the length of the vertical line is $T - \tau$. The value of the integral over t at any τ is $C^2(\tau)$ times the length of the line of integration since $C^2(\tau)$ does not depend on t. Thus

$$\int_0^T \int_{-t}^{T-t} C^2(\tau) \, d\tau \, dt = \int_{-T}^0 (T + \tau) C^2(\tau) \, d\tau + \int_0^T (T - \tau) C^2(\tau) \, d\tau$$

$$= \int_{-T}^T (T - |\tau|) C^2(\tau) \, d\tau$$

The absolute value signs have been used in order to write two integrals (the first over $\tau < 0$ and the second over $\tau > 0$) as one.

Using this result,

$$\langle \hat{C}_x^2(0) \rangle = C_x^2(0) + \frac{2}{T^2} \int_{-T}^T (T - |\tau|) C_x^2(\tau) \, d\tau$$

Consider the case $T \gg \tau$ but not infinite; then

$$\frac{2}{T^2} \int_{-T}^T (T - |\tau|) C_x^2(\tau) \, d\tau = \frac{2}{T} \int_{-T}^T \left(1 - \frac{|\tau|}{T}\right) C_x^2(\tau) \, d\tau$$

$$\simeq \frac{2}{T} \int_{-\infty}^\infty C_x^2(\tau) \, d\tau$$

The symbol ∞ in the last integral implies only $T \gg \tau$. In this case

$$\langle \hat{C}_x^2(0) \rangle \simeq C_x^2(0) + \frac{2}{T} \int_{-\infty}^\infty C_x^2(\tau) \, d\tau$$

The mean square error at $\tau = 0$ of a signal $x(t)$ having true correlation function $C_x(\tau)$, due to finite time estimation, is approximately

$$\varepsilon^2 \simeq \frac{\langle \hat{C}_x^2(0) \rangle - C_x^2(0)}{C_x^2(0)} = \frac{2}{T} \frac{\int_{-\infty}^\infty C_x^2(\tau) \, d\tau}{C_x^2(0)}$$

In this expression, the double-sided correlation function is implied:

$$\int_{-\infty}^\infty C_x^2(\tau) \, d\tau = 2 \int_0^\infty C_x^2(\tau) \, d\tau$$

Since $C_x(\tau)$ is symmetric (Section 43), the equation for $C_x(\tau)$ is the same for the double-sided or the single-sided function. For example, Lorentzian noise has

$$S(f) = \frac{K\theta}{1 + \omega^2\theta^2}, \quad f > 0; \qquad C(\tau) = \frac{K}{4} e^{-\tau/\theta}, \quad \tau \geq 0$$

and

$$SS(f) = \frac{1}{2}\frac{K\theta}{1 + \omega^2\theta^2}, \quad -\infty < f < \infty; \quad C(\tau) = \frac{K}{4} e^{-|\tau|/\theta}; \quad -\infty < \tau < \infty$$

In the last expression $|\tau|$ implies a decaying function on either side of $\tau = 0$.

The final equation for the variance of $x(t)$ is

$$C_x(0) \pm \varepsilon$$

where $C_x(\tau)$ is the true correlation function and

$$\varepsilon = \left[\frac{2}{TC_x^2(0)} \int_{-\infty}^{\infty} C_x^2(\tau)\, d\tau \right]^{1/2}$$

The Case $\tau \neq 0$. We have calculated the error due to finite time measurement in the first point of the correlation function. Bendat and Piersol (1971) give the following expression for the mean square error expected at any τ [pp. 181–184; Equation (6.72)]:

$$\varepsilon^2(\tau) \simeq \frac{1}{T} \int_{-\infty}^{\infty} [C^2(u) + C(u - \tau)C(u + \tau)]\, du$$

When $\tau = 0$, this expression reduces to the previous formula. For large τ (but $\tau \ll T$),

$$\varepsilon \simeq \frac{1}{T} \int_{-\infty}^{\infty} C^2(\tau)\, d\tau$$

or one half of the value of the mean square error at $\tau = 0$.

PROBLEM. Consider Lorentzian noise with true spectral density $K\theta/(1 + \omega^2\theta^2)$. Show that the expected error in an estimate of the variance measured over time interval T is $(2\theta/T)^{1/2}$. Let $f_\theta = 100$ Hz and $T = 1$ sec. What is the percent error?

EXERCISE. Calculate the mean square error in the correlation function of Lorentzian noise at any value of τ. (See Bendat, 1958, Section 7.3-1.)

Show that the error in the mean value of $x(t)$ is given by

$$\varepsilon^2 \simeq \frac{1}{T} \int_{-\infty}^{\infty} C(\tau) \, d\tau$$

[see Bendat and Piersol, (1971), Equation (6.19)].
    We previously assumed the mean value of $x(t)$ to be zero.

b. Integral Spectra

    To calculate the expected error in the integral spectrum we need the correlation function of L-filtered noise. Examples of correlation functions of RC-filtered noise are given in Section 52; the error formulas for integral spectra are obtained by using the results of Section 52 and by letting $f_1 = f_2 = l$ be a continuous variable.

    White Noise. Consider the spectral density $S(f) = A$. The correlation function of white noise passed through an L filter of center frequency l is

$$C(\tau, l) = \frac{A\pi}{4} l(1 - 2\pi\tau l)e^{-2\pi\tau l}$$

The integral spectrum is the variance at the output of each L filter against its center frequency l; thus

$$I(l) = C(0, l) = \frac{A\pi}{4} l$$

    The mean square error in the integral spectrum at any l is given by

$$\varepsilon^2 = \frac{\langle \hat{I}^2(l) \rangle - I^2(l)}{I^2(l)} = \frac{\langle \hat{C}^2(0, l) \rangle - C^2(0, l)}{C^2(0, l)}$$

In Section 55.a we derived a formula for the error in the correlation function at $\tau = 0$; this formula may be applied to the present problem.
    It is required to calculate $\langle \hat{I}^2(l) \rangle = \langle \hat{C}^2(0, l) \rangle$. From (a), the mean square error of the correlation function at $\tau = 0$ is

$$\varepsilon^2 = \frac{2}{T} \int_{-\infty}^{\infty} \frac{C^2(\tau, l)}{C^2(0, l)} \, d\tau$$

Let $x = 2\pi\tau$; then

$$\varepsilon^2 = \frac{2}{T} \int_{-\infty}^{\infty} (1 - 2\pi\tau l)^2 e^{-4\pi\tau l} \, d\tau$$

$$= \frac{4}{2\pi T} \int_{0}^{\infty} (1 - xl)^2 e^{-2xl} \, dx$$

Now

$$\int_0^\infty (1 - xl)^2 e^{-2xl}\, dx = \int_0^\infty (e^{-2xl} - 2xle^{-2xl} + x^2 l^2 e^{-2xl})\, dx$$

$$= \left[\frac{1}{2l} - \frac{2l}{(2l)^2} + \frac{2l^2}{(2l)^3} \right] = \frac{1}{4l}$$

Therefore

$$\varepsilon^2 = \frac{1}{2\pi Tl}$$

$\varepsilon = (1/2\pi Tl)^{1/2}$ is the expected error in the integral spectrum of white noise at every point along the frequency axis l. The final expression for white noise is

$$I(l) = \frac{A\pi}{4} l \pm \frac{1}{(2\pi lT)^{1/2}}$$

PROBLEM. Calculate the percent error expected in the integral spectrum of white noise at 1, 10, and 100 Hz for estimates derived over the time interval 1 sec.

PROBLEM. Consider band-limited white noise where $f_1 \neq f_2$. In Section 52 the correlation function of white noise passed through a low-pass/high-pass RC filter with $f_1 \neq f_2$ was shown to be

$$C(\tau) = \frac{A\pi}{2} f_2^2 \frac{f_2 e^{-2\pi \tau f_2} - f_1 e^{-2\pi \tau f_1}}{f_2^2 - f_1^2}$$

Derive an expression for the mean square error at $\tau = 0$ expected from an estimate of $C(\tau)$ over time interval T for all f_1 and f_2. Consider the limiting value of ε^2 for $f_1 = 0$, $f_2 \to \infty$, and $f_1 = f_2$.

1/f Noise. The correlation function for L-filtered noise with spectral density B/f is

$$C(\tau, l) = \frac{B}{2} [1 - \pi \tau l \bar{E}(2\pi \tau l)]$$

(Section 52). The integral spectrum is

$$I(l) = C(0, l) = \frac{B}{2}$$

since $\bar{E}(0) = 0$ (Section 53).

It is required to calculate

$$\varepsilon^2 = \frac{2}{T} \int_{-\infty}^{\infty} \frac{C(\tau, l)}{C(0, l)} \, d\tau = \frac{4}{T} \int_{0}^{\infty} (1 - \pi\tau l \bar{E})^2 \, d\tau$$

Let $u = 2\pi\tau l$, then

$$\varepsilon^2 = \frac{4}{2\pi T l} \int_{0}^{\infty} \left[1 - \frac{u}{2} \bar{E}(u) \right]^2 du$$

where

$$\bar{E}(u) = e^{-u} \text{Ei}(u) - e^u \text{Ei}(-u)$$

and Ei are the exponential integrals defined in Section 53. The integration required for the evaluation of L-filtered $1/f$ noise is difficult. Even after expansion of the integrand, the solution is not in the *Table of Integrals of the Exponential Integral*, by Geller and Ng (1969), the most comprehensive list available.

Numerical integration based on tabulated values of $\bar{E}$ [DeFelice and Sokol, Equation (29), 1976] gave the result

$$\varepsilon^2 = \frac{2.4652 \cdots}{2\pi l T}$$

or

$$\int_{0}^{\infty} \left[1 - \frac{u}{2} \bar{E}(u) \right]^2 du = \frac{1}{4} (2.4652 \cdots) = 0.6163$$

It was also shown that the integral is bounded. The analytical result has been kindly provided by M. Geller (June 7, 1976, personal communication). The answer is

$$\int_{0}^{\infty} \left[1 - \frac{u}{2} \bar{E}(u) \right]^2 du = \frac{\pi^2}{16} = 0.6168 \cdots$$

The following is a sketch of the derivation of this formula, courtesy M. Geller.

The Evaluation of the Error Integral for $1/f$ Noise. Let

$$A = \int_{0}^{\infty} \left[1 - \frac{u}{2} \bar{E}(u) \right]^2 du$$

where

$$\bar{E}(u) = e^{-u} \text{Ei}(u) - e^u \text{Ei}(-u)$$

From Geller and Ng (1969), # 3.3.22,

$$\bar{E}(u) = 2u \int_0^\infty \frac{\sin t}{t^2 + u^2} dt$$

Let

$$A(a) = \int_0^\infty e^{-au} \left[1 - \frac{u}{2} \bar{E}(u)\right]^2 du$$

and

$$A = \lim_{a \to 0} A(a)$$

Then

$$A(a) = \int_0^\infty e^{-au} \left[1 - u\bar{E}(u) + \frac{u^2}{4} \bar{E}^2(u)\right] du \qquad (55.1)$$

$$= \int_0^\infty e^{-au} \left[1 - 2u^2 \int_0^\infty \frac{\sin t}{t^2 + u^2} dt + u^4 \int_0^\infty \frac{\sin t \, dt}{t^2 + u^2} \int_0^\infty \frac{\sin t \, dt}{t^2 + u^2}\right] du$$

Now

$$\frac{u^2}{t^2 + u^2} = 1 - \frac{t^2}{t^2 + u^2}$$

therefore

$$u^2 \int_0^\infty \frac{\sin t}{t^2 + u^2} dt = \int_0^\infty \sin t \, dt - \int_0^\infty \frac{t^2 \sin t \, dt}{t^2 + u^2}$$

Also, since

$$\frac{u^4}{(t^2 + u^2)(v^2 + u^2)} = 1 + \frac{1}{v^2 - t^2} \left(\frac{t^4}{t^2 + u^2} - \frac{v^4}{v^2 + u^2}\right)$$

then

$$u^4 \int_0^\infty \frac{\sin t \, dt}{t^2 + u^2} \int_0^\infty \frac{\sin t \, dt}{t^2 + u^2}$$

$$= u^4 \int_0^\infty \int_0^\infty \frac{\sin t \sin v \, dt \, dv}{(t^2 + u^2)(v^2 + u^2)}$$

$$= \int_0^\infty \int_0^\infty \sin t \sin v \, dt \, dv$$

$$+ \int_0^\infty \int_0^\infty \frac{1}{v^2 - t^2} \left(\frac{t^4}{t^2 + u^2} - \frac{v^4}{v^2 + u^2}\right) \sin t \sin v \, dt \, dv$$

$$= \int_0^\infty \int_0^\infty \sin t \sin v \, dt \, dv$$

$$+ \int_0^\infty \int_0^\infty \frac{t^4 \sin t \sin v \, dt \, dv}{(v^2 - t^2)(t^2 + u^2)} - \int_0^\infty \int_0^\infty \frac{v^4 \sin t \sin v \, dt \, dv}{(v^2 - t^2)(v^2 + u^2)}$$

Now evoke the integral

$$\int_0^\infty e^{-bt} \sin t \, dt = \frac{1}{b^2 + 1}$$

As $b \to 0$,

$$\int_0^\infty \sin t \, dt \to 1$$

Similarly

$$\int_0^\infty t \sin t \, dt \to 0$$

and

$$\int_0^\infty t^2 \sin t \, dt \to -2$$

Therefore, Equation (55.1) becomes

$$A(a) = \int_0^\infty e^{-au} \, du - 2 \int_0^\infty e^{-au} \, du \int_0^\infty \sin t \, dt$$

$$+ 2 \int_0^\infty \int_0^\infty \frac{e^{-au} t^2 \sin t \, dt \, du}{t^2 + u^2} + \int_0^\infty e^{-au} \, du \int_0^\infty \int_0^\infty \sin t \sin v \, dt \, dv$$

$$+ \int_0^\infty \int_0^\infty \int_0^\infty \frac{e^{-au} t^4 \sin t \sin v \, dt \, dv \, du}{(v^2 - t^2)(t^2 + u^2)} - \int_0^\infty \int_0^\infty \int_0^\infty \frac{e^{-au} v^4 \sin t \sin v \, dt \, dv \, du}{(v^2 - t^2)(v^2 + u^2)}$$

Since

$$\int_0^\infty u^n e^{-au} \, du = \frac{n!}{a^{n+1}}$$

then

$$A(a) = \frac{1}{a} [1 - 2(1) + 1(1)(1)] + 2 \int_0^\infty t^2 \sin t \int_0^\infty \frac{e^{-au}}{t^2 + u^2} \, du \, dt$$

$$+ \int_0^\infty \sin v \int_0^\infty \frac{t^4 \sin t}{v^2 - t^2} \int_0^\infty \frac{e^{-au}}{t^2 + u^2} \, du \, dt \, dv$$

$$- \int_0^\infty \sin t \int_0^\infty \frac{v^4 \sin v}{v^2 - t^2} \int_0^\infty \frac{e^{-au}}{v^2 + u^2} \, du \, dv \, dt \qquad (55.2)$$

Since

$$\int_0^\infty \frac{du}{t^2 + u^2} = \frac{1}{t} \int_0^\infty \frac{d(u/t)}{1 + (u/t)^2} = \frac{\pi}{2t}$$

taking the limit $a \to 0$, Equation (55.2) becomes

$$
A = \pi \int_0^\infty t \sin t \, dt + \frac{\pi}{2} \int_0^\infty \sin v \int_0^\infty \frac{t^3 \sin t}{v^2 - t^2} \, dt \, dv - \frac{\pi}{2} \int_0^\infty \sin t \int_0^\infty \frac{v^3 \sin v}{v^2 - t^2} \, dv \, dt
$$

$$
= \pi(0) + \frac{\pi}{2} \int_0^\infty \sin y \int_0^\infty \frac{x^3 \sin x \, dx \, dy}{y^2 - x^2} - \frac{\pi}{2} \int_0^\infty \sin y \int_0^\infty \frac{x^3 \sin x \, dx \, dy}{x^2 - y^2}
$$

$$
= \pi \int_0^\infty \sin y \int_0^\infty \frac{x^3 \sin x \, dx \, dy}{y^2 - x^2}
$$

$$
= \pi \int_0^\infty x^3 \sin x \int_0^\infty \frac{\sin y \, dy}{y^2 - x^2} \, dx
$$

Now

$$
\int_0^\infty \frac{\sin \mu x \, dx}{a^2 - x^2} = \frac{1}{a} \left[\sin \mu a \, \mathrm{Ci}(\mu a) - \cos \mu a \, \mathrm{Si}(\mu a) \right]
$$

where

$$
\mathrm{Si}(y) = \int_0^y \frac{\sin t}{t} \, dt
$$

$$
\mathrm{Ci}(y) = - \int_y^\infty \frac{\cos t}{t} \, dt
$$

$$
\mathrm{si}(y) = - \int_y^\infty \frac{\sin t}{t} \, dt
$$

$$
\mathrm{Si}(y) - \frac{\pi}{2} = \mathrm{si}(y)
$$

Then

$$
\int_0^\infty \frac{\sin y \, dy}{y^2 - x^2} = - \frac{1}{x} \left[\sin x \, \mathrm{Ci}(x) - \cos x \, \mathrm{si}(x) - \frac{\pi}{2} \cos x \right]
$$

Since

$$
\sin x \, \mathrm{Ci}(x) - \cos x \, \mathrm{si}(x) = \int_0^\infty \frac{e^{-xt}}{1 + t} \, dt
$$

then

$$
A = -\pi \int_0^\infty x^3 \sin x \left\{ \frac{1}{x} \left[\sin x \, \mathrm{Ci}(x) - \cos x \, \mathrm{si}(x) \right] \right\} dx
$$

$$
+ \frac{\pi^2}{2} \int_0^\infty x^3 \sin x \left(\frac{1}{x} \cos x \right) dx
$$

$$
= -\pi \int_0^\infty x^2 \sin x \int_0^\infty \frac{e^{-xt} \, dt}{t^2 + 1} \, dx + \frac{\pi^2}{2} \int_0^\infty x^2 \sin x \cos x \, dx
$$

Now

$$\int_0^\infty x^2 \sin x \cos x \, dx = \int_0^\infty \frac{x^2 \sin 2x}{2} dx$$

$$= \int_0^\infty \frac{y^2}{4} \frac{\sin y}{2} d\left(\frac{y}{2}\right)$$

Therefore

$$A = \frac{\pi^2}{2}\left(-\frac{1}{8}\right) - \pi \int_0^\infty x^2 \sin x \int_0^\infty \frac{e^{-xt} \, dt}{t^2 + 1} dx$$

$$= -\frac{\pi^2}{16} - \pi \int_0^\infty \frac{1}{t^2 + 1} \int_0^\infty x^2 e^{-xt} \sin x \, dx \, dt$$

Now (BH # 361-2)

$$\int_0^\infty x^2 e^{-xt} \sin x \, dx = \frac{6t^2 - 2}{(t^2 + 1)^3}$$

Since

$$\int_0^\infty \frac{t^2 \, dt}{(t^2 + 1)^4} = \frac{\pi}{32}$$

$$\int_0^\infty \frac{dt}{(t^2 + 1)^4} = \frac{5\pi}{32}$$

then

$$A = -\frac{\pi^2}{16} - \pi\left[\frac{6\pi}{32} - 2\left(\frac{5\pi}{32}\right)\right]$$

$$= -\frac{\pi^2}{16} - \pi\left(-\frac{4\pi}{32}\right) = \frac{\pi^2}{16}$$

Therefore

$$\int_0^\infty \left[1 - \frac{u}{2}\bar{E}(u)\right]^2 du = \frac{\pi^2}{16}$$

The $1/f$ noise error for the integral spectrum becomes

$$\varepsilon^2 = \frac{4}{2\pi l T} \frac{\pi^2}{16}$$

$$= \frac{(\pi/2)^2}{2\pi l T}$$

$$= \frac{2.467\cdots}{2\pi l T}$$

The final expression for $1/f$ noise is

$$I(l) = \frac{B}{2} \pm \frac{\pi/2}{(2\pi lT)^{1/2}}$$

PROBLEM. Calculate the percent error expected in the integral spectrum of $1/f$ noise at 1, 10, and 100 Hz for estimates derived from the time interval 1 sec.

PROBLEM. The correlation function of $1/f$ noise filtered by a low-pass/high-pass RC filter with $f_1 \neq f_2$ is

$$C(\tau) = \frac{B f_2^2}{2(f_2^2 - f_1^2)} [E(2\pi\tau f_2) - E(2\pi\tau f_1)]$$

(Section 52). Derive an expression for the error at $\tau = 0$ for all f_1 and f_2. Consider the limiting values of ε^2 as $f_1 = 0$, $f_2 \to \infty$ and $f_1 = f_2$.

PROBLEM. Show that

$$\left[1 - \frac{u}{2} \bar{E}(u) \right]^2 = \left(\frac{u}{2} \frac{dE}{du} \right)^2$$

where $\bar{E}$ and E are defined in Section 53. This equality offers another possibility for evaluating the $1/f$ noise error integral.

PROBLEM. Plot $\varepsilon(l)$ for $T = 1$ sec and $1 \leq l \leq 1000$ Hz for white and $1/f$ noise. Compare this graph with $\varepsilon(l)$ derived for Lorentzian noise immediately below; $\varepsilon(l)$ depends on θ for Lorentzian noise.

Lorentzian Noise. Consider noise with spectral density $K\theta/(1 + \omega^2\theta^2)$ passed through an L filter. In Section 52 it was shown that the correlation function of the output noise is

$$C(\tau, l) = \frac{K}{4} f_\theta l^2 \left[\frac{l e^{-2\pi\tau l} - f_\theta e^{-2\pi\tau/f_\theta}}{(l^2 - f_\theta^2)^2} - \frac{(1 - 2\pi\tau l)e^{-2\pi\tau l}}{2l(l^2 - f_\theta^2)^2} \right]$$

Let $x = 2\pi\tau$, $m = f_\theta$, and $c = K\theta$. Recall that $K\theta\pi/2 = K/(4f_\theta)$; then

$$C(x) = \frac{c\pi}{2} m^2 l^2 \left[\frac{l e^{-xl} - m e^{-xm}}{(l^2 - m^2)^2} - \frac{(1 - xl)e^{-xl}}{2l(l^2 - m^2)} \right]$$

The normalized mean square error is given by

$$\varepsilon^2 = \frac{(2/T) \int_{-\infty}^{\infty} C^2(\tau) \, d\tau}{C^2(0)}$$

It can be shown that (Section 52)

$$C(0) = \frac{c\pi}{4} \frac{f_0^2 f_2}{(f_2 + f_0)^2} = \frac{c\pi}{4} \frac{m^2 l}{(m + l)^2}$$

Therefore

$$\varepsilon^2 = \frac{(c\pi/2)^2 (m^2 l^2)^2}{(c\pi/4)^2 (m^2 l)^2 / (m + l)^4} \frac{4}{T} \int_0^\infty [\ ]^2 \frac{dx}{2\pi}$$

which reduces to

$$\varepsilon^2 = \frac{8l^2 (l + m)^4}{\pi T} \int_0^\infty [\ ]^2 \, dx$$

where

$$[\ ]^2 = \left[\frac{l e^{-xl} - m e^{-xm}}{(m^2 - l^2)^2} - \frac{(1 - xl) e^{-xl}}{2l(l^2 - m^2)} \right]^2$$

Expanding, we have

$$[\ ]^2 = \frac{l^2 e^{-2xl} + m^2 e^{-2xm} - 2lm e^{-x(l+m)}}{(m^2 - l^2)^4}$$

$$+ \frac{(1 + x^2 l^2 - 2xl) e^{-2xl}}{4l^2 (l^2 - m^2)^2} - 2 \left[\frac{(l e^{-xl} - m e^{-xm})(e^{-xl} - xl e^{-xl})}{2l(l^2 - m^2)(m^2 - l^2)^2} \right]$$

We make use of the definite integral

$$\int_0^\infty x^n e^{-ax} \, dx = \frac{n!}{a^{n+1}}$$

to evaluate $\int_0^\infty [\ ]^2 \, dx$:

$$\int_0^\infty [\ ]^2 \, dx = \frac{\dfrac{l^2}{2l} + \dfrac{m^2}{2m} - \dfrac{2lm}{l + m}}{(m^2 - l^2)^4} + \frac{\dfrac{1}{2l} + \dfrac{2l^2}{(2l)^3} - \dfrac{2l}{(2l)^2}}{4l^2 (l^2 - m^2)^2}$$

$$- 2 \left[\frac{\dfrac{l}{2l} - \dfrac{m}{m + l} - \dfrac{l^2}{(2l)^2} + \dfrac{m}{(m + l)^2}}{2l(l^2 - m^2)^3} \right]$$

$$= \frac{\frac{1}{2}(l + m) - 2lm/(l + m)}{(m^2 - l^2)^4} + \frac{1/4l}{4l^2 (l^2 - m^2)^2}$$

$$- 2 \left[\frac{\dfrac{1}{4} - \dfrac{m}{l + m} \left(1 - \dfrac{l}{l + m}\right)}{2l(l^2 - m^2)^3} \right]$$

$$= \frac{\frac{1}{2}(l + m)^2 - 2lm}{(l + m)(m^2 - l^2)^4} + \frac{1}{16l^3 (l^2 - m^2)^2} - 2 \left[\frac{\frac{1}{4} - m^2/(l + m)^2}{2l(l^2 - m^2)^3} \right]$$

Continuing the reduction:

$$\int_0^\infty [\ ]^2\, dx = \frac{(l+m)^2 - 4lm}{2(l+m)(m^2 - l^2)^4} + \frac{1}{16l^3(l^2 - m^2)^2} - 2\left[\frac{(l+m)^2 - 4m^2}{8l(l^2 - m^2)^3(l+m)^2}\right]$$

$$= \frac{(l-m)^2}{2(l+m)(l^2 - m^2)^4} + \frac{1}{16l^3(l^2 - m^2)^2} - \left[\frac{l^2 + 2lm - 3m^2}{4l(l^2 - m^2)^3(l+m)^2}\right]$$

$$= \frac{(l-m)^2}{2(l+m)(l-m)^4(l+m)^4} + \frac{1}{16l^3(l+m)^2(l-m)^2}$$
$$- \left[\frac{(l+3m)(l-m)}{4l(l+m)^3(l-m)^3(l+m)^2}\right]$$

Therefore, the expression for mean square error becomes

$$\varepsilon^2 = \frac{8l^2(l+m)^4}{\pi T}\left[\frac{1}{2(l+m)^5(l-m)^2} + \frac{1}{16l^3(l+m)^2(l-m)^2}\right.$$
$$\left. - \frac{(l+3m)}{4l(l+m)^5(l-m)^2}\right]$$

$$= \frac{8l^2}{\pi T}\left[\frac{1}{2(l+m)(l-m)^2} + \frac{(l+m)^2}{16l^3(l-m)^2} - \frac{l+3m}{4l(l+m)(l-m)^2}\right]$$

$$= \frac{4l^2}{\pi T}\frac{1}{(l-m)^2}\left[\frac{1}{(l+m)} + \frac{(l+m)^2}{8l^3} - \frac{l+3m}{2l(l+m)}\right]$$

$$= \frac{4l^2}{\pi T}\frac{1}{(l-m)^2}\left[\frac{8l^3 + (l+m)^3 - (l+3m)4l^2}{8l^3(l+m)}\right]$$

or

$$\varepsilon^2 = \frac{1}{2\pi Tl(l+m)}\left[\frac{5l^3 + m^3 + 3m^2 l - 9l^2 m}{(l-m)^2}\right]$$

Let $p = l/m = l/f_\theta$; the expression for the mean square error becomes

$$\varepsilon^2 = \frac{1}{2\pi lT}\left[\frac{5p^3 + 1 + 3p - 9p^2}{(p+1)(p-1)^2}\right]$$

which is equivalent to DeFelice and Sokol [1976, Equation (34)].

PROBLEM. Show that for

$$l \ll f_\theta \qquad \varepsilon^2 \to \frac{1}{2\pi lT}$$

$$l = f_\theta \qquad \varepsilon^2 = \frac{3}{2\pi f_\theta T}$$

$$l \gg f_\theta \qquad \varepsilon^2 \to \frac{5}{2\pi lT}$$

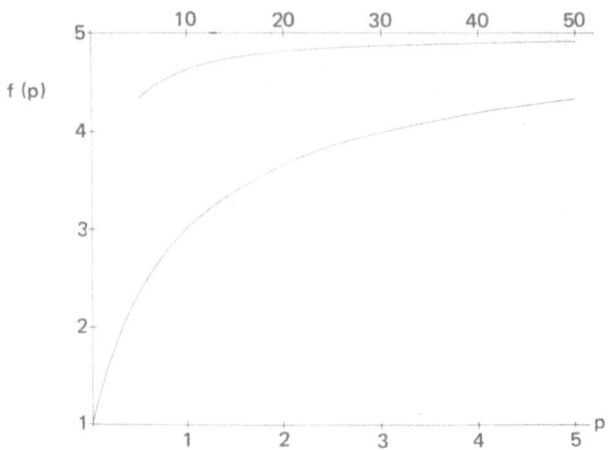

Figure 55.2. The function $f(p)$, defined in the text, for p between 0 and 50.

Figure 55.2 shows the function

$$f(p) = \frac{5p^3 + 1 + 3p - 9p^2}{(p + 1)(p - 1)^2}$$

plotted against p for values between $p = 0$ and 50. $f(p)$ increases mono-tonically with p.

PROBLEM.   Calculate the slope of $f(p)$. At low values of p, $f(p)$ increases more quickly than $f(p) \propto p$ while at higher values the reverse is true. Investigate the slope of $f(p)$ at $p = 0$ and as $p \to \infty$. Find the crossover point, i.e., the value of p where the slope is equal to unity. Call this value (p_1).

Using the result of the above problem, when $p = p_1$, $f(p) = p = l/f_\theta$. At this point the mean square error is

$$\varepsilon^2 = \frac{1}{2\pi lT} f(p_1) = \frac{1}{2\pi T f_\theta} = \frac{\theta}{T}$$

Below $p = p_1$, the error increases as l increases; above $p = p_1$, the error decreases as l increases.

PROBLEM.   Start with the equation for the correlation function of L-filtered Lorentzian noise. Show that when $l = f_\theta$, the correlation function is

$$C(\tau) = \frac{K}{32} e^{-2\pi\tau l}[1 + 2\pi\tau l - (2\pi\tau l)^2]$$

c. Spectral Density

An approximate formula for the mean square error expected in spectral density estimates is given in Bendat and Piersol (1971, Section 6.5).

Consider a signal $x(t)$ with zero mean averaged over time interval T. Let $x(t)$ be passed through a set of narrow-band filters of constant bandwidth Δf. The symbol Δf is used to define a specific bandwidth for a realizable filter, e.g.,

$$\Delta f = \int_0^\infty |Q|^2 \, df$$

where Q is the transfer function of a Q filter (Section 39). $|Q|^2$ is used since it appears in the definition of the practical measure of spectral density at the frequency f_0 (Section 34):

$$S(f_0) = \frac{\int_0^\infty |Q|^2 S(f) \, df}{\int_0^\infty |Q|^2 \, df}$$

f_0 is the center frequency of the Q filter and $S(f)$ is the true spectral density of $x(t)$ before filtering.

Let $x_0(t)$ be the output of the filter. Consider an average of $x_0(t)$ over time interval T. Then the above expression is an estimate of $S(f_0)$ called $\hat{S}(f_0)$. For finite T,

$$\hat{S}(f_0) = \frac{1}{\Delta f} \int_0^\infty |Q|^2 S(f) \, df$$

This expression is equivalent to Bendat and Piersol [their equation (6.82)], who, in order to evaluate the right-hand side, replace a realizable filter by an idealized rectangular filter. With this assumption

$$\int_0^\infty |Q|^2 S(f) \, df \simeq \int_{f-\delta}^{f+\delta} S(f) \, df$$

where $\delta = \Delta f/2$. This equation is equivalent to Bendat and Piersol's Equation (6.84). To obtain expressions for the mean square error in estimates of spectral densities, Bendat and Piersol expand $S(f)$ in a Taylor series about the center frequency of the idealized filter and substitute the integral limits shown above.

These assumptions, which are roughly equivalent to assuming that $S(f)$ does not vary over Δf, lead to the following expression for the mean square error expected in spectral density estimates [Bendat and Piersol,

1971, Equation (6.98)]

$$\varepsilon^2(f) \simeq \frac{1}{T\,\Delta f} + \left(\frac{\Delta f^2}{24}\right)^2 \left[\frac{S''(f)}{S(f)}\right]^2$$

where $S(f)$ is the true spectral density and

$$S''(f) = \frac{d^2}{df^2}\,[S(f)]$$

As expected, the error depends on the type of noise being analyzed. To emphasize this point, several examples are given below.

A Pure Lorentzian. In this case

$$S(f) = \frac{c}{1 + (f/f_\theta)^2}$$

$$= \frac{cf_\theta^2}{f_\theta^2 + f^2}$$

where $c = K\theta = K/2\pi f_\theta$.

It is required to calculate

$$\frac{dS(f)}{df} = cf_\theta^2\,\frac{-2f}{(f_\theta^2 + f^2)^2}$$

and

$$\frac{d^2 S(f)}{df^2} = cf_\theta^2\left[\frac{-2}{(f_\theta^2 + f^2)^2} + \frac{2(2f)(2f)}{(f_\theta^2 + f^2)^3}\right]$$

$$= cf_\theta^2\left[\frac{-2(f_\theta^2 + f^2) + 8f^2}{(f_\theta^2 + f^2)^3}\right]$$

$$= 2cf_\theta^2\left[\frac{3f^2 - f_\theta^2}{(f_\theta^2 + f^2)^3}\right]$$

Therefore

$$\frac{S''(f)}{S(f)} = \frac{2(3f^2 - f_\theta^2)}{(f_\theta^2 + f^2)^3/(f_\theta^2 + f^2)}$$

$$= 2\,\frac{3f^2 - f_\theta^2}{(f_\theta^2 + f^2)^2}$$

The mean square normalized error for Lorentzian noise is

$$\varepsilon^2 \simeq \frac{1}{T\,\Delta f} + \frac{4\,\Delta f^4}{576}\left[\frac{3f^2 - f_\theta^2}{(f_\theta^2 + f^2)^2}\right]^2$$

Notice that when $f = (1/3)^{1/2}$, f_0, $\varepsilon^2 = 1/T \, \Delta f$ at this point, for $\Delta f = 1$ Hz and $T = 1$ sec, $\varepsilon = 1$, which implies a 100% error. Define the frequency dependent term as

$$\varepsilon_f^2 = \frac{\Delta f^4}{144} \frac{(3f^2 - f_\theta^2)^2}{(f_\theta^2 + f^2)^4}$$

Then, for

$$f = 0.01 f_\theta, \qquad \varepsilon_f^2 = \frac{0.9990 \, \Delta f^4}{144 f_\theta^4}$$

$$f = 0.1 f_\theta, \qquad \varepsilon_f^2 = \frac{0.9042 \, \Delta f^4}{144 f_\theta^4}$$

$$f = 0.5 f_\theta, \qquad \varepsilon_f^2 = \frac{0.0256 \, \Delta f^4}{144 f_\theta^4}$$

$$f = 0.6 f_\theta, \qquad \varepsilon_f^2 = \frac{0.0019 \, \Delta f^4}{144 f_\theta^4}$$

$$f = 1.0 f_\theta, \qquad \varepsilon_f^2 = \frac{0.2500 \, \Delta f^4}{144 f_\theta^4}$$

$$f = 2.5 f_\theta, \qquad \varepsilon_f^2 = \frac{0.1140 \, \Delta f^4}{144 f_\theta^4}$$

$$f = 10.0 f_\theta, \qquad \varepsilon_f^2 = \frac{0.0009 \, \Delta f^4}{144 f_\theta^4}$$

The error is critical near f_θ, though of little practical significance compared to $1/(T \, \Delta f)$.

Band-Limited White Noise. Suppose the Lorentzian is limited in frequency by a simple low-pass/high-pass filter; the resultant spectral density is

$$S(f) = A \left[\frac{f_2^2 f^2}{(f_1^2 + f^2)(f_2^2 + f^2)} \right]$$

This expression is subject to the same analysis as above; the region where $S(f) \simeq \text{const}$, $\varepsilon_f \simeq 0$. The mean square normalized error is given simply by

$$\varepsilon^2 = \frac{1}{T \, \Delta f}$$

Band-Limited 1/f Noise. An analogous expression exists, with B/f replacing A:

$$S(f) \approx \frac{B}{f} \left[\frac{f_2^2 f^2}{(f_1^2 + f^2)(f_2^2 + f^2)} \right]$$

Rather than evaluate this expression, consider a frequency window between F_1 and F_2. In this range

$$S(f) = B/f, \qquad F_1 \leq f \leq F_2$$

Then

$$\frac{d^2S}{df^2} = \frac{2B}{f^3}$$

and

$$\left(\frac{d^2S/df^2}{S}\right)^2 = \left(\frac{2}{f^2}\right)^2$$

The mean square normalized error is

$$\varepsilon^2 = \frac{1}{T\,\Delta f} + \frac{\Delta f^4}{144 f^4}, \qquad F_1 \leq f \leq F_2$$

Let $F_1 = 1$ Hz and $F_2 = 1000$ Hz. Then for $f = 1$ Hz, $\Delta f = 1$ Hz, and $T = 1$ sec:

$$\varepsilon^2 = 1 + \frac{1}{144} = 1.007$$

which implies about 100% error. The calculation for $T = 100$ sec implies about 1.7% error.

To a good approximation, the final expression for the expected error in spectral density estimates using constant bandwidth Δf filters and averaging the output over time interval T is

$$\hat{S}(f) = S(f) \pm \frac{1}{(T\,\Delta f)^{1/2}}$$

where $S(f)$ is white, $1/f$, or Lorentzian noise.

56. Comparison of Errors in Correlation Functions, Integral Spectra, and Spectral Densities

We have derived expressions for the mean square error ε^2 in the previous section. The expected error ε is the rms deviation from the true value of the function if one could measure over all time. One way of expressing this is to write

$$\hat{S}(f) = S(f) \pm \varepsilon(f)$$
$$\hat{I}(l) = I(l) \pm \varepsilon(l)$$
$$\hat{C}(\tau) = C(\tau) \pm \varepsilon(\tau)$$

In this section we discuss the idealizations on which the formulas for ε are based.

An important difference between $S(f)$ and the functions $I(l)$ and $C(\tau)$ is that, in principle, each point of the spectral density is statistically independent of every other point. [The proof of this statement lies in the definition of spectral density (Section 34)

$$\hat{S}(f) = \frac{2}{T} \mid \bar{x}(f) \mid^2$$

$\bar{x}(f)$ is the Fourier transform of $x(t)$; the spectral density of $x(t)$ is $S(f)$. The terms of a Fourier expansion are orthogonal.] The integral spectrum and the correlation function, on the other hand, are each composed of highly correlated points. Another way of expressing this is to say that each point of $S(f)$ derives only from new information about $x(t)$ but each point in $I(l)$ or $C(\tau)$ derives from new information plus information already contained in other points.

Measured spectral density points are independent only if the analyzing narrow-band filters do not overlap. In principle, the frequency range may be divided into space-filling, rectangular windows that represent the bank of analyzing filters. In practice, of course, this is impossible. Narrow filters may be constructed, but such filters have tails that extend on either side of the center frequency (see Figure 39.1). To approximate statistically independent points, the overlap of adjacent filters must be small. This is achieved by maintaining an appropriate spacing between filters. However, spacing the filters to reduce overlap leaves areas of the frequency range uncovered and leads to a loss of available data. One solution is to make the filters sharper and to space them more closely. This provides more points in the spectrum and makes better use of the data. However, each point has a larger error since

$$\varepsilon(f) \simeq \frac{1}{(T \, \varDelta f)^{1/2}}$$

Adjacent points in a correlation function are not statistically independent. A simple example is the value $C(0)$; $C(0)$ is the variance of the noise:

$$\sigma^2 = \int_0^\infty S(f) \, df = C(0)$$

(Section 34 and 43). Evidently $C(0)$ contains information from the entire frequency range. The general expression for $C(\tau)$ is

$$C(\tau) = \int_0^\infty S(f) \cos \omega\tau \, df$$

Therefore, every point of the correlation function uses the same range. If independent data points derive only from nonoverlapping frequency windows, each point in $C(\tau)$ must be correlated with every point.

It is natural to ask how the inverse operation

$$S(f) = 4 \int_0^\infty C(\tau) \cos \omega\tau \, d\tau$$

uncorrelates the points. There is no simple answer. Note also that although $S(f)$ is positive definite $C(\tau)$ may assume both positive and negative values. A trivial example is the delta function; if $S(f) = A \, \delta(f - f_0)$ then $C(\tau) = A \cos \omega_0\tau$. However, it makes no sense to state that $C(\tau) = A \, \delta(\tau - \tau_0)$ implies $S(f) = 4A \cos \omega_0\tau$ for the correlation function must have its largest value at $\tau = 0$. The correct statement is that $C(\tau) = A \, \delta(\tau)$ implies $S(f) = 4A$. In such questions it is perhaps simplest to recall how $S(f)$ and $C(\tau)$ are measured. $C(\tau)$ is derived from a time-shifting operation on $x(t)$ before multiplying $x(t)$ by itself; it is easy to imagine how a negative product could result. $S(f)$, on the other hand, only filters $x(t)$ before multiplying, so that the self-product must always be positive.

We now compare the error formulas for $S(f)$ and $C(\tau)$. If $S(f)$ is composed of uncorrelated data points at each f, then the $\varepsilon(f)$ are also uncorrelated. By similar reasoning, the $\varepsilon(\tau)$ must be highly correlated. The practical implication of this is that, for the same length of data $x(t)$, correlation functions are smoother than spectral densities.

For example, consider Lorentzian noise. The spectral density and the correlation function are

$$S(f) = \frac{K\theta}{1 + \omega^2\theta^2}$$

and

$$C(\tau) = \frac{K}{4} e^{-\tau/\theta}$$

Let $T = 1$ sec and $\theta = 1$ msec. $f_\theta = 160$ Hz is the cutoff frequency of the noise. The relative error in $S(f)$, neglecting the small frequency-dependent term (Section 55.c) is independent of f and of θ. Assume an ideal constant bandwidth analyzer with $\Delta f = 5$ Hz. The expected error at every f is

$$\varepsilon \simeq \frac{1}{(T\,\Delta f)^{1/2}} = \frac{1}{5^{1/2}} \simeq 0.447$$

or 45%. The relative errors in the correlation function are easily given for $\tau = 0$ and for large τ (Section 55.a). They are

$$\varepsilon(\tau = 0) \simeq \left(\frac{2}{T}\int_{-\infty}^{\infty} e^{-2\tau/\theta}\,d\tau\right)^{1/2}$$

and

$$\varepsilon \text{ (large } \tau) \simeq \left(\frac{1}{T}\int_{-\infty}^{\infty} e^{-2\tau/\theta}\,d\tau\right)^{1/2}$$

Evaluating these expressions

$$\frac{2}{T}\int_{-\infty}^{\infty} e^{-2\tau/\theta}\,d\tau = \frac{2\theta}{T}\int_{0}^{\infty} e^{-x}\,dx = \frac{2\theta}{T}$$

where $x = 2\tau/\theta$. Thus,

$$\varepsilon(\tau = 0) \simeq \left(\frac{2\theta}{T}\right)^{1/2} = \frac{1}{(500)^{1/2}} \simeq 0.0447$$

or 4.5%. Also,

$$\varepsilon \text{ (large } \tau) \simeq \left(\frac{\theta}{T}\right)^{1/2} = \frac{1}{(1000)^{1/2}} \simeq 0.0316$$

or 3%. Roughly speaking, two Lorentzians viewed as a correlation function and differing in either K or θ by about 5% would be resolved; the expected distribution of data points in the two cases would be clearly separate. Plotted on semi-logarithmic paper (log C–linear τ), the curves could be fit with a straight line for rapid estimates of K and θ.

Two Lorentzians differing by a few percent in their parameters would be practically indistinguishable as spectral densities if the expected error for each curve is 45%. This does not imply that nonlinear curve-fitting procedures and statistical tests on the data could not resolve the two curves. It merely points to the practical differences between using correlation function analysis or spectral density analysis on the same data.

We note that increasing the length of time of the experiment affects $\varepsilon(f)$ and $\varepsilon(\tau)$ in the same way. One way to decrease the expected error in the amplitude of $S(f)$ is to increase Δf. One should then calculate the effect of filter overlap and take into account how $S(f)$ may change over the domain of any filter. This method results in the integral spectral analysis.

The integral spectrum, like the correlation function, is composed of highly correlated data points. In this case, the reason for the correlation is obvious since every point of the integral spectrum intentionally shares a broad frequency range with its neighbors (see Section 45).

As an example, consider $1/f$ noise analyzed by the integral spectra and the spectral density method. If $S(f) = B/f$, then (as before)

$$\varepsilon(f) \simeq \frac{1}{(T \, \varDelta f)^{1/2}}$$

If $\varDelta f = 5$ Hz and $T = 1$ sec, the expected error is again about 45%. The expected error in $I(l) = B/2$ is (Section 55.b)

$$\varepsilon(l) \simeq \frac{1}{2} \left(\frac{\pi}{2l T} \right)^{1/2}$$

For 5 Hz and $T = 1$ sec,

$$\varepsilon(l = 5 \text{ Hz}) \simeq \frac{1}{2} \left(\frac{\pi}{10} \right)^{1/2} \simeq 0.28$$

or 28%. For 500 Hz

$$\varepsilon(l = 1000 \text{ Hz}) \simeq \frac{1}{2} \left(\frac{\pi}{1000} \right)^{1/2} \simeq 0.028$$

or 2.8%. Consider an experiment on B/f noise in which the effect of an external parameter on B is sought. For example, does temperature affect the magnitude of $1/f$ noise? In the above example, a 28% change is resolved in the integral spectrum at low frequencies; the data points cluster closer and closer about the line $I(l) = B/2$ as l becomes larger. In the spectral density, a 28% difference would practically be lost in the 45% error expected for each curve.

In effect, the integral spectrum smooths the spectral density and increases amplitude resolution at the expense of frequency resolution.

The smoothing effect of integral spectral analysis is explained by considering the process as a convolution. The relationship between the integral spectrum and the spectral density is

$$I(l) = \int_0^\infty S(f) K(l, f) \, df$$

where $K(l, f)$ is the square modulus of an L filter (Section 48). Under a variable transform, the relationship becomes

$$I(y) = \frac{1}{2} \int_{-\infty}^{\infty} e^{x/2}\, S(x)K(y - x)\, dx$$

where $f = \exp(x/2)$ and $l = \exp(y/2)$.

Let the transformed spectral density be $s(x)$, where

$$s(x) = \tfrac{1}{2}e^{x/2}S(x)$$

The convolution operation may be written (Section 41)

$$I(y) = s(y) * K(y)$$

The function $s(x)$ is drawn in x space $(-\infty < x < \infty)$. The filter is drawn in x space at some particular y, e.g., at $y = 0$, the function

$$K(y - x) = \frac{e^{x-y}}{(1 + e^{x-y})^2} = K(x - y)$$

becomes

$$K(-x) = K(x) = \frac{e^{x}}{(1 + e^{x})^2}$$

Imagine that $s(x)$ is stationary and that $K(x)$ shifts along the x axis. The area of the product $s(x)K(x)$ is plotted as a function of the shift. If $s(x)$ has some general trend about which it fluctuates, this process results in a smoother version of $s(x)$. An excellent discussion of the smoothing effect of convolution is given in Bracewell (1965, Chapter 3).

The spectral density may also be thought of as a convolution integral. Consider the double-sided spectral density $SS(f)$. By definition of the delta function (Section 24)

$$SS(f) = \int_{-\infty}^{\infty} SS(u)\, \delta(u - f)\, du$$

$SS(u)$ is playing the role of $s(x)$ and $\delta(u - f)$ the role of $K(y - x)$. The filter is now infinitely sharp and has no smoothing effect.

PROBLEM. To illustrate the features of convolution, a simple numerical example is suggested. Consider a "skyline" function of 20 steps separated by fixed Δx. Let

$$s(x) = 1, 4, 1, 5, 9, 2, 6, 5, 3, 5, 8, 9, 7, 9, 3, 2, 3, 8, 4, 6$$

Each value of $s(x)$ is constant over the interval $\Delta x = 1$. Consider a rectangular function $K(x)$ of height $1/2$ and width 2. Convolve $s(x)$ with $K(x)$. Hint: The first few values of the convolution are $2\frac{1}{2}$, $2\frac{1}{2}$, 3, 7, $5\frac{1}{2}$, etc. Repeat the process when $K(x)$ has height $\frac{1}{3}$ and width 3. The first few values are now 2, $3\frac{1}{3}$, 5, $5\frac{1}{3}$, $5\frac{2}{3}$, $4\frac{1}{3}$, etc. What happens if the filter is as narrow as the "graininess" of the function itself, i.e., $K(x)$ of height 1 and width 1?

57. Correcting Correlation Functions for ac Coupling

An advantage of the spectral density is the simplicity of certain operations in the frequency domain compared to the time domain. For example, multiplication in the frequency domain is equivalent to convolution in the time domain.

Consider noise passed through a high-pass filter. Filtering is necessary if the noise is superimposed on a large dc signal. If the noise fluctuations are small compared with the dc component, the dc component must be removed before the signal is amplified for analysis. One solution is to subtract the dc component: however, if the dc component varies in time some sort of dynamic subtraction is necessary. This procedure may be complicated and could itself introduce frequency components into the noise. In principle, however, subtraction does avoid the problem treated below.

AC coupling also removes the dc component of the noise signal. In the frequency domain correction for ac coupling is easily performed. A simple high-pass filter (Section 38) has the transfer function

$$Y = \frac{i\omega\tau_1}{1 + i\omega\tau_1} = \frac{i(f/f_1)}{1 + i(f/f_1)}$$

where $\tau_1 = C_1 R_1$ is the time constant of the circuit. If the spectrum to be measured, $S_i(f)$, is passed through such a filter, the output spectrum is simply

$$S_0(f) = S_i(f) \mid Y \mid^2$$

Dividing the output spectrum by $\mid Y \mid^2$ retrieves the original spectrum; since $\mid Y \mid^2$ is known, this is easily done. (Some workers prefer to use very sharp high-pass filters rather than the CR circuit of Section 38. In this case, the exact shape of $\mid Y \mid^2$ is ignored; $\mid Y \mid^2$ is assumed to be a perfect step below which all frequency components are perfectly removed and above which all are left undistorted. This is an approximation.)

Consider high-pass filtering in the time domain. Let the correlation function of the input noise be $K(\tau)$ and let $C(\tau)$ be the correlation function of the same noise passed through a filter. $C(\tau)$ will be corrected to give $K(\tau)$.

In the time domain, the corrected function is

$$K(\tau) = \overline{S_0/|\,Y\,|^2}$$

where the bar signifies the minus-*i* Fourier transform (Section 25). $K(\tau)$ may be derived from $C(\tau)$ by convolution with a function $k(\tau)$, thus

$$K(\tau) = k(\tau) * C(\tau)$$

By the convolution theorem (Section 41)

$$\bar{K} = \bar{k}\bar{C}$$

or

$$S_i = \bar{k}S_0$$

Solving for $\bar{k}$,

$$\bar{k} = \frac{S_i}{S_0} = \frac{1}{|\,Y\,|^2}$$

Taking the plus-*i* Fourier transform of both sides of this equation gives

$$k = \overrightarrow{1/|\,Y\,|^2}$$

Once the filter is known, $k(\tau)$ may be calculated. The corrected correlation function is then $k(\tau)$ convolved with $C(\tau)$.

For example, let Y be a simple CR high-pass filter. Then

$$\frac{1}{|\,Y\,|^2} = \frac{1 + (f/f_1)^2}{(f/f_1)^2} = 1 + \left(\frac{f_1}{f}\right)^2$$

Evaluating the plus-*i* transform of this expression

$$k(\tau) = \int_{-\infty}^{\infty} \left(1 + \frac{f_1^2}{f^2}\right) e^{i2\pi f \tau}\, df$$

$$= \delta(\tau) - 4\pi^2 f_1^2 \int_{-\infty}^{\infty} \frac{1}{(2\pi f)^2}\, e^{i2\pi f \tau}\, df$$

Let $\theta_1 = 1/2\pi f_1$. The integral is of the form CF # I-416 (in CF # I-416, let $\lambda = \mu = 0$):

$$\int_{-\infty}^{\infty} \frac{1}{(i2\pi f)^2} e^{i2\pi f\tau} \, df = \frac{1}{2} |\tau|$$

Therefore

$$k(\tau) = \delta(\tau) - \frac{1}{2\theta_1{}^2} |\tau|$$

Since

$$C(\tau) * \delta(\tau) = C(\tau)$$

the corrected correlation function $K(\tau)$ is related to the measured correlation function $C(\tau)$ and the ac coupling time constant θ by

$$K(\tau) = C(\tau) - \frac{1}{2\theta_1{}^2} |\tau| * C(\tau)$$

Double-sided correlation functions are implied.

To test the equation above, consider a process with a Lorentzian spectrum. Let $c = K\theta$. Then

$$S(f) = \frac{c}{1 + (f/f_\theta)^2}$$

The measured (ac-distorted) correlation function is (Section 52)

$$C(\tau) = \frac{K}{4} \frac{f_\theta}{f_\theta{}^2 - f_1{}^2} (f_\theta e^{-2\pi\tau f_\theta} - f_1 e^{-2\pi\tau f_1})$$

$$= \frac{c}{4} \frac{\theta_1{}^2}{\theta_1{}^2 - \theta^2} \left(\frac{e^{-\tau/\theta}}{\theta} - \frac{e^{-\tau/\theta_1}}{\theta_1} \right)$$

where $\theta_1 = 1/2\pi f_1$, $\theta = 1/2\pi f_\theta$ are the coupling and the process time constants, respectively. As $f_1 \to 0$, the correct correlation function $K(\tau)$ is obtained since the filter is effectively removed. The equation above reduces to the well-known relaxation formula:

$$K(\tau) = C(\tau) = \frac{c}{4\theta} e^{-\tau/\theta}$$

which is the time domain version of the Lorentzian.

In order to perform the appropriate transformations, the double-sided frequency and time domain expressions will be used. This was explicit in the evaluation of $k(\tau)$. The correlation functions shall all be the double-sided functions. In the present example

$$K(\tau) = \frac{c}{4\theta}\, e^{-|\tau|/\theta}$$

and so on.

It is required to evaluate $K(\tau)$, where $C(\tau)$ is the ac-distorted correlation function. The result should agree with $K(\tau)$ directly above. Omitting the subscripts, only convolutions of the form

$$|\tau| * \frac{1}{\theta}\, e^{-|\tau|/\theta}$$

are required. This equation may be written

$$\frac{1}{\theta}\, [\tau H(\tau) - \tau H(\tau)] * [H(\tau)e^{-\tau/\theta} + H(-\tau)e^{\tau/\theta}]$$

where $H(\tau)$ is the Heaviside step function (Section 26). The evaluation reduces to the sum of four integrals:

For $\tau > 0$:

$$\frac{1}{\theta} \int_{-\infty}^{\infty} (\tau - u)H(\tau - u)H(u)e^{-u/\theta}\, du = \frac{1}{\theta} \int_{0}^{\tau} (\tau - u)e^{-u/\theta}\, du$$

$$\frac{1}{\theta} \int_{-\infty}^{\infty} (\tau - u)H(\tau - u)H(-u)e^{u/\theta}\, du = \frac{1}{\theta} \int_{-\infty}^{0} (\tau - u)e^{-u/\theta}\, du$$

$$\frac{1}{\theta} \int_{-\infty}^{\infty} (-\tau + u)H(-\tau + u)H(u)e^{-u/\theta}\, du = \frac{1}{\theta} \int_{\tau}^{\infty} (-\tau + u)e^{-u/\theta}\, du$$

$$\frac{1}{\theta} \int_{\infty}^{\infty} (-\tau + u)H(-\tau + u)H(-u)e^{u/\theta}\, du = 0$$

Evaluating the three nonzero terms requires the integrals

$$\frac{1}{\theta} \int e^{-u/\theta}\, du = -e^{-u/\theta}$$

$$\frac{1}{\theta} \int e^{u/\theta}\, du = e^{u/\theta}$$

$$\frac{1}{\theta} \int u e^{-u/\theta}\, du = -(u + \theta)e^{-u/\theta}$$

$$\frac{1}{\theta} \int u e^{u/\theta}\, du = (u - \theta)e^{u/\theta}$$

In order of sequence, the three nonzero terms are

$$\frac{1}{\theta} \int_0^\tau (\tau - u)e^{-u/\theta} = \tau(-e^{-u/\theta})_0^\tau - [-(u + \theta)e^{-u/\theta}]_0^\tau$$
$$= \tau(-e^{-\tau/\theta} + 1) - [-(\tau + \theta)e^{-\tau/\theta} + \theta]$$
$$= \tau - \theta + \theta e^{-\tau/\theta}$$

and

$$\frac{1}{\theta} \int_{-\infty}^0 (\tau - u)e^{u/\theta} \, du = \tau + \theta$$

and

$$\frac{1}{\theta} \int_\tau^\infty (-\tau + u)e^{-u/\theta} \, du = \theta e^{-\tau/\theta}$$

Adding these three gives the value of $|\tau| * (1/\theta)e^{-|\tau|/\theta}$ for $\tau > 0$, namely

$$2\tau + 2\theta e^{-\tau/\theta}$$

PROBLEM. Show that for $\tau < 0$, $|\tau| * (1/\theta)e^{-|\tau|/\theta}$ has the value

$$-2\tau + 2\theta e^{\tau/\theta}$$

This is a special case of the general result that the convolution of two (real) even functions is also an even function.

It has therefore been shown that

$$|\tau| * \frac{1}{\theta} e^{-|\tau|/\theta} = 2|\tau| + 2\theta e^{-|\tau|/\theta}$$

We may now evaluate

$$K(\tau) = C(\tau) - \frac{1}{2\theta_1^2} |\tau| * C(\tau)$$

when

$$C(\tau) = \frac{c}{4} \frac{\theta_1^2}{\theta_1^2 - \theta_2} \left(\frac{1}{\theta} e^{-|\tau|/\theta} - \frac{1}{\theta_1} e^{-|\tau|/\theta_1} \right)$$

Namely,

$$\frac{4(\theta_1^2 - \theta^2)}{c\theta_1^2} K(\tau) = \left(\frac{e^{-\tau/\theta}}{\theta} - \frac{e^{-\tau/\theta_1}}{\theta_1} \right) - \frac{1}{2\theta_1^2} (2\tau + 2\theta e^{-\tau/\theta} - 2\tau - 2\theta_1 e^{-\tau/\theta_1})$$

$$= \left(\frac{1}{\theta} - \frac{\theta}{\theta_1^2} \right) e^{-\tau/\theta}$$

Rearranging the above expression gives

$$K(\tau) = \frac{c}{4} \left(\frac{\theta_1{}^2}{\theta_1{}^2 - \theta^2} \right) \left(\frac{\theta_1{}^2 - \theta^2}{\theta \theta_1{}^2} \right) e^{-\tau/\theta}$$

$$= \frac{c}{4} \frac{1}{\theta} e^{-\tau/\theta}$$

which agrees with the answer for $f_1 \to 0$ ($\tau > 0$).

PROBLEM. Show that for $\tau < 0$,

$$K(\tau) = \frac{c}{4} \frac{1}{\theta} e^{\tau/\theta}$$

and therefore

$$K(\tau) = \frac{c}{4} \frac{1}{\theta_0} e^{-|\tau|/\theta}$$

PROBLEM. Show by direct integration that the area under the product of $|\tau - \theta|$ and $e^{-|\tau|/\theta}$ agrees with $2|\tau| + 2\theta e^{-|\tau|/\theta} = |\tau| \cdot (1/\theta)e^{-|\tau|/\theta}$. Graph the functions $|\tau - \theta|$ and $e^{-|\tau|/\theta}$.

SOLUTION. Evaluate the three integrals:

$$\int_{-\infty}^{0} (-\tau + \theta)e^{\tau/\theta} \, d\tau$$

$$\int_{0}^{\theta} (-\tau + \theta)e^{-\tau/\theta} \, d\tau$$

$$\int_{\theta}^{\infty} (\tau - \theta)e^{-\tau/\theta} \, d\tau$$

Compare the sum of these with the convolution integral. Repeat for $\tau < 0$.

5

Noise Sources

Previous chapters have presented methods available for the analysis of noise. Examples included white, $1/f$, and Lorentzian noise, whose properties in three modes of analysis are given in Table 58.1.

These are one-sided functions and correspond to the usual conditions of measurement. If A is in units of V^2/Hz, then $S(f)$ is a function expressed in V^2/Hz vs. Hz, $I(l)$ in V^2 vs. Hz, and $C(\tau)$ in V^2 vs. time.

Some of the functions in Table 58.1 have no physical meaning. For example,

$$\sigma^2 = \int_0^\infty S(f)\,df = A \int_0^\infty df \to \infty$$

TABLE 58.1. *A Comparison of the Theoretical Expressions for White Noise, $1/f$ Noise, and Lorentzian Noise Analyzed as a Spectral Density $S(f)$, an Integral Spectrum $I(l)$, or a Correlation Function $C(\tau)$*

Noise	$S(f)$	$I(l)$	$C(\tau)$	
White	A	$\dfrac{A\pi}{4}l$	$\dfrac{A}{2}\delta(\tau)$	
$1/f$	$\dfrac{B}{f}$	$\dfrac{B}{2}$	$-B\,\mathrm{Ci}(2\pi f_0\tau)$	$f > f_0$
Lorentzian	$\dfrac{K(f/f_0)}{1+(f/f_0)^2}$	$\dfrac{K}{8}\dfrac{(l/f_0)}{(1+l/f_0)^2}$	$\dfrac{K}{4}e^{-\tau/\theta}$	

231

for white noise and

$$\sigma^2 = \int_0^\infty \frac{B}{f}\, df = B \ln f \Big|_0^\infty \to \infty$$

for $1/f$ noise. The variance of white noise is infinite because of the high frequency limit; the variance of $1/f$ noise is infinite at both limits of integration. These are nonphysical results since σ represents the expected deviation of the noise signal from some average value and σ^2 is proportional to the signal energy.

The singularities that occur in white and $1/f$ noise are removed by band limiting the noise. But the question remains whether or not there is such a thing as $S(f) = A$ noise, internal to some physical object, that we observe only in band-limited form? This problem is discussed below.

This chapter deals with the physical origin of the noise sources introduced in previous chapters. These sources are the most important ones for the interpretation of biological membrane noise.

A. WHITE NOISE

The term "white noise" simply expresses the property that the spectral density of the noise contains all frequencies equally:

$$S(f) = A$$

where A is independent of frequency. The term is analogous to the term "white light," loosely defined as light that contains all colors. Band-limiting white noise is analogous to passing white light through colored glass. The two phenomena are more than just analogous since both are electromagnetic processes; however, they usually occur in entirely different frequency ranges.

Nothing is implied about the source or the type of noise in the term "white noise." One could be referring to any fluctuation, not necessarily electrical noise. As used in this book, and by most workers in electrical noise, the term refers to a specific process first measured by Johnson in the 1920s.

58. Johnson Noise

J. B. Johnson was the first to measure electrical fluctuations due to thermal agitation of charge in conductors. The following paragraph was published in 1927, one year before his more detailed paper appeared in *Physical Review* (Johnson, 1927, pp. 367–368):

> Ordinary electrical conductors are sources of random voltage fluctuations, as the result of thermal agitation of the electrical charges in the conductor. The average effect of the fluctuations has been measured by means of a vacuum tube amplifier, where it manifests itself as a component of the phenomenon commonly called "tube noise." A part of the "tube noise" arises in the first tube and other elements of the apparatus; the remainder in the input resistance, with a mean square voltage fluctuation $(V^2)_m$ which is proportional to the resistance R of that conductor. The ratio $(V^2)_m/R$, of the order of 10^{-18} watt at room temperature, is independent of the material and shape of the conductor, but is proportional to the absolute temperature. In the range of audio frequencies, at least, the noise contains all frequencies at equal amplitudes. The noise of an input resistance of only 5000 ohms may exceed that of the rest of the circuit, so that the limit of useful amplification is at times set by the thermal agitation of charges in the input resistance of the amplifier.

Johnson studied tube noise before he studied thermal noise. Tube noise may be reduced as technology improves. Thermal noise is a fundamental lower limit of noise expected from any conductor; it may be reduced only by lowering the temperature or decreasing the bandwidth.

Johnson expressed his results in terms of power, namely,

$$\frac{(V^2)_m}{R} \simeq 10^{-18} \text{ W} \qquad \text{(room temperature)}$$

for any conductor. This emphasizes the independence of thermal noise on the material and the shape of the conductor. In our notation

$$(V^2)_m = \sigma_V^2$$

is the variance of the voltage noise across a resistor R. Johnson (1928) showed the constancy of voltage noise variance per unit resistor per unit bandwidth for both electronic and ionic conductors.

Spontaneous electrical fluctuations were predicted long before Johnson's experiments. Fürth (1956) has given a history of early thinking in fluctuation theory. Johnson was the first experimentalist to measure electrical fluctuations for which the origin was identified as the thermal agitation of electrical charge.

PROBLEM. Using Nyquist's formula for Johnson noise (see below) show that the 10^{-18} W quoted by Johnson implies a bandwidth of about 60 Hz.

The universal nature of Johnson's finding apparently inspired H. Nyquist to explain the experimental results from first principles. In Nyquist's words (Nyquist, 1927, p. 614):

> At the December, 1926, meeting of the Am. Phys. Soc., J. B. Johnson reported the discovery and measurement of an e.m.f. due to the thermal agitation in conductors. The present paper outlines a theoretical derivation of this effect. A nondissipative transmission line is brought into thermodynamic equilibrium with conductors of a definite temperature. The line is then isolated and its energy investigated statistically. The resultant formula is $E^2_\nu\, d\nu = 4kTR\, d\nu$ for the r.m.s. e.m.f. E_ν contributed in a frequency range one cycle wide by a network whose resistance component at the frequency ν is R. T and k are the absolute temperature and the Boltzmann constant. Experimental data are available for the audible range and there the agreement between the formula and the data is good. It will be observed that neither the charge nor the mass nor any other property of the carrier of electricity enters the formula explicitly. They enter indirectly through R. The formula above is based on the equipartition law. If the quantum distribution law is used the expression becomes
>
> $$E_\nu^2\, d\nu = [4h\nu R/(e^{h\nu/kT} - 1)]\, d\nu$$
>
> The two expressions are indistinguishable in the range of measurements.

Johnson noise is often referred to as Nyquist noise since the theoretical basis of Johnson noise is the Nyquist theorem.

Nyquist interprets Johnson's experiments by defining a source of electromotive force (emf) present in all conductors and due to the thermal agitation of charge. In effect, a real resistor is replaced by a noiseless resistor R in series with a voltage generator $e(t)$. This voltage generator (a random noise signal) has the property that its true spectral density is

$$S_e(f) = 4kTR = \text{const}$$

PROBLEM. Show that $4kTR$ has the units of V^2/Hz. Nyquist uses the symbol E_ν^2 for $S(f)$.

The open-circuit voltage from R and $e(t)$ in series (see Section 31) is

$$V(t) = e(t)$$

Therefore

$$S_V = S_e = 4kTR$$

The equivalent circuit of the noise source is convenient since it allows a rapid calculation of the open-circuit noise expected from any network.

PROBLEM.   Show that the Johnson noise measured from two resistors r_1 and r_2 in parallel has spectral density

$$S_v = 4kT \frac{r_1 r_2}{r_1 + r_2}$$

SOLUTION (see Sections 31 and 35).   By the cross-product rule, the open-circuit voltage is

$$V = \frac{e_1 r_2 + e_2 r_1}{r_1 + r_2} \quad \text{or} \quad \bar{V} = \frac{\bar{e}_1 r_2 + \bar{e}_2 r_1}{r_1 + r_2}$$

in the frequency domain. The spectral density is

$$S_V = \frac{S_{e_1} r_2{}^2 + S_{e_2} r_1{}^2}{(r_1 + r_2)^2}$$

Since $S_{e_1} = 4kTr_1$ and $S_{e_2} = 4kTr_2$, then

$$S_V = 4kT \frac{r_1 r_2{}^2 + r_2 r_1{}^2}{(r_1 + r_2)^2} = 4kT \frac{r_1 r_2}{r_1 + r_2}$$

PROBLEM.   Show that the open-circuit Johnson noise from a parallel RC circuit has spectral density

$$S_V = \frac{4kTR}{1 + \omega^2 \tau^2}, \quad \tau = RC$$

(The capacitor in the RC circuit is considered noiseless.)

SOLUTION.   If there is no noise source in C the cross-product rule gives an open-circuit voltage

$$\bar{V} = \frac{\bar{e}_1 (1/i\omega C)}{R + (1/i\omega C)} = \frac{\bar{e}_1}{1 + i\omega \tau}$$

Therefore

$$S_V = \frac{S_{e_1}}{1 + \omega^2 \tau^2} = \frac{4kTR}{1 + \omega^2 \tau^2}$$

PROBLEM.   Calculate the spectral density of Johnson noise expected from an *RrLC* circuit. The circuit has three branches: *R* in parallel with *C* parallel with *r* and *L* (*r* and *L* in series). The capacitor and inductor are considered noiseless.

PROBLEM.   Show that the variance of the Johnson noise from a parallel *RC* circuit is

$$\sigma_V{}^2 = \frac{kT}{C}$$

The variance is finite because higher frequencies from the noise source in *R* are shorted to ground through *C* (see Section 35). The variance does not depend on *R*; as *R* becomes larger, the noise source ($4kRT$) increases but the roll-off frequency ($1/2\pi RC$) decreases. The two effects cancel and leave the area under $S_V(f)$ constant for all *R*. On the other hand, a change in *C* affects only the roll-off.

The simplification introduced by Johnson and by Nyquist is the use of equivalent circuits to describe electrical noise. Their method of replacing real devices by noiseless elements in series or parallel with noise generators is used extensively in the analysis of biological membrane noise.

59. Derivation of the Nyquist Formula

The following derivation follows Nyquist (1928), although some steps have been expanded upon and the notation has been changed to conform to our own.

The derivation depends on two principles from equilibrium thermodynamics. The first is the principle of detailed balance. According to Bridgman (1928, pp. 90–102) this may be stated as follows:

> No system in thermal equilibrium in an environment at constant temperature spontaneously and of itself arrives in such a condition that any of the processes taking place in the system by which energy may be extracted, run in a preferred direction, without a compensating reverse process.

A complete review of the principle in its earliest forms is found in R. C. Tolman (1925, pp. 436–439).

Consider two resistors connected in parallel. The principle of detailed balance says that if both resistors are in thermal equilibrium and at the same constant temperature, the electrical fluctuations in them may not cause one resistor to cool or to heat at the expense of the other.

According to Nyquist, each resistor may be regarded as a noiseless resistor R in series with a noise generator $e(t)$. Let R_1 and e_1 stand for one resistor and let R_2 and e_2 stand for the other. The total current in the circuit formed by placing R_1 and R_2 in parallel is

$$i = \frac{e_1 + e_2}{R_1 + R_2}$$

Let $R_1 = R_2 = R$. The current due to R_1 is

$$i_1 = \frac{e_1}{2R}$$

and the current due to R_2 is

$$i_2 = \frac{e_2}{2R}$$

When a current i flows through a resistor R, the resistor heats up. The actual rise in temperature depends on the size and material of the resistor; however, the energy dissipated by the current per unit time (i.e., the power) is always given by $i^2 R$. The energy dissipated per unit time in R_1 due to the noise current from R_2 is

$$w_1 = (i_2)^2 R = \frac{e_2^2}{4R}$$

The energy dissipated per unit time in R_2 due to the noise current from R_1 is

$$w_2 = (i_1)^2 R_2 = \frac{e_1^2}{4R}$$

The principle of detailed balance states that on the average these two powers must be equal.

Nyquist's derivation of the formula for Johnson noise uses an infinitely long, nondissipative transmission line separating two resistors. This is a special case of the Heaviside line discussed in Section 88. The nondissipative line has zero longitudinal resistance and infinite shunt resistance, i.e.,

$$r'' = 0 \quad \text{and} \quad r_m \rightarrow \infty$$

Let l' be the inductance per unit length (l' is in units of H/cm) and c_m the capacitance per unit length (c_m is in F cm); the resistance of the nondissipative transmission line of length l is

$$l(l'/c_m)^{1/2}$$

If the transmission line is perfectly matched to the two terminal resistors there will be no reflection of signals at the ends of the line. The condition for a matched line separating R_1 and R_2 is

$$R_1 = R_2 = R = l(l'/c_m)^{1/2}$$

Signals on the line are transmitted as solutions to a wave equation at constant velocity θ (Section 88). The time required to transmit power between R_1 and R_2, or R_2 to R_1, is

$$t = l/\theta$$

Consider the state of the transmission line joining R_1 and R_2 for time t. The energy states of the line may be analyzed as one would a vibrating string drawn between two points. The vibrations are broken down into their individual frequency components indicated by the index n.

Let $e_1(t)$ be the signal on the line due to R_1. The modes of vibration of the transmission line are described by

$$e_1{}^n(t) = a_n \sin \omega_n t$$

where a_n is the amplitude of the vibration, $\omega_n = 2\pi f_n$, f_n is the natural frequency, and $t = l/\theta$. The lowest frequency that will fit on the line has a node at each end, i.e., one-half cycle of the sine wave. The condition which determines this lowest fundamental is

$$\omega_1 t = \pi \quad \text{or} \quad f_1 = \theta/2l$$

The higher fundamental frequencies on the line are determined by

$$\omega_2 t = 2\pi \quad \text{or} \quad f_2 = \theta/l$$
$$\omega_3 t = 3\pi \quad \text{or} \quad f_3 = 3\theta/2l$$
$$\omega_4 t = 4\pi \quad \text{or} \quad f_4 = 2\theta/l$$

and so on.

In classical physics the number of modes of vibration on the transmission is unbounded. However, in a limited frequency range Δf, the number of allowed modes is finite. The number of allowed modes in Δf is given by

$$\text{number of modes} = \frac{\Delta f}{f_1} = \frac{2l}{\theta} \Delta f$$

PROBLEM. Verify that the above formula is reasonable by working out a few cases, e.g., consider the number of modes between the fourth and first:

$$\text{number of modes} = \frac{f_4 - f_1}{f_1} = \frac{2l}{\theta}\left(\frac{2\theta}{l} - \frac{\theta}{2l}\right) = 3$$

Draw the first four modes of vibration on the line *l*. Derive a general expression for the number of modes between the *m*th and the *n*th ($m > n$).

From classical thermodynamics, the average energy associated with each degree of freedom is $\frac{1}{2}kT$, where k is Boltzmann's constant and T is the absolute temperature. The principle is true only for the kinetic energy of systems, although in special cases it applies to the potential energy as well. [A careful discussion may be found in Guggenheim (1967, Chapter 7).] One special case is the harmonic oscillator, and another is the *lc* transmission line discussed above.

A mode of vibration on the transmission line may be identified with a degree of freedom in the classical sense. For each mode of vibration there is an average electrical energy stored in the electric field of the capacitor and an average magnetic energy stored in the magnetic field of the inductor. The classical principle of the equipartition of energy states (for the non-dissipative transmission line) is

$$\frac{\text{average electrical energy}}{\text{mode of vibration}} = \frac{1}{2}kT$$

$$\frac{\text{average magnetic energy}}{\text{mode of vibration}} = \frac{1}{2}kT$$

The available energy is distributed equally among the various modes.

The total average energy on the line per mode of vibration is kT. Since we know the number of modes in the frequency range Δf, we may calculate the average energy in this range:

$$\langle \text{energy} \rangle_{\Delta f} = \frac{\text{total average energy}}{\text{mode}} \text{(number of modes)}$$

$$= kT\frac{\Delta f}{f_1} = \frac{2kTl}{\theta}\,\Delta f$$

The average energy delivered to the line during the time l/θ is the average power. Thus

$$\langle\text{power}\rangle_{\Delta f} = \frac{\langle\text{energy}\rangle_{\Delta f}}{l/\theta} = 2kT\,\Delta f$$

Of this power, one-half is due to R_1 and one-half to R_2, i.e.,

$$\langle w_1 \rangle_{\Delta f} = \frac{\langle e_2{}^2 \rangle_{\Delta f}}{4R} = kT\,\Delta f$$

$$\langle w_2 \rangle_{\Delta f} = \frac{\langle e_1{}^2 \rangle_{\Delta f}}{4R} = kT\,\Delta f$$

Therefore

$$\langle e_1{}^2 \rangle_{\Delta f} = \langle e_2{}^2 \rangle_{\Delta f} = 4kTR\,\Delta f$$

which is one form of the Johnson noise formula. Notice that

$$\frac{\langle e_1{}^2 \rangle_{\Delta f}}{\Delta f} = 4kTR$$

is the spectral density $S(f)$ if Δf is small (Section 39). The numerator is the variance of the noise source $e_1(t)$ in the frequency range Δf. In the limit of small Δf

$$S(f) = 4kTR$$

Since the location of Δf in the frequency domain was arbitrary, the above expression is valid for all frequencies. That $S(f)$ is frequency independent is a direct consequence of the equipartition of energy.

I have purposely followed Nyquist's 1928 derivation in detail to point out some of its features. The approach is satisfying since it ties a basic physical process to a fundamental law of thermodynamics. It may appear, however, that the factor 4 is a result of the particular conditions of the problem, viz., two resistors, $R_1 = R_2 = R$, connected by a matched, "lossless" transmission line. The factor appears more convincingly in Section 60. Also, the use of the equipartition theorem depends on there being a large number of modes of vibration in Δf (in order to speak of an average energy per mode) and yet our final step was to let Δf become arbitrarily small; since

$$\text{number of modes} = \frac{2l}{\theta}\,\Delta f$$

the number of modes also becomes small unless l is large (since θ is constant; Section 88). And l must be large to accommodate low frequencies. Is there a size limitation imposed by the derivation? Nyquist says only that one may consider R_1 internal to R_2 in order to avoid radiation losses from the line.

EXERCISE. A formula for Johnson noise may be derived directly from theories of electrical conduction. An example is found in Goldman (1948, p. 395). Goldman uses the theoretical expression of Drude for the resistivity of a metal; the resistivity is given by

$$\varrho = \frac{6kT}{ne^2 D\bar{v}}$$

where D and $\bar{v}$ are the mean free path and the mean thermal velocity of free electrons in the metal. Compare this expression with ϱ in Section 13. Drude's formula is derived by Jeans elsewhere (Jeans, 1925, pp. 226 and 303). This approach arrives at the expression

$$S(f) = 3kTR$$

instead of $4kTR$. Nyquist's method does not depend on any particular theory of electrical resistance.

In our discussion of Nyquist's formula, the spectral density was the voltage spectral density. To emphasize this a subscript V is used:

$$S_V = 4kTR \ominus \text{V}^2/\text{Hz}$$

Instead of visualizing a voltage noise generator $e(t)$ in series with R, one may picture a current noise generator $i(t)$ in parallel with R (see Section 31). The current generator is related to the voltage generator by

$$i = \frac{e}{R}$$

Therefore,

$$S_i = \frac{S_e}{R^2}$$

Since the open-circuit voltage $V = e$, and since the closed-circuit current $I = i$,

$$S_I = \frac{4kT}{R} \ominus \text{A}^2/\text{Hz}$$

The power spectral density (Section 37) is given by

$$S_W = 4kT \ominus \text{W/Hz}$$

The unit W/Hz $\ominus$ energy; S_W emphasizes the universal nature of Johnson noise and its independence of the type of resistor being considered.

An interesting account of early work on Johnson noise is found in E. B. Moullin (1938). In this book Nyquist's ingenious proof is severely criticized; Moullin sees no reason to replace a real resistor by a noiseless one in series with a noise emf. Other points of the derivation are also questioned, although the result is accepted by Moullin "as true and of immense convenience in calculation" (Moullin, 1938, p. 39). Moullin describes experiments by F. C. Williams in which circuits at different temperatures are connected. Johnson noise has been measured from a circuit in which the capacitor and resistor were held at different temperatures. Varying the temperature of the capacitance did not affect the noise and eliminated it as the source of thermal noise (Moullin, 1938, p. 33).

PROBLEM. Show that the maximum power a voltage source $e(t)$ in series with R can deliver to an external resistor is

$$w^{\text{max}} = \frac{e^2}{4R}$$

Consider two resistors in parallel where $R_1 \neq R_2$. The power delivered to R_2 from R_1 is

$$w_2 = i_1{}^2 R_2 = \frac{e_1{}^2}{(R_1 + R_2)^2} R_2$$

Since the maximum power R_1 may deliver is $e_1{}^2/4R_1$, the reflected power is

$$\frac{e_1{}^2}{4R_1} - \frac{e_1{}^2 R_2}{(R_1 + R_2)^2} = \frac{e_1{}^2}{4R_1} \frac{(R_2 - R_1)^2}{(R_1 + R_2)^2}$$

The coefficient of reflection is defined as

$$r = \frac{R_2 - R_1}{R_2 + R_1}$$

By this definition the reflected power is the delivered power times $|r|^2$ and the absorbed power is the delivered power times $1 - |r|^2$. The absolute value signs are used to allow for complex impedances. This case is treated in Section 60.

60. Nyquist Formula for an Arbitrary Impedance

In Section 59 only pure resistors were considered. Nyquist (1928) also derived a formula for the Johnson noise expected from an arbitrary impedance Z. The formula is

$$S(f) = 4kT\mathrm{Re}Z$$

where

$$Z = \mathrm{Re}Z + \mathrm{Im}Z$$

and ReZ is the real part and ImZ is the imaginary part of Z.

EXAMPLE. Consider an *RC* circuit, *R* and *C* in parallel. The real and imaginary parts of its impedance are given by

$$Z = \frac{R}{1 + i\omega\tau} = \frac{R}{1 + \omega^2\tau^2} + i\,\frac{-\omega R}{1 + \omega^2\tau^2}$$

where $\tau = RC$. The Nyquist formula gives

$$S(f) = 4kT\,\frac{R}{1 + \omega^2\tau^2}$$

Compare this approach with the problems given at the end of Section 58 in which $S(f)$ was obtained directly from circuit theory for specific impedances.

To demonstrate Nyquist's formula consider a resistance *R* in parallel with an impedance Z. Part of the energy transferred from *R* to Z and from Z to *R* will be reflected, but each must deliver (on average) the same power to the other if the circuit is at constant temperature and in equilibrium.

In this case the coefficient of reflection (Section 59) is

$$\varrho = \frac{Z - R}{Z + R}$$

PROBLEM. Show that

$$1 - |\varrho|^2 = \frac{4R(\mathrm{Re}\,Z)}{(\mathrm{Re}\,Z + R)^2 + (\mathrm{Im}\,Z)^2}$$

The absorbed power at Z due to R is

$$\frac{e_R^2}{4R}(1 - |\varrho|^2) = \frac{e_R^2 \text{Re } Z}{(\text{Re } Z + R)^2 + (\text{Im } Z)^2}$$

From Section 59 the mean square value of e_R in the frequency band Δf is

$$\langle e_R \rangle_{\Delta f} = 4kTR \ \Delta f$$

Therefore, the average absorbed power at Z due to R in the frequency band Δf is

$$\langle w_Z \rangle_{\Delta f} = \frac{4kTR(\text{Re } Z)}{(\text{Re } Z + R)^2 + (\text{Im } Z)^2} \ \Delta f$$

This must equal (on the average) the absorbed power at R due to Z

$$\langle w_R \rangle_{\Delta f} = \frac{4kT(\text{Re } Z)R}{(\text{Re } Z + R)^2 + (\text{Im } Z)^2} \ \Delta f$$

The power absorbed in R due to Z is

$$w_R = i_R^2 R = \frac{e_Z^2}{|R + Z|^2} R = \frac{e_Z^2 R}{(\text{Re } Z + R)^2 + (\text{Im})^2}$$

and therefore

$$\langle w_R \rangle_{\Delta f} = \frac{\langle e_Z^2 \rangle_{\Delta f} R}{(\text{Re } Z + R)^2 + (\text{Im } Z)^2}$$

Comparing the last two expressions,

$$\langle e_Z^2 \rangle_{\Delta f} = 4kT \text{ Re } Z \ \Delta f$$

For Δf arbitrarily small,

$$S(f) = 4kT\text{Re}Z$$

PROBLEM. Show that a corollary to the above result is that the maximum power that can be extracted from an impedance Z in series with a voltage generator e_Z is

$$e_Z^2/4\text{Re}Z$$

PROBLEM. In the derivation of $S(f) = 4kTR$ for a pure resistor and $S(f) = 4kT\text{Re}Z$ for an arbitrary impedance, the distinction between $e(t)$ and $\bar{e}(f)$ was dropped. Argue from Rayleigh's theorem (Section 34) that this is legitimate.

Johnson noise from an arbitrary impedance may be calculated in two ways. The simplest is to calculate ReZ directly and use Nyquist's formula. The second method associates a noise generator with spectral density $4kTR$ to each resistive component and one calculates the open-circuit voltage noise from circuit theory. All capacitors and inductors are considered noiseless. These methods are equivalent. Some interesting questions arise when the impedance being considered is the phenomenological impedance of the excitable membrane. These are discussed in Section 82.

61. Quantum Theory Formulation of Nyquist's Equation

Nyquist's equation $S(f) = 4kTR$ conflicts with the definition of the variance

$$\sigma^2 = \int_0^\infty S(f)\, df$$

since it implies an infinite variance. Nyquist (1928) replaced the classical equipartition law by the substitution

$$kT \to \frac{hf}{e^{hf/kT} - 1}$$

where $h = 6.6256 \times 10^{-34}$ J sec is Planck's constant. The expansion

$$e^u \simeq 1 + u + \cdots, \qquad u \ll 1$$

implies

$$\frac{hf}{e^{hf/kT} - 1} \simeq kT$$

for low frequencies. The condition for which the approximation is valid is

$$\frac{hf}{kT} \ll 1$$

or

$$f \ll \frac{k}{h} T$$

Since

$$\frac{k}{h} = \frac{1.38 \times 10^{-23} \text{ J/}^\circ\text{K}}{6.63 \times 10^{-34} \text{ J sec}} = 0.2 \times 10^{11}/^\circ\text{K sec}$$

the condition on frequency at room temperature $(T \simeq 300°K)$ is

$$f \ll 0.2 \times 10^{11} \times 300 \quad \text{sec}^{-1}$$

or

$$f \ll 6 \times 10^{12} \text{ Hz}$$

One would have to work at very low temperatures or unusually high frequencies to measure the quantum effect on Johnson noise. To the author's knowledge, no experimental verification of the quantum formulation of Nyquist's equation has been published.

The equipartition of kinetic energy in quantum theory states that the average energy per degree of freedom is actually

$$\frac{1}{2} hf + \frac{hf}{e^{hf/kT} - 1}$$

where $\frac{1}{2}hf$ is called the zero point energy. It is a point of controversy whether or not the zero point energy should be included in Nyquist's formula for Johnson noise. If it is included, the problem of an infinite variance arises once again, for

$$\int_0^\infty (\tfrac{1}{2} hf) \, df \to \infty$$

The second term is integrable. Let $u = hf/kT$; then

$$\int_0^\infty \frac{hf \, df}{e^{hf/kT} - 1} = \frac{(kT)^2}{h} \int_0^\infty \frac{u \, du}{e^u - 1}$$

The integral on the right-hand side has the value

$$\int_0^\infty \frac{du}{e^u - 1} = \frac{\pi^2}{6}$$

Excluding the zero point energy, the quantum theory variance for Johnson noise from a pure resistor R is

$$\sigma^2 = 4kTR \, \frac{kT}{h} \, \frac{\pi^2}{6}$$

The total Johnson noise power is

$$\frac{\sigma^2}{R} = \frac{1}{6h} \, (2\pi kT)^2 = 1.7 \times 10^{-7} \text{ W}$$

at room temperature.

PROBLEM. Show that at room temperature, the total open band variance from a 10-MΩ resistor is about 1.7 V², which implies an rms deviation of about 1.3 V.

EXERCISE. Compare the classical distribution laws with those for two classes of quantum particles. The classical Maxwell–Boltzmann distribution laws apply to identical but distinguishable particles. Maxwell–Boltzmann statistics predict that the average number of particles having energy ε is proportional to

$$e^{-\varepsilon/kT}$$

This is the most probable distribution of energy for classical particles. The constant of proportionality is $\exp(\mu/kT)$, where μ is the chemical potential (the free energy per molecule).

The distribution law that applies to identical, indistinguishable particles having integral spin is called the Bose–Einstein distribution law. Such particles are called bosons, the most common example being a photon of light. The average number of bosons having energy ε is

$$\langle N \rangle = \frac{1}{e^{(\varepsilon-\mu)/kT} - 1}$$

For photons $\mu = 0$ and $\varepsilon = hf$. The average number of photons with the energy hf is

$$\langle N \rangle = \frac{1}{e^{hf/kT} - 1}$$

The average energy associated with this set of photons is the energy per photon (hf) times the average number of photons, or

$$\langle E \rangle = \frac{hf}{e^{hf/kT} - 1}$$

This is the quantum theory form of the equipartition theorem. The equipartition theorem is a theorem about kinetic energy. Only in special cases (e.g., the harmonic oscillator) may an equivalent generalization be made about potential energy (see Guggenheim, 1967, Chap. 7). The above expression for $\langle E \rangle$ refers to the average total energy; it reduces to the average total energy (kinetic plus potential) of the classical harmonic oscillator ($\frac{1}{2}kT + \frac{1}{2}kT$) in the limit of low frequencies or high temperatures (Guggenheim, 1955, p. 36). The quantum form of the equipartition theorem for photons was used by Nyquist (1928); the photon is the quantized unit of the electromagnetic field set up on the hypothetical transmission line between the two resistors.

The distribution law that applies to identical, indistinguishable particles having half-integral spin is called the Fermi–Dirac distribution law. Such particles are called fermions, the most common example being an electron. The average number of fermions having energy ε is

$$\langle N \rangle = \frac{1}{e^{(\varepsilon - \mu)/kT} + 1}$$

which differs from bosons only in the sign in the denominator (Guggenheim, 1967, p. 36).*

It is interesting to approach Johnson noise from the point of view of the resistor. Landau and Lifshitz (1959, Chap. 5) give the distribution of free electrons in metals. At very low temperatures the electrons will occupy the lowest energy levels in the metal; N electrons may occupy the lowest $N/2$ levels (two per level). In the classical theory of an electron gas, the average kinetic energy of an electron at temperature T is

$$\langle E \rangle = \frac{3}{2} kT$$

($\frac{1}{2}kT$ per degree of freedom). At $T = 0$ this energy is zero. All classical electrons try to crowd into the lowest level; this is forbidden in the quantum theory. An excellent discussion of this is given in Smith (1961) which shows that the maximum energy of the lowest of electron configurations is

$$E_m = \left(\frac{3}{\pi} \right)^{2/3} \left(\frac{h}{8m} \right) n^{2/3}$$

where m is the electron mass and n is the concentration of free electrons. The average kinetic energy per electron is

$$\langle E \rangle = \frac{3}{5} E_m$$

Even in this lowest energy condition, $\langle E \rangle$ is quite large. Smith gives E_m in the range 2.1 to 7.2 eV for some common metals (Smith 1961, Chap. 2).

---

* A careful discussion of the boson and fermion distribution laws is given in Landau and Lifshitz (1959, Chap. 5).

PROBLEM. Calculate kT at room temperature in electron volts.

A more difficult problem is to calculate $\langle E \rangle$ for electrons at an arbitrary temperature. This is also done by Smith (1961, p. 38). The average kinetic energy per electron is approximately

$$\langle E \rangle = \frac{3}{5} E_m \left[1 + \frac{5\pi^2}{12} \left(\frac{kT}{E_m} \right)^2 + \cdots \right]$$

(cf. Eq. 88 in Smith, 1961, Chap. 2). The first term is the average kinetic energy of the lowest state, i.e., the zero point energy. (Recall that the zero point energy for photons is $hf/2$). The second term is the approximate average thermal energy per electron; the second term can be rewritten as

$$\frac{3}{2} kT \frac{kT}{E_m} \frac{\pi^2}{6} \ominus \frac{\text{energy}}{\text{electron}}$$

The classical expression is simply $3kT/2$.

EXERCISE. Calculate the average thermal energy of an electron in silver [see Table 2.1 in Smith (1961) for the value of E_m]. Compare this energy with the classical value.

According to the quantum theory, the Johnson noise power from any resistor is

$$4kT \frac{kT}{h} \frac{\pi^2}{6} \ominus \frac{\text{energy}}{\text{time}}$$

The ratio of the last two formulas defines a fundamental time for an electron in a metal with maximum zero point energy E_m, viz.,

$$\Delta t = \frac{3}{8} \frac{h}{E_m} \simeq 0.3 \times 10^{-15} \text{ sec}$$

for $E_m = 5$ eV.

PROBLEM. What is the physical meaning of Δt?

Suggested Reading. For an excellent discussion of the relationship between classical and quantum radiation laws and their relationship to noise theory see Chap. VIII of Born's *Natural Philosophy of Cause and Chance* (1949); Chap. IX deals with the concept of probability as it appears in quantum mechanics. Highly recommended.

62. *Johnson Noise and the Nernst Equation*

Consider a perfect membrane permeable to one ionic species whose concentration on one side of the membrane is n'' and on the other side is n'. The Nernst potential (Section 14) is

$$V'' - V' = -\frac{kT}{ze} \ln \frac{n''}{n'}$$

Let $V = V'' - V'$. Then

$$V = \frac{kT}{ze} (\ln n' - \ln n'')$$

This is an equilibrium potential. Consider deviations in V due to deviations in n from their mean values n' and n''. Formally,

$$\delta V = \frac{kT}{ze} \left(\frac{\delta n'}{n'} - \frac{\delta n''}{n''} \right)$$

The mean square value of these deviations is

$$\langle \delta V^2 \rangle = \left(\frac{kT}{ze} \right)^2 \left[\left\langle \left(\frac{\delta n'}{n'} \right)^2 \right\rangle + \left\langle \left(\frac{\delta n''}{n''} \right)^2 \right\rangle - 2 \left\langle \frac{\delta n' \, \delta n''}{n' n''} \right\rangle \right]$$

If the total number of particles is constant,

$$n' + n'' = \text{const}$$

then

$$\delta n' = -\delta n''$$

In the limit that $n' = n'' = n$,

$$\langle \delta V^2 \rangle = \left(\frac{kT}{ze} \right)^2 \left[4 \left\langle \frac{\delta n^2}{n^2} \right\rangle \right]$$

Let N be the total number of particles and v be the volume of the sample. Therefore $n = N/v$. If

$$\langle \delta N^2 \rangle = N$$

then

$$\langle \delta n^2 \rangle = \left\langle \delta \left(\frac{N}{v} \right)^2 \right\rangle = \left(\frac{1}{v^2} \right) N$$

and

$$\left\langle \frac{\delta n^2}{n^2} \right\rangle = \frac{1}{N}$$

Substituting this expression into the last formula for $\langle \delta V^2 \rangle$ gives

$$\langle \delta V^2 \rangle = 4 \left(\frac{kT}{ze} \right)^2 \frac{1}{N}$$

Since the system is in equilibrium, the average fluctuation just calculated must equal that predicted by the Nyquist formula for Johnson noise, viz.,

$$\langle \delta V^2 \rangle = 4kTR \, \Delta f$$

Equating these expressions gives

$$R \, \Delta f = \frac{kT}{z^2 e^2 N}$$

This may be termed the fundamental resistance of a conductor with N charge carriers of valence z. Define $G = 1/R$; then

$$G = N \left(\frac{z^2 e^2}{kT} \right) \Delta f$$

is the fundamental conductance.

PROBLEM.   Show that at room temperature

$$\frac{e^2}{kT} \simeq 6 \times 10^{-18} \text{ F}$$

Let

$$\Delta f = \frac{kT}{h} \frac{\pi^2}{6} \simeq 0.1 \times 10^{14} \text{ Hz}$$

be the fundamental bandwidth of a sample at room temperature. This bandwidth is imposed by the quantum theory of Johnson noise (Section 61). The fundamental conductance per charge carrier of valence ± 1 is

$$\frac{G}{N} = \left(\frac{e^2}{kT} \right) \Delta f \simeq 0.6 \times 10^{-4} \, \Omega^{-1}$$

at room temperature.

PROBLEM. Calculate the fundamental resistance of 1 cm³ of Cu ($n = 8.7 \times 10^{22}/\text{cm}^3$).

PROBLEM. Show that if the conductivity of a material is given by

$$\sigma = z^2 e^2 n u$$

where u is the mobility of the charge carrier, then

$$u = \frac{l^2}{kT} \Delta f$$

where l is the length of the conductor. Calculate $\Delta f/kT$ in J sec. By this definition, the fundamental mobility depends on the length of the conductor.

63. *Measurement of Johnson Noise*

The original paper by Johnson (1928) gives a detailed account of the technical problems he had to solve. Today low-noise amplifiers are available for which the measurement of Johnson noise is routine. A popular choice is the Princeton Applied Research model 113 (PAR 113). All of the data presented in this section were measured using this amplifier.

One advantage of the PAR 113 is that it is inherently an integral spectrometer (Section 45). The high and low bandpass limits of the amplifier are simple RC filters. Although the high- and low-frequency filters are not perfectly uncoupled, they do approximate a transfer function of the form Y, where

$$| Y |^2 = \frac{f_2^2 f^2}{(f^2 + f_1^2)(f^2 + f_2^2)}$$

$f_1 = 1/2\pi R_1 C_1$ is the lower corner frequency and $f_2 = 1/2\pi R_2 C_2$ is the upper corner frequency. When $f_1 = f_2 = f_0$,

$$| Y |^2 = | L |^2 = \frac{f_0^2 f^2}{(f^2 + f_0)^2}$$

f_0 is the center frequency of an L filter (see Figure 45.2). When f_0 is replaced by the symbol l, it is considered a continuous variable (filters infinitely close). Here we consider only point spectra. Figure 63.1 shows the observed noise from a PAR 113 amplifier with a resistor R at its input. In this figure $f_1 = 0.3$ Hz and $f_2 = 3000$ Hz. R varies between 0 and 100 MΩ.

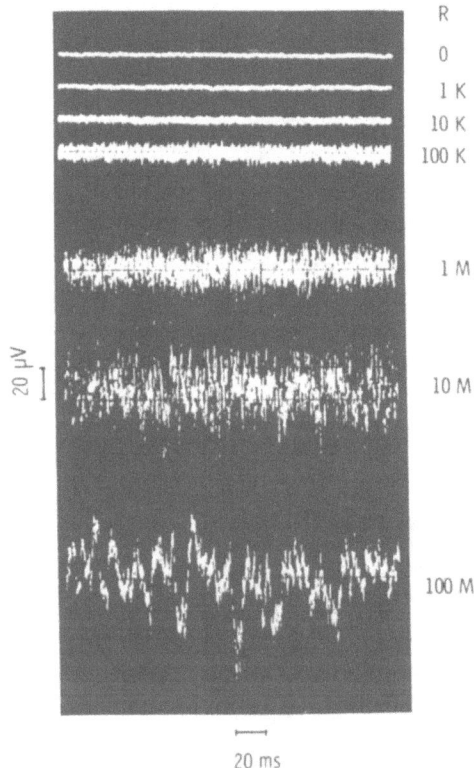

Figure 63.1. Johnson noise from resistors between 0 and 100 MΩ. Bandwidth is constant for all traces (limited by a CR filter at 0.3 Hz at the low end and an RC filter at 3000 Hz at the high end). The trace at 100 MΩ shows low-frequency fluctuations due to filtering by stray capacitance at the input and a contribution of amplifier current noise. (Courtesy D. L. Ypey.)

A rough calculation shows that the observed noise agrees with the Nyquist formula. Consider $R = 1$ MΩ in Figure 63.1. The expected variance of the noise is

$$\sigma^2 = 4kTR \int_0^\infty \frac{f_2^2 f\, df}{(f^2 + f_1^2)(f^2 + f_2^2)}$$

The integral is of standard form; from Section 52

$$\sigma^2 = 4kTR \frac{\pi}{2} \frac{f_2^2}{f_1 + f_2}$$

For $R = 10^6\ \Omega$, $f_1 = 0.3$ Hz, $f_2 = 3000$ Hz, and room temperature,

$$4kT \simeq 1.6 \times 10^{-20}\ \text{J}$$

and

$$\sigma^2 \simeq 1.6 \times 10^{-14}\left(\frac{\pi}{2}(3000)\right)$$

$$\simeq 7.2 \times 10^{-11}\ \text{V}^2$$

The expected peak-to-peak noise is about 2.8σ (Section 33), therefore, the peak-to-peak Johnson noise from the 1-MΩ resistor in Figure 63.1 should be about

$$pp \simeq 2.8(72 \times 10^{-12})^{1/2} \simeq 24 \; \mu V$$

Increasing R by 10 increases the peak-to-peak noise by $(10)^{1/2}$. The calculated Johnson noise agrees well with the noise traces shown in Figure 63.1 for R between 100 KΩ and 10 MΩ.

In Figure 63.2 R is held constant and the bandwidth is varied. $R = 0.6$ MΩ in all cases. The top trace shows the output noise for $f_1 = 3$ Hz and $f_2 = 1000$ Hz. The bottom six traces have $f_1 = f_2 = f_0$. f_0 varies between 1000 Hz and 3 Hz.

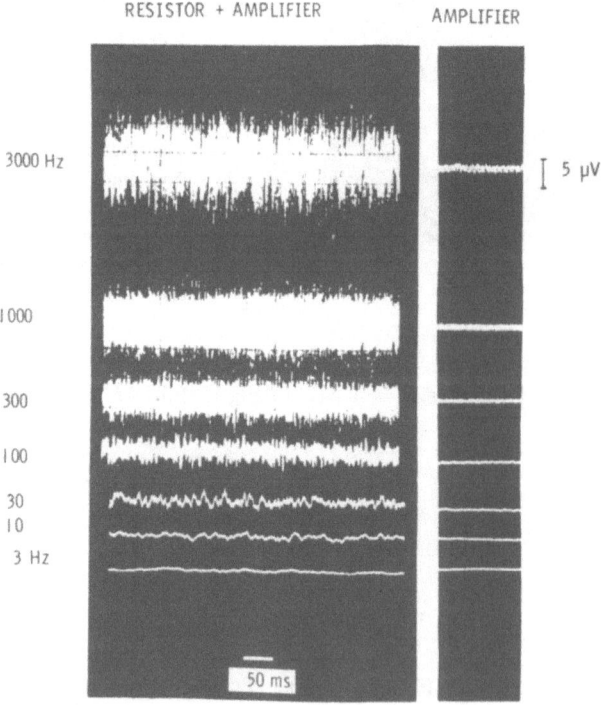

Figure 63.2. Johnson noise from a 0.6-MΩ resistor. The top trace shows broadband noise limited by a *CR* filter at the lower end (3 Hz) and an *RC* filter at the high end (3000 Hz). The bottom six traces show Johnson noise band limited by *L* filters (Section 45) centered at 3, 10, 30, 100, 300, and 1000 Hz. Zero net current flows through the resistor. Amplifier voltage noise (shorted input) is shown to the right. (Courtesy D. L. Ypey.)

The effective bandwidth of an L filter is

$$\int_0^\infty \frac{f_0^2 f^2}{f^2 + f_0^2} \, df = \frac{\pi}{2} f_0$$

PROBLEM. Calculate the expected peak-to-peak Johnson noise from a 0.6-MΩ resistor if the noise is passed through an L filter having $f_0 = 100$ Hz. Compare this result with Figure 63.2.

Increasing f_0 by 10 increases the peak-to-peak noise by $(10)^{1/2}$. In Figure 63.2 the noise traces for $f_0 = 10$, 100, and 1000 Hz do increase roughly by factors of 3.

The right-hand traces shown in Figure 63.2 were measured by grounding the input ($R = 0$) for each bandpass. These traces represent the amplifier's inherent voltage noise; in addition to voltage noise, amplifiers may have current noise sources at their input. The close agreement between the measured noise and the expected Johnson noise indicates that the total contribution of the amplifier is small.

The voltage fluctuations shown in Figures 63.1 and 63.2 result directly from the thermal motion of electrons in the resistor. The movement of an electron is a current and this current, dropped across the resistor, appears as a voltage. At any particular moment, all the electronic currents do not exactly cancel one another; this results in a net voltage fluctuation about a mean value of zero. Figures 63.1 and 63.2 show, so to speak, the movement of electrons in conductors.

In Figure 63.3 (bottom two traces) the variance of the amplifier voltage noise and the measured resistor noise are plotted against the center frequency of six L filters (Section 45). Schematic diagrams are shown to indicate the experimental setup. For the amplifier noise (bottom trace) the input of the PAR 113 is grounded. For the resistor noise (middle trace) the PAR has two 1.2-MΩ resistors at its input. Thus $R = 0.6$ MΩ. (The reason for using two parallel resistors will become apparent when we allow current to flow in the circuit.)

Figure 63.3 is a point integral spectrum (Section 45). Theoretically, the integral spectrum for Johnson noise is

$$L(l) = \int_0^\infty S(f) \, | \, L \, |^2 \, df$$

where

$$S(f) = 4kTR$$

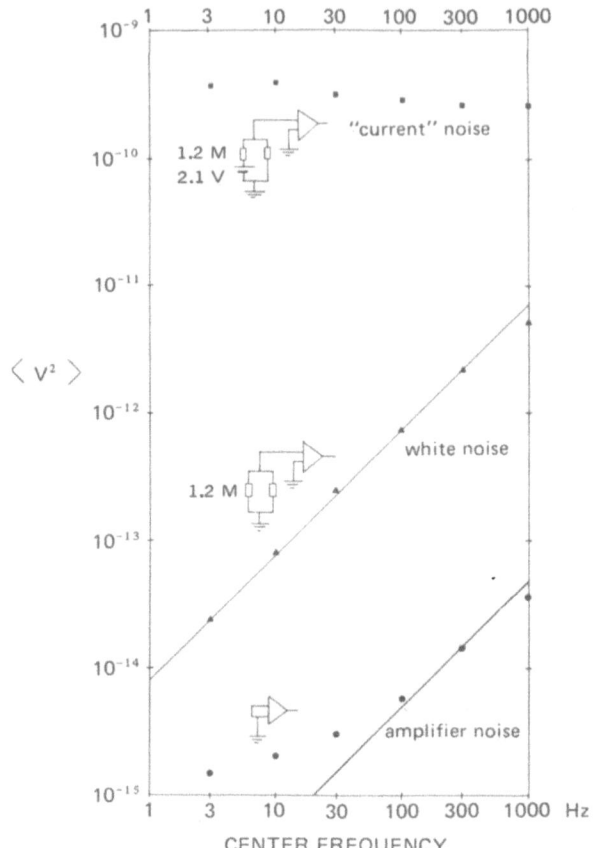

Figure 63.3. The variance of voltage noise against the center frequency of six L filters used to analyze the noise. Three experiments are shown. In the bottom experiment, the input of the amplifier is shorted; the straight line represents white noise. In the middle experiment, the input is a 0.6-MΩ resistor with zero net current; the straight line represents the theoretical Johnson noise expected from this resistor. In the top experiment, the equivalent input is a 0.6-MΩ resistor; a battery is introduced in the circuit to provide a net current flow. The experimental points (solid squares) approximate $1/f$ noise. (Courtesy D. L. Ypey.)

and

$$| L |^2 = \frac{l^2 f^2}{f^2 + l^2}$$

From Section 47, the integral spectrum for Johnson noise is

$$I(l) = 4kTR\left(\frac{\pi}{4} l\right) = \pi kTRl$$

The straight line through the "white noise" curve in Figure 63.3 is $I(l)$ with $R = 0.6$ MΩ. The fit is good except near $f_0 = 1000$ Hz. The low value is due to a decreased bandwidth of the L filter at 1000 Hz caused by stray capacitance at the input.

Amplifier noise is not white over the entire frequency range. Recall that B/f noise (Section 47) has the integral spectrum

$$I(l) = \frac{B}{2}$$

The amplifier noise may be approximated by a sum of $1/f$ plus white noise, with the $1/f$ component dominating at the lower frequencies. Johnson noise is one order of magnitude above background noise at $f_0 = 3$ Hz and two orders of magnitude above background at 300 Hz.

So far we have considered only those cases with zero net current through R. The top trace in Figure 63.3 was measured with a net current flowing through the resistor; this results in an extra noise component discussed in the next section.

B. 1/f NOISE

$1/f$ noise is literally noise with spectral density

$$S(f) = B/f$$

where B is constant for a particular steady-state experiment. The name has come to apply to noise of the form

$$S(f) = B/f^\alpha$$

where α is close to one. Other names are in use, for example, flicker noise. In this book, I have used the term "$1/f$ noise" rather loosely to mean excess noise in a resistor held away from equilibrium by passing a steady current.

$1/f$ noise may be obtained by filtering white noise. For example, the infinite RC cable (Section 29) is an appropriate filter (see Verveen and DeFelice, 1974). The infinite cable may be approximated by a cascaded linear network (Barnes and Jarvis, 1971). Such devices are useful in the construction of $1/f$ noise generators. In the following paragraphs, however, $1/f$ noise is generated by a physical process, not by filtering.

64. *Excess Noise in Carbon Resistors*

In contrast to Johnson noise, excess noise does depend on the resistor type. I have selected a common laboratory device, the carbon composition resistor, to illustrate the generation of excess noise. Even among carbon composition resistors some types are noisier than others and every case must be treated individually.

Compare the top diagram in Figure 63.3 (labeled "current noise") with the center diagram (labeled "white noise"). In both the top and center diagrams the input resistance to the amplifier is 0.6 MΩ. In the top circuit diagram, the average current is

$$I = \frac{2.1 \text{ V}}{2.4 \text{ M}\Omega} = 0.875 \ \mu\text{A}$$

In the center diagram the average current is

$$I = 0$$

The average voltage across the resistors in the top circuit is

$$V = \frac{(2.1 \text{ V})(1.2 \text{ M}\Omega)}{1.2 \text{ M}\Omega + 1.2 \text{ M}\Omega} = 1.05 \text{ V}$$

In the center diagram the average voltage is

$$V = 0$$

The average energy dissipated in the top circuit is *I–V* for each resistor, or

$$W = 2(0.875 \ \mu\text{A})(1.05 \text{ V})$$
$$= 1.8375 \ \mu\text{W}$$

Note that the last expression is also the square of the open-circuit voltage over the input resistance (0.6 MΩ) of the circuit. The average energy dissipated in the central circuit of Figure 63.3 is

$$W = 0$$

This is the condition of thermal equilibrium.

PROBLEM. Calculate the rms Johnson current noise, voltage noise, and power noise associated with two 1.2-MΩ resistors in parallel. Assume the maximum bandwidth (0.1×10^{14} Hz at room temperature, Section 61). Compare these with I, V, and W calculated above.

Excess noise from carbon composition resistors is not white. To illustrate this, compare Figure 63.2 with Figure 64.1. In Figure 64.1 a 1.0-V battery in the circuit gives 0.4 μA. Compare the 3–1000-Hz band in Figure 63.2 with the same band in Figure 64.1. The latter is over 10 times as large and has marked low-frequency components compared with the even, grassy appearance of white noise. (Note the tenfold voltage scale difference between the two figures.)

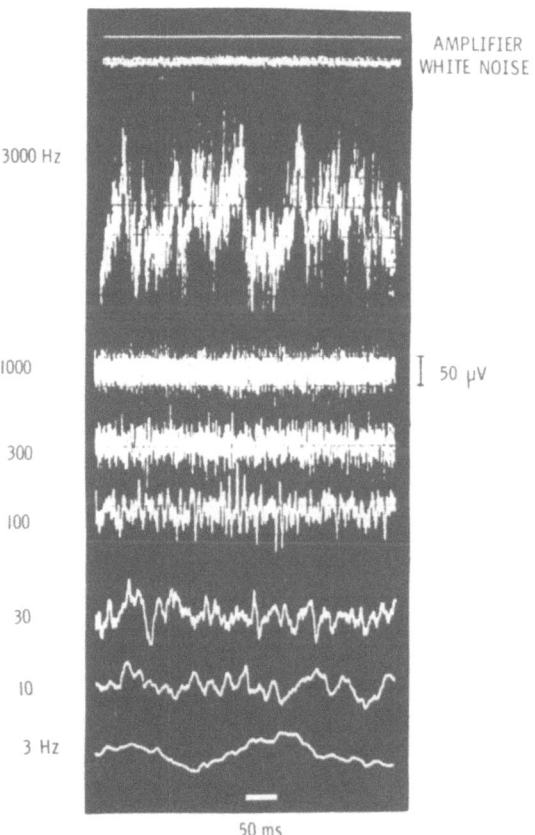

Figure 64.1. Johnson noise plus excess noise from two 1.2-MΩ resistors in parallel. A 1-V battery in the circuit causes a steady current to flow through the resistors. The bandwidths used are the same as in Figure 63.2, but note that the voltage scale is 10 times as large. Amplifier voltage noise (shorted input) and 3–1000 Hz Johnson noise from a 0.6-MΩ resistor are shown at the top. (Courtesy D. L. Ypey.)

Compare the bottom six traces in Figures 63.2 and 64.1. In both figures a bank of L filters is used with center frequencies f_0 between 3 and 1000 Hz. The bandwidths are

$$\Delta f = \frac{\pi}{4} f_0$$

The bandwidth of the analyzing L filter increases proportionally with the filter center frequency (Section 45). Johnson noise increases with f_0 (Figure 63.2) but excess noise is roughly independent of f_0 (Figure 64.1). The variance of excess noise against f_0 is approximately flat, as shown in Figure 63.3 (top); this is $1/f$ noise.

PROBLEM.   The noise in Figure 64.1 has a spectral density of the form

$$S(f) = B/f$$

Estimate B from the data shown.

In summary, $1/f$ noise has the same average amplitude in the bandwidth Δf centered at f_0 as it does in the bandwidth $10 \, \Delta f$ centered at $10 f_0$, etc. White noise would have one-third the amplitude.

65. Excess Noise Depends on Current

If the circuit voltage in Figure 65.1 is varied, different currents flow in the loop and different amplitudes of excess noise occur. The band width in Figure 65.1 is held constant by two RC filters with $f_1 = 10$ and $f_2 = 100$ Hz. A plot of the variance against the current gives a parabola for moderate currents; thus

$$\sigma_V^2 \propto I^2$$

The direction of current is unimportant.
Since

$$V = IR$$

then

$$\sigma_V^2 \propto V^2$$

for constant R. And since

$$W = I^2 R = V^2/R$$

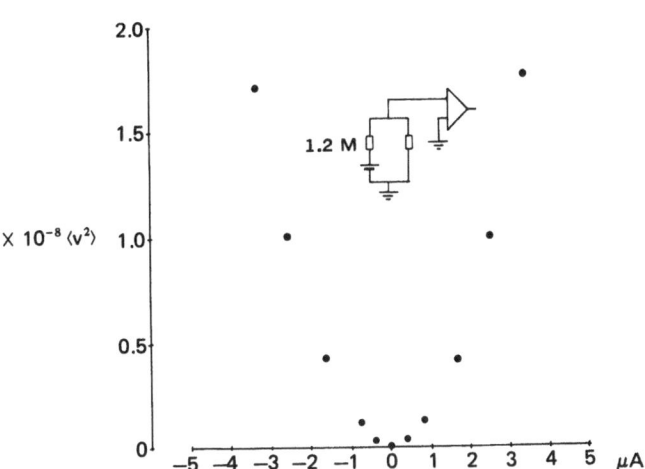

Figure 65.1. Excess voltage noise from two 1.2-MΩ resistors in parallel. The battery in the circuit is varied to cause clockwise (positive) and counterclockwise (negative) steady currents to flow. The variance of the voltage noise across the resistors (limited by a *CR* filter at 10 Hz and a *RC* filter at 100 Hz) is plotted against the current through the resistors. Johnson noise is negligible. (Courtesy D. L. Ypey.)

then

$$\sigma_V^2 \propto W$$

The amplitude of the excess noise is directly proportional to the energy dissipation in the circuit.

PROBLEM. Assume $S(f) = B/f$. Calculate B from Figure 65.1; calculate the proportionality constants in the last three equations for σ_V^2.

SOLUTION. The general relationship between variance and band-limited spectral density is

$$\sigma^2 = \int_0^\infty S(f) \, | \, Y \, |^2 \, df$$

where Y is the voltage transfer function of the band-limiting filter. In this case

$$\sigma_V^2 = \int_0^\infty \frac{B}{f} \frac{f^2 f_2^2 \, df}{(f + f_1)^2 (f + f_2)^2}$$

From Section 52

$$\sigma_V^2 = B \frac{f_2^2}{f_2^2 - f_1^2} \ln \frac{f_2}{f_1}$$

Select a value of σ_V^2 at a particular I (e.g., $\sigma_V^2 \simeq 10^{-8}$ V² for $I \simeq 2.7$ μA). $f_1 = 10$ Hz and $f_2 = 100$ Hz. Find B. Assume

$$B = bI^2$$

Find b. Find the proportionality constants when σ_V^2 is plotted against V^2; against W. Check the units.

PROBLEM. Show that if $f_2/f_1 = $ constant, the same variance is obtained for $1/f$ noise for all f_1 and f_2. Compare with Figure 65.1 for $f_1 = 10, f_2 = 100$ Hz; $f_1 = 100, f_2 = 1000$ Hz; etc.

66. The Effect of Resistor Size on Excess Noise

Consider a resistor R of length l, cross-sectional area A, and resistivity ϱ. Then

$$R = \varrho \frac{l}{A}$$

The resistor volume is

$$V = lA$$

If l and A are doubled, R stays the same:

$$R = \varrho \frac{2l}{2A} = \varrho \frac{l}{A}$$

but the resistor volume is quadrupled:

$$V = 4lA$$

The spectral density of Johnson noise is

$$S_V(f) = 4kTR \ \text{V}^2/\text{Hz}$$

and the variance is

$$\sigma_V^2 = \int_0^\infty S_V(f) \, df$$

Since Johnson noise depends only on temperature, bandwidth, and the ohmic value of the resistance, Johnson noise from a resistor of length l and cross-sectional area A should be the same as one of length $2l$ and area $2A$.

Johnson noise is fundamentally an intensive (rather than extensive) property of the resistor. This is seen easily from the power spectral density

$$S_W(f) = 4kT \, \text{W/Hz}$$

which depends only on temperature.

The intensive nature of Johnson noise may be demonstrated from circuit analysis. According to Nyquist, if any real resistor is replaced by a noiseless R in series with a voltage noise generator $e(t)$, the spectral density of $e(t)$ is

$$S_e(f) = 4kTR$$

Consider the parallel series circuit shown in Section 36.b. Two resistors in series and two such branches in parallel have the open-circuit voltage

$$V = \frac{(ete)2R + (ete)2R}{2R + 2R}$$

and the open-circuit voltage spectral density:

$$S_V(f) = \frac{8S_e R^2 + 8S_e R^2}{(4R)^2} = S_e$$

(Do not combine terms in V before squaring. See Sections 35 and 36 for a full discussion.) In the parallel series array, the open-circuit voltage spectral density is equal to the voltage spectral density of an individual component of the array

$$S_V(f) = S_e(f)$$

This assumes each component has the same average properties; the form of the noise source has not been assumed.

For Johnson noise

$$S_V(f) = 4kTR$$

The equivalence of S_V and S_e is obtained immediately by writing down the equivalent circuit for the parallel series array:

$$\frac{(2R)(2R)}{2R + 2R} = R$$

In effect, the parallel series array doubles the length and doubles the cross-sectional area of the unit resistor of which it is composed. Johnson noise measurements, like resistance measurements, cannot distinguish between scaled resistors. Excess noise does not scale; the larger the volume the smaller the excess noise for the same external conditions. To show this the following experiment was done.

Resistors of value $R = 1.20$ MΩ $\pm$ 0.1% were selected with a Keithley 602 electrometer from a batch of TRW 10%, $\frac{1}{2}$-W mil-R-11 style RC20 carbon composition resistors. A single resistor of this sort will be referred to as a $1N$ resistor.

Place two $1N$ resistors in series and two such branches in parallel. This is a $4N$ resistor; it has the same value R as a $1N$ resistor but four times the volume. In a similar fashion, construct $9N$, $16N$, $25N$, ..., resistors, all of the same ohmic value but of 9, 16, 25, ..., times the volume of the original.

Figure 66.1 shows data from two $1N$ resistors in parallel (left) and two $16N$ resistors in parallel (right). The emf in these circuits (see the inset in Figure 65.1) was one cell of a large automotive battery dropped through a

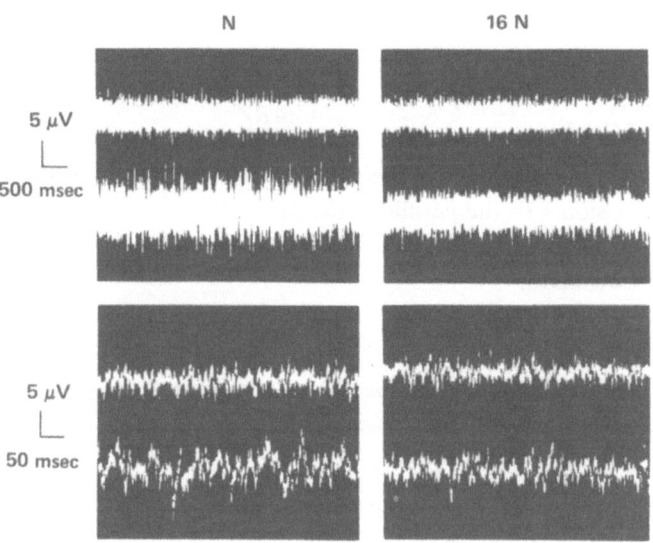

Figure 66.1. *Upper left panel.* Voltage noise from two resistors in parallel, without current (upper trace) and with current (lower trace). The lower left panel shows the same data on an expanded time axis. *Upper right panel.* Voltage noise from 32 resistors arranged in two groups of 16 in parallel. The resistors are combined such that they have the same ohmic value as those used on the left, but having 16 times the volume (see text). The upper trace has zero net current and the lower trace has the same current as that passing through the smaller resistor to the left. The lower right panel shows the same data on an expanded time axis. Larger resistors have less excess noise.

precision attenuator. The combined internal resistance of this voltage source was less than 1000 Ω.

Let E be the emf in the circuit. The current is

$$I = \frac{E}{2R}$$

and the total energy dissipation is

$$W = 2\left(\frac{E}{2R}\right)^2 R = \frac{E^2}{2R}$$

Figure 66.1 (top left) shows two noise traces. The upper trace is for zero current, or

$$W = 0$$

and represents Johnson noise from 0.6 MΩ in the band limit set by $f_1 = 10$ Hz and $f_2 = 100$ Hz. The lower trace is the same circuit with a current of 91.3 nA. The energy dissipation for each resistor is 10^{-8} W, for a total of

$$W = 20 \text{ nW}$$

The fluctuations are due to Johnson noise plus excess noise due to holding the resistors in a steady state away from equilibrium. Figure 66.1 (bottom left) shows the same data on an expanded time scale.

Figure 66.1 (top right) repeats the experiment, except that two 16N resistors replace the 1N resistors. This is the only change. The top trace has $W = 0$ and the bottom trace has $W = 20$ nW. Figure 66.1 (bottom right) shows the same data on an expanded time scale.

First compare the two upper photographs in Figure 66.1. The top traces are roughly equal, thus a 1N resistor and a 16N resistor have the same Johnson noise. The two lower traces are unequal. Presumably Johnson noise is still the same (we are not far from equilibrium). The excess noise added in the 1N case is larger than in the 16N case.

PROBLEM.   The value of the 1N and the 16N resistor is 1.2 MΩ. The noise is band limited by a simple low-pass, high-pass RC filter with $f_1 = 10$ Hz and $f_2 = 100$ Hz. Show that the expected average peak-to-peak value for the Johnson noise at the input of the amplifiers is 3.3 μV. Compare with Figure 66.1.

SOLUTION. The variance of the band-limited noise from R is (Section 52)

$$\sigma_V^2 = 4kTR\left(\frac{\pi}{2}\,\frac{f_2^2}{f_1+f_2}\right)$$

of which we expect to see one-half for two resistors R in parallel.

Excess noise in the $16N$ case would equal the $1N$ case if the energy being dissipated in the larger resistor were increased by a factor of 16. This is shown in Figure 66.2. Figure 66.2 shows two correlation functions of Johnson plus excess noise. In one case (solid line) the noise is generated by two $1N$ resistors in parallel, each dissipating 6.23 nW. In the second case (dashed line) the noise is from two $16N$ resistors in parallel, each dissipating 98.7 nW. The Johnson noise is equal in the two cases. Since the two correlation functions nearly superimpose, the excess noise generated in the two cases must be the same.

The rule we have exemplified is this: Excess noise in two resistors near equilibrium will be the same if their power-to-volume ratio is equal. It is implied that we are speaking of the same sort of resistors simply scaled in size.

PROBLEM. Show that the currents flowing in the two cases shown in Figure 66.2 are $I = 72$ nA ($1N$ case) and $I = 287$ nA ($16N$ case).

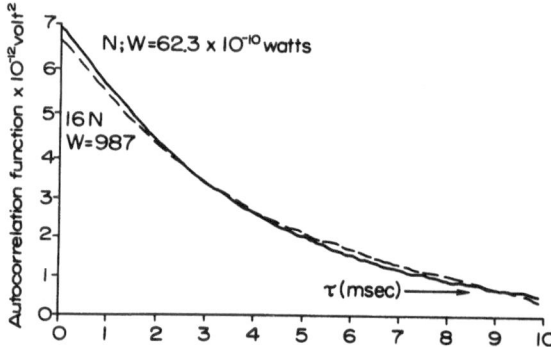

Figure 66.2. Correlation functions of voltage noise measured from two resistor groups. One group is composed of two resistors (N) in parallel and the second of two larger resistors ($16N$) in parallel. The $16N$ resistor has the same ohmic value as the N resistor by parallel-series arrangement. The current passing through both groups of resistors is selected such that the power being dissipated in the resistors with the larger volume is about 16 times that of the power being dissipated in the resistors with the smaller volume. The graphs show that the variance and the spectral density of the measured noise are the same in the two cases.

EXERCISE. In DeFelice (1976) it was shown that the shape of the correlation functions in Figure 66.2 fit a power spectral density from a $1N$ resistor of the form

$$S_W = 4kT + gW/f$$

The excess noise is $1/f$ noise, and its variance is proportional to W. One may also write

$$S_V = 4kTR + gV^2/f$$

or

$$S_I = 4kT/R + gI^2/f$$

since $W = I^2R = V^2/R$. g is a dimensionless quantity and depends on the type of resistor.

(a) Derive an expression for the theoretical correlation function $C_V(\tau)$ for the band-limited noise shown in Figures 66.1 and 66.2; assume Johnson plus $1/f$ noise is the correct form for the spectral density (see Section 52 and 53).

(b) Plot $C_V(\tau)$ and compare it with the measured curve shown in Figure 66.2.

(c) Estimate the expected error in the correlation functions shown in Figure 66.2 at $\tau = 0$. The measured $C_V(\tau)$ were calculated from 32.77 sec of data (see Section 55). The exact calculation is rather lengthy; estimates may be obtained by making suitable approximations.

Since scaling leaves Johnson noise the same, we ignore Johnson noise and concentrate on the variance of the excess noise. Let W be the same for a $1N$, $4N$, $9N$, and $16N$ resistor. All resistors have the same R, but have volumes in the ratio $1:4:9:16$. Then (excess noise variance only)

$$\sigma_1^2 = 4\sigma_4^2 = 9\sigma_9^2 = 16\sigma_{16}^2$$

Data supporting this result are shown in DeFelice (1976, Figure 3). If the spectral density of the excess noise is proportional to gW, then at constant W and constant bandwidth

$$g_1 = 4g_4 = 9g_9 = 16g_{16}$$

Thus the g value for a resistor scaled to N times its size is $1/N$ times the original value; $1/g$ is an extensive property of a resistor.

Consider the following thought experiment. Place a $1N$ resistor of value R in one box, and a $16N$ resistor of value R in another box. The I–V curves, the energy dissipation, and the Johnson noise will be the same for the two cases. No measurement of I, V, W, or noise at equilibrium will distinguish between the two boxes. However, a measurement of excess noise will reveal which box has the $16N$ resistor. For the same external conditions, the larger resistor will be quieter.*

An average value for g_1 obtained from many of the TRW resistors referred to above is

$$g_1 = 7.2 \times 10^{-10}$$

Compare one of these resistors dissipating energy at the rate W, and four in a parallel series arrangement dissipating energy $4W$. Each one of the four is dissipating W and is, therefore, in the same state as the single resistor. Since

$$S_V = S_e$$

for the parallel series arrangement (Section 36), the noise from the single resistor and the group of four is the same. Another way to express this is to set

$$g_4 = 1.8 \times 10^{-10}$$

so that the product gW is the same in both cases.

The formula $S_V = S_e$ has assumed that the elementary noise generators e are statistically independent (Sections 35 and 36). Thus, within the experimental error of the scaling experiments, the excess noise generators (like the Johnson noise generators) are independent.

Suppose that an arbitrary resistor is subdivided into many equal-size volumes separated by imaginary boundaries. Let v be the elementary volume and V be the total volume. If W is the total energy being dissipated, then in v

$$w = \frac{v}{V} W$$

---

* An interesting point is raised by considering the temperature. For the same W, the $16N$ resistor would not get as hot as the $1N$ resistor; thus the term $4kRT$ would actually be different in the two cases; also, R is usually temperature dependent. These kinds of effects may be controlled for. The experiments discussed in this section have been purposely held as close to equilibrium as possible, in order to avoid this sort of complication.

Also,

$$g_v = \frac{V}{v} g_V$$

and therefore

$$g_v w = g_V W$$

Measuring $g_V W$ for a given resistor R is a measure of the same quantity for an arbitrarily small resistor of a similar material.

Suppose the elementary volume v is reduced to the dimension of a single charge carrier in the resistor. Keeping to the example shown in Figure 66.2,

$$g_v w = 44.9 \times 10^{-19} \text{ J/sec}$$

as an average value for each charge carrier in the resistor.

PROBLEM. Compare the variance of Johnson noise power with the excess noise power for the case shown in Figure 66.2.

SOLUTION. Show that for Johnson noise

$$\sigma_W{}^2 = 4kT \frac{f_2^2}{f_1 + f_2} \frac{\pi}{2}$$

and for excess noise

$$\sigma_W{}^2 = g_v w \frac{f_2^2}{f_1 + f_2} \frac{\ln(f_2/f_1)}{f_2 - f_1}$$

Substitute in the values of the parameters and compare the two values of $\sigma_W{}^2$. Note that the bandwidth of the noise for the excess noise power includes the factor $1/\langle f \rangle$, where $\langle f \rangle$ is the logarithmic average of f_1 and f_2 (Section 19).

PROBLEM. Four experimental situations are shown in Figure 66.3, along with the measured values of the current flowing in the loop, the input resistance at the amplifier input, and the variance of the voltage noise in a band set by $f_1 = 3$ and $f_2 = 30$ Hz. Show that cases B, C and D are consistent with A if each component resistor (all have the value 1.2 MΩ) is an independent noise source. Calculate the g value for a component resistor.

EXERCISE. $1/f$ noise from carbon composition resistors is often placed in a separate category because of the special construction of these resistors. Carbon composition resistors consist of compact grains of conducting

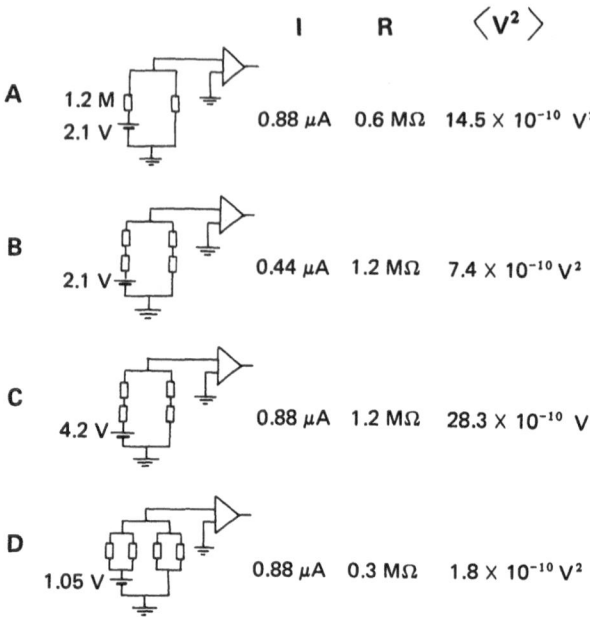

Figure 66.3. Four experimental situations for voltage noise measurement. I is the current flowing in the circuit in each case, and R is the equivalent input resistance. $\langle V^2 \rangle$ is the voltage noise variance measured in a broadband RC filter with $f_1 = 3$ Hz and $f_2 = 30$ Hz. (Courtesy D. L. Ypey.)

material; van der Ziel (1976, Section 7.4) has written that the $1/f$ noise from carbon resistors is due to resistance fluctuations between grains. If a steady current is passed through the carbon resistor, a fluctuation in voltage is seen, viz.,

$$\delta V = I\,\delta R$$

and

$$S_V(f) = I^2 S_R(f)$$

If $S_R(f)$ has a $1/f$ spectrum, so does $S_V(f)$. Carbon resistors differ in the way the compact grains make contact with the external metal wire. This is another possible source of resistance fluctuation. Consider Figure 66.4. Two $1N$ 1.2-MΩ resistors in parallel have a battery inserted in the loop of voltage 1,2, and 4 V. The potential fluctuations across the parallel combination are measured in a band set by $f_1 = 3$ Hz and $f_2 = 3000$ Hz. In this particular case, "burst noise" is seen. These bursts are small rectangular pulses whose amplitude is proportional to the mean current flowing in the loop. Calculate δR from Figure 66.4. What is the percent change in R that δR represents? We show in Section 72 that a random telegraph signal (rectangular pulses with random durations) has a Lorentzian spectral density, i.e.,

$$S(f) = \frac{K\theta}{1 + \omega^2\theta^2}$$

where $1/\theta$ is related to the average rate of flipping from one state to another. Assume there are ten values of θ in a sample, related by

$$\theta_1 = 2\theta_2 = 3\theta_3$$

Add these ten Lorentzians graphically on log–log paper. What is the approximate slope of the composite spectrum between $f_1 = 1/2\pi\theta_1$, and $f_2 = 1/2\pi\theta_2$. Let $\theta_1 = 100$ msec.

Note in Figure 66.4 that the base-line noise increases with the mean current flowing in the loop. This is the $1/f$ noise already discussed in Sections 64–66. An *ad hoc* theory of $1/f$ noise may be developed, assuming that excess noise results from resistance fluctuations with a wide distribution of time constants.

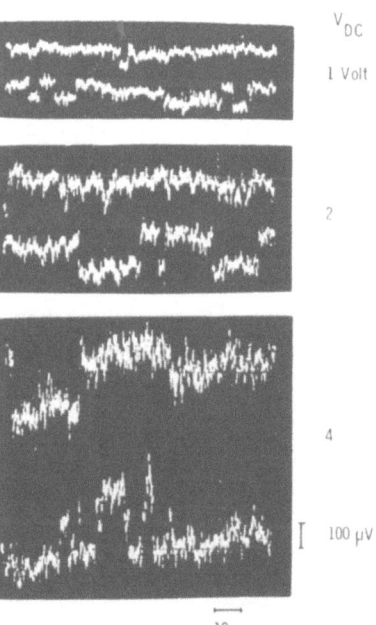

Figure 66.4. Examples of burst noise from carbon composition resistors. Two 1.2-MΩ resistors in parallel with 1, 2, and 4 V in the loop. Note that the size of a rectangular burst is roughly proportional to the current through the resistor. (Courtesy D. L. Ypey.)

V_{DC}

1 Volt

2

4

100 μV

10 ms

It may be added that the phenomena shown in Sections 64–66 also exist for low-noise wire-wound resistors. In these resistors the *g* values are much lower than in comparable ohmic value carbon resistors; however, the qualitative features shown in Figures 64.1, 65.1, 66.1, and 66.2 are the same. Discrete steps are rarely seen in wire-wound resistors.

67. Resistor Noise: A Selected Chronological Bibliography

I have selected resistor noise from ordinary laboratory resistors as a first example of excess noise since it is so easy to observe. The experiments shown in Sections 64–66 may be repeated with relative ease; the conclusion that excess noise near equilibrium is $1/f$ noise may be reached using only oscilloscope and simple filter theory (Figure 64.1).

There is more to learn from electronic conductors; although we now leave them, the following list of references may be of interest to some readers. It was prepared with D. Ypey in 1977; the list is not exhaustive but it does give a fairly complete picture of the literature on $1/f$ noise up to early 1976.

After some of the references, a key feature of the article is identified.

Volume.   This refers to studies of the effect of resistor volume on excess noise. Many of the observations are explained by the assumption of independent internal sources discussed in Section 66.

Temperature.   The effect of temperature on $1/f$ noise is still an open question. Some theorists have excluded the possibility that $1/f$ noise depends on temperature in principle. However, scattered experimental evidence suggests such a dependence. Generally, the temperature range available to biologists is too narrow for the question to be of practical importance.

Burst.   This refers to the type of noise seen in Figure 66.4.

Index.   Many authors have suggested a noise index to characterize the expected excess noise from a particular resistor. One such index is the *g* value referred to in Section 66.

Otto, R. (1935). Das Rauschen von Kohlemikrophonen. *Hochfrequens. Techn. Elek.* **45**, 187–198. *Volume, Temperature.*

Christensen, C. J. and G. L. Pearson (1936). Spontaneous resistance fluctuations in carbon microphones and other granular resistances. *Bell. Syst. Tech. J.* **15**, 197–223. *Volume, Temperature.*

Harris, E. J. (1948). Circuit and current noise. *Electron. Eng.* **20**, 145–148.

Blackburn, John F. (1949). *Components Handbook. MIT Radiation Laboratory Series.* New York: McGraw-Hill Book Co. *Index, Temperature.*

Ehrenfried, G. (1949). Fixed composition resistors. In: *Components Handbook* (ed. J. Blackburn), *MIT Radiation Laboratory Series*, Chap. 2, pp. 33–64. New York: McGraw-Hill Book Co.

Fagen, M. D. and G. Ehrenfried (1949). Fixed wire-wound and miscellaneous resistors. In: *Components Handbook* (ed. J. Blackburn), *MIT Radiation Laboratory Series*, Chap. 3, pp. 65–114. New York: McGraw-Hill Book Co.

Campbell, R. H., Jr., and R. A. Chipman (1949). Noise from current-carrying resistors 20 to 500 kc. *Proc. IRE—Waves and Electrons Section* **37**, 938–942. *Burst.*

Hettich, Alfred (1950). Geometrische Dimensionen und Widerstandsrauchen. *Frequenz* **4**, 14–25. *Volume.*

van der Ziel, A. (1950). On the noise spectra of semi-conductor noise and of flicker effect. *Physica* **16**, 359–371.

MacFarlane, G. G. (1950). A theory of contact noise in semiconductors. *Proc. Phys. Soc. (London)* **63B**, 807–814. *Temperature.*

Bell, D. A. (1952). Current noise in semi-conductors: A re-examination of Bernamont's data. *Philos. Mag.* **43**, 1107–1111. *Burst.*

Templeton, I. M. and D. K. C. MacDonald (1953). The electrical conductivity and current noise of carbon resistors [at different temperatures]. *Proc. Phys. Soc. (London)* **B66**, 680–687. *Temperature.*

Bell, D. A. (1953). Current noise in semiconductors. *Wireless Engineer* **30**, pp. 23–24.

Rollin, B. V. and I. M. Templeton (1953). Noise in semiconductors at very low frequencies. *Proc. Phys. Soc. (London)* **66B**, 259–261.

Bell, D. A. and K. Y. Chong (1954). Current noise in composition resistors. *Wireless Engineer* **31**, 142–144. *Volume.*

Bell, D. A. (1954). Phenomenological approach to "current noise." *Br. J. Appl. Phys.* **6**, 284–287. *Volume.*

van der Ziel, Aldert (1954). Excess noise in semiconductors and vacuum tubes. In: *Noise*, Chap. 8, pp. 190–232. New York: Prentice-Hall, Inc.: *Volume.*

Conrad, George T. (1954). Noise measurements of composition resistors. I. The method and equipment. *IRE Trans. Component Parts* **PGCP4**, 61–78.

Conrad, George T. (1954). Noise measurements of composition resistors. II. Characteristics and comparison of resistors. *IRE Trans. Component Parts.* **CP4**, 79–92: *Index, Volume, Temperature.*

Conrad, G. T. (1956). A proposed current noise index for composition resistors. *IRE Trans. Component Parts* (March 1956) **CP-3**, 14–20.

Bell, D. A. (1958). Semiconductor noise as a queuing problem. *Proc. Phys. Soc. (London)* **72**, 27–32.

van der Ziel, A. (1959). Flicker noise in semiconductor material. In: *Fluctuation Phenomena in Semi-Conductors*, Chap. 3, pp. 46–64. London: Butterworths Scientific Publications. *Temperature.*

Bell, D. A. (1959). The $1/f$ spectrum of current noise; noise in metal films rectifiers and transistors, and "bursts." In: *Electrical Noise*, Chaps. 10 and 11, pp. 210–264. London: D. Van Nostrand Company. *Volume, Temperature*, and *Burst.*

Conrad, G. T., N. Newman, and A. P. Stansbury (1960). A recommended standard resistor-noise test system. *IRE Trans. Component Parts* (September 1960) **CP-7**, 71–88: *Index.*

Bennet, William R. (1960). 1/*f* noise (from Chap. 5: Noise in semiconductors). In: *Electrical Noise*. New York: McGraw-Hill Book Co., Inc.

Stansbury, Alan (1961). Measuring resistor current noise. *Elec. Equip. Eng.* (June 1961) **9**, 17–18.

Brophy, J. J. (1966). Noise (from chapter on amplifiers). In: *Basic Electronics for Scientists*, pp. 268–275. New York: McGraw-Hill Book Co.

King, Robert (1966). Introduction; current noise. In: *Electrical Noise*, pp. 1–7, 87–94. London: Chapman and Hall, Ltd. *Burst.*

Brophy, J. J. (1968). Statistics of 1/*f* noise. *Phys. Rev.* **166**, 827–831.

Brophy, J. J. (1969). Zero-crossing statistics of 1/*f* noise. *J. Appl. Phys.* **40**, 567–69.

van der Ziel, A. (1970). Flicker noise. In: *Noise*, Section 6–4, pp. 106–118. Englewood Cliffs, New Jersey: Prentice-Hall, Inc.. *Burst.*

Brophy, J. J. (1970). Low-frequency variance noise. *J. Appl. Phys.* **41**, 2913–2916.

Brophy, J. J. (1972). *The Radio-Electronic Master*. Official catalog of electronic parts, instruments and equipment. Garden City, New York: United Technical Publications.

Purcell, W. E. (1972). Variance noise spectra of 1/*f* noise. *J. Appl. Phys.* **43**, 2890–2895.

Motchenbacher, C. D. and F. C. Fitchen (1973). Noise in passive components. In: *Low-Noise Electronic Design*, Chap. 9, pp. 171–190. New York: John Wiley and Sons. *Index.*

Dell, R. A., M. Epstein, and C. R. Kannewurf (1973). Experimental study of 1/*f* noise stationarity by digital techniques. *J. Appl. Phys.* **44**, 472–476.

Mimaki, T. (1973). Zero-crossing intervals of Gaussian processes. *J. Appl. Phys.* **44**, 477–1485.

Bell, T. H. (1974). Representation of random noise by random pulses. *J. Appl. Phys.* **45**, 1902–1904.

Moore, W. J. (1974). Statistical studies of 1/*f* noise from carbon resistors. *J. Appl. Phys.* **45**, 1896–1901. *Burst.*

Jones, B. K. and J. D. Francis (1975). Direct correlation between 1/*f* and other noise sources. *J. Phys. D: Appl. Phys.* **8**, 1172–1176.

Anderson, A. C., J. H. Anderson, and M. P. Zaitlin (1976). Some observations on resistance thermometry below 1 K. *Rev. Sci. Instrum.* **47**, 407–411. *Temperature, Volume.*

van der Ziel, A. (1976). Flicker noise and generation—recombination noise. In: *Noise in Measurements*. Chap. 7, pp. 77–87. New York: Wiley-Interscience.

DeFelice, Louis J. (1976). 1/*f* resistor noise. *J. Appl. Phys.* **47**, 350–352. *Volume, Index.*

68. Excess Noise in Ionic Conductors

Less work has been done on excess noise from ionic conductors than on electronic conductors although there are many similarities between the two. As a first example, we consider excess noise from a collodion membrane separating two aqueous salt solutions.

Ionic solutions are relatively poor conductors compared to metals, but they are comparable to semiconductors. To estimate the relative conductance of ionic resistors and carbon composition resistors, assume that the volume of the packed carbon grains of a 1-MΩ resistor have dimensions $A = 10^{-3}$ cm^2 and $l = 1$ cm (typical values). Then

$$\varrho = \frac{A}{l} R = 1000 \ \Omega \ \text{cm}$$

This resistivity is about equal to a 0.01 M concentration of KCl (Section 12).

In practice, excess noise in ionic conductors is easily observed if the resistance is the order of megohms. One way of achieving a high resistance is to force the ions to flow through small openings; another is to lower the concentration. A collodion membrane separating two aqueous salt solutions provides such a system: The density and amount of collodion may be varied to produce a network of micropores through which the ions must flow, and the concentration of the bathing solution may be varied to adjust the resistance (DeFelice and Michalides, 1972, have described this system in detail).

Collodion is a nitrated cellulose with weak cation exchange properties. The membranes discussed below were bathed in equal concentrations of salt solution on both sides of the membrane; for zero current, the potential across the membrane is zero.

Consider a membrane bathed in nM KCl and in parallel with a resistor R_0. A battery E is placed in the loop to provide an average current through the membrane. This configuration is identical to that present in Sections 64–66, except that now one of the resistors is ionic.

If R is the resistance of the collodion membrane, the current in the loop is

$$I = \frac{E}{R + R_0}$$

If V is the potential across the membrane, the current is

$$I = \frac{E - V}{R_0}$$

A plot of I vs. V characterizes the membrane's resistance.

In general the I–V relationship of a collodion membrane is nonlinear. The resistance decreases at higher current densities (Figure 4, DeFelice and Michalides, 1972). R is the slope resistance (the steady-state resistance; Section 23).

Let $e(t)$ be the equivalent noise generator in the collodion membrane and $e_0(t)$ the equivalent noise generator in R_0. By the cross-product rule (Section 31), the open-circuit voltage from the parallel combination of R and R_0 is

$$V = \frac{eR_0 + e_0R}{R_0 + R}$$

From Section 35 the voltage spectral density is

$$S_V = \frac{S_eR_0{}^2 + S_{e_0}R^2}{(R + R_0)^2}$$

Suppose R_0 is a low-noise resistor with negligible excess noise for the currents we are passing. Then

$$S_{e_0} = 4kTR_0$$

We wish to test the hypothesis that the noise from the ionic resistor obeys the same empirical formula as carbon composition resistors, viz.,

$$S_e = (4kT + gW/f)R$$

where $W = I^2R$ is the energy dissipated in the collodion membrane.

Substitution of the last two equations into that for $S_V(f)$ gives

$$S_V(f) = 4kT\frac{RR_0}{R + R_0} + \frac{gI^2}{f}\left(\frac{RR_0}{R + R_0}\right)^2$$

Introduce the symbol $\mathbb{R}$ for the parallel combination of R and R_0:

$$\mathbb{R} = \frac{RR_0}{R + R_0}$$

If the noise is passed through a filter with voltage transfer function Q (Section 38), the variance of the noise measured across the membrane is

$$\sigma_V{}^2 = \int_0^\infty S_V(f)\, df$$

or

$$\sigma_V{}^2 = 4kT\mathbb{R}\int_0^\infty |Q|^2\, df + gI^2\mathbb{R}^2\int_0^\infty \frac{1}{f}|Q|^2\, df$$

(Note that $\mathbb{R}$ goes to R as R_0 goes to infinity.)

We then test this expectation near equilibrium. It is convenient to introduce an index that compares the excess noise to the Johnson noise.

Define

$$\varepsilon = \frac{\text{excess noise}}{\text{Johnson noise}}$$

From the above expression

$$\varepsilon = \frac{gI^2 R}{4kT} \frac{\int_0^\infty (1/f) \mid Q \mid^2 df}{\int_0^\infty \mid Q \mid^2 df}$$

This is the excess noise index for the parallel RR_0 circuit. It must be determined separately that the excess noise from R_0 and the battery are negligible.

Suppose Q represents an ideal narrow filter (Section 39); then

$$\mid Q \mid^2 = F \delta(f - f_0)$$

where f_0 is the pass frequency and F is a constant.

Taking this ideal character of a Q filter as an approximation gives

$$\varepsilon = \frac{gI^2 R}{4kT} \frac{1}{f_0}$$

PROBLEM. Above we assumed that R and R_0 are pure resistors. Show that the same expression for ε results if R and R_0 are complex impedances Z and Z_0, if Z has the property

$$\mid Z \mid^2 = R \times ReZ$$

This holds for a parallel RC circuit.

PROBLEM. Calculate the effect of finite bandwidth on the factor $1/f_0$ in the expression for ε.

SOLUTION. In Sections 39 and 40 we derived, for a Q filter with $\mathcal{Q} > 1/2$,

$$\int_0^\infty \frac{1}{f} \mid Q \mid^2 df = \frac{1}{(4\mathcal{Q}^2 - 1)^{1/2}} \left(1 + \frac{2}{\pi} \tan^{-1} \frac{2\mathcal{Q}^2 - 1}{(4\mathcal{Q}^2 - 1)^{1/2}} \right)$$

and

$$\int_0^\infty \mid Q \mid^2 df = \frac{\pi}{2} \frac{f_0}{\mathcal{Q}}$$

Compare the ratio of these two expressions with $1/f_0$.

The dependence of ε on f_0 requires only a relative measurement of noise. For arbitrary bandpass (either by external circuits or the membrane's own impedance) measure the total noise away from equilibrium and divide it by the noise at equilibrium: then by the definition

$$\frac{\text{total noise away from equilibrium}}{\text{noise at equilibrium}} - 1 = \varepsilon$$

This is repeated for a series of bandpass filters with different values of f_0. DeFelice and Michalides (1972) show that for constant membrane conditions (membrane current I and bathing concentration n) the product

$$\varepsilon f_0 = \text{const}$$

for f_0 between 8 and 5000 Hz. In other words, excess noise from collodion membranes is $1/f$ noise. These experiments, like those on carbon composition resistors, were done near equilibrium.

Since excess noise falls off with frequency and noise in the equilibrium state is white, the ratio of total noise to equilibrium noise approaches 1 (ε approaches zero) as f_0 increases. This occurred around 1000 Hz in the collodion membranes studied.

We now turn to the dependence of the noise amplitude on the membrane conditions.

The resistance of the membrane (see Figure 68.1) may be varied by changing the bathing concentration of ions (n). Figure 68.1 shows the dependence of R (taken as the slope of an I–V curve at $I = 0$) on the concentration of KCl in the bath. The simplest theory of resistance as a function of ionic concentration (Section 13) has

$$R \propto \frac{1}{n}$$

Even water does not obey this simple relationship. In Section 12 we showed that as n is increased, the decrease in R is less than that expected from an inverse dependence on n. This is even more striking in the watery micropores of the collodion membrane. The actual concentration of n in the micropores is unknown; however, the concentration profile must change since R does. Figure 68.1 shows that R increases steadily as n decreases, but far below the increase expected from the above equation.

It appears there is a fixed concentration of ions in the membrane below which one may not go even in very weak external solutions. The

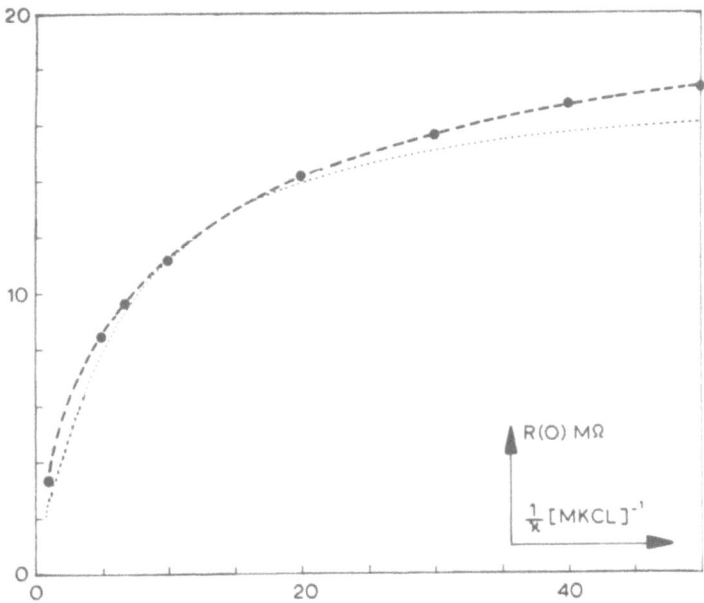

Figure 68.1. The dc slope resistance of a collodion membrane evaluated at zero current (solid dots) as a function of inverse concentration in the solution (moles^{-1}). The dashed line is simply a smooth curve drawn through the experimental points. The dotted line is the curve $R \propto 1/(n + c)$, where c is a constant. (From De Felice and Michalides, 1972.)

dotted line in Figure 68.1 is the equation

$$R \propto \frac{1}{n + c}$$

where c is constant.

PROBLEM.   Calculate the value of c used to fit the above equation to the experimental points in Figure 68.1. Compare the dependence of R on n with the integral resistance discussed in Section 19.

Does excess noise depend on n? To test this, measure ε at different values of n and for different currents I flowing through the membrane. Since n changes the membrane resistance (Figure 68.1), this effect was unfolded from the measurements by dividing each value of ε by the measured value of R. Figure 68.2 shows that for n between 0.02 M and 1 M KCl, and I between about 5 and 100 nA,

$$\frac{\varepsilon}{R} = \text{const } I^2$$

PROBLEM.   Compare the data of Figure 68.2 with the theoretical expression

$$\frac{\varepsilon}{R} = \frac{gI^2}{4kT}\frac{1}{f_0}$$

Figure 68.2 has $f_0 = 16$ Hz. Show that $g \simeq 3 \times 10^{-10}$. What is the percent error made in determining g by ignoring the effect of finite bandwidth of the Q filter ($\mathcal{Q} = 16$)?

Figure 68.2 shows that the ratio of excess to Johnson noise is constant for constant I^2R regardless of the value of R. For example, suppose n is increased such that the resistance is reduced by a factor of 4. If I is

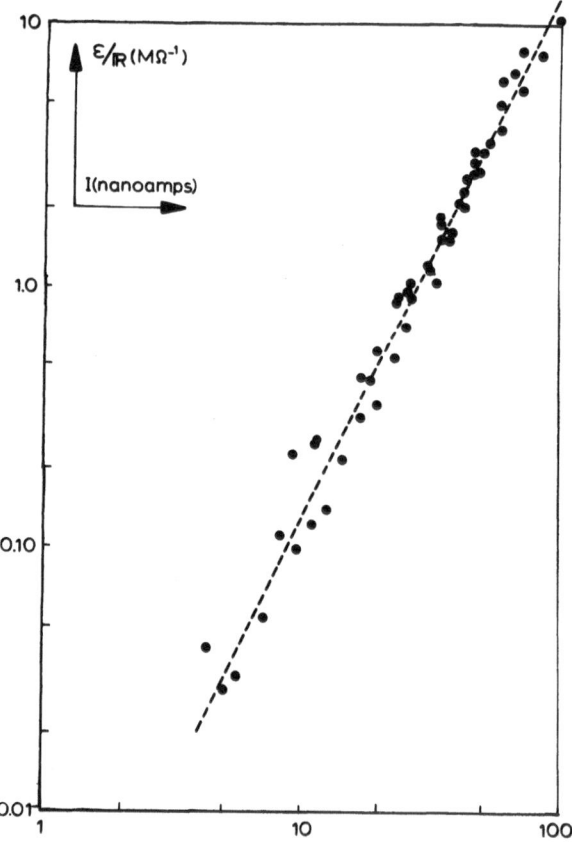

Figure 68.2. Excess noise per unit equivalent resistance of the noise source against mean current through the collodion membrane. The points are from the same membrane shown in Figure 68.1 bathed in eight different KCl concentrations between 0.02 and 1 M KCl. The dashed line is the curve $\varepsilon/R \propto I^2$. (From De Felice and Michalides, 1972.)

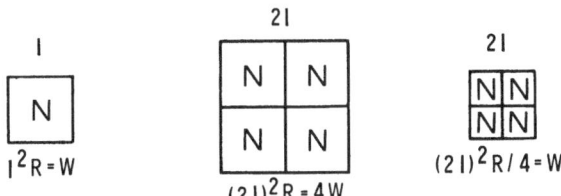

Fig. 68.3. Schematic drawing of three resistors. Under the conditions of carrier density and current shown, all three resistors have the same ratio of excess to Johnson noise.

doubled, ε stays the same because

$$\varepsilon_{N/4} = \frac{g(2I)^2 R/4}{4kT} = \frac{gI^2 R}{4kT} = \varepsilon_N$$

The subscript on ε signifies the relative effective density of charge carriers in the membrane. In Section 66 the same symbol N was used to index resistor volume at constant carrier density. Figure 68.3 will help compare these two cases.

In this figure the first resistor has a standard volume and an effective charge carrier number N; the second resistor has four times the volume of the first but the same effective carrier density; the third resistor has the same volume as the first but four times the carrier density. The current passing through each resistor is shown above each diagram; the rate of energy dissipation is shown below each diagram. Under the conditions shown, all three resistors have the same ratio of excess to Johnson noise, i.e., the same value of ε.

Case (a). The total resistance is R, the current through the resistor is I, and the power is W. N effective carriers dissipate energy at the rate W in a resistor of value R.

Case (b). The total resistance is R and the resistance of each compartment is also R. If the current is such that the total power is $4W$, then each compartment will share one-fourth this value. Thus, any one of the four compartments in (b) is in exactly the same state as (a). Since the open-circuit voltage spectral density of the entire resistor is identical to any component in a parallel-series circuit (Sections 36 and 66) the excess noise in (b) has the same average properties as the excess noise in (a). Since the Johnson noise in (a) and (b) are the same, ε is also the same.

Case (c). The total resistance is $R/4$. This is indicated by keeping the volume in (c) the same as it was in (a), but raising the effective number of charge carriers by 4. As before, the resistance of each compartment in (c) also has the value $R/4$. The experimental result is this: if the total power is W, so that each compartment dissipates energy at the rate $W/4$, the ratio of excess to Johnson noise is the same in (c) as it is in (a). In (c) N carriers dissipate energy at the rate $W/4$ in a resistor of value $R/4$.

In cases (a) and (b), comparing ε is the same as comparing excess noise since the Johnson noise is the same. In case (c), however, Johnson noise is reduced to one-fourth. Since ε in (c) has the same value as ε in (a), the excess noise in (c) must also be reduced by one-fourth. Focusing on the excess noise alone, case (c) is generating less; to raise (c) to the level of (a) or (b), a current of $4I$ is necessary. This will give a total power of $4W$, or a rate of energy dissipation of W for each compartment.

The following rule has evolved from our experiments on excess noise near equilibrium: If the same number of effective charge carriers dissipate energy at the same average rate, the variance of the excess noise is the same. It appears that the constant feature of excess noise is the energy dissipated per carrier in any particular medium.

Suggested Reading. For additional information on excess noise from model ionic conductors see Green and Yafuso (1969), Hooge (1970, 1972, 1976), Hooge and Gaal (1971), Yafuso and Green (1971), Michalides *et al.* (1973) and Green (1974). For additional information on the relationship between entrophy production, irreversible processes, and nonequilibrium thermodynamics see Prigogine (1955), Katchalsky and Curran (1965), and de Groot (1966).

69. Excess Noise in Lipid Bilayers

Synthetic membranes may be formed from phospholipids forming a layer two molecules thick. This bilayer may separate two salt solutions; certain proteins placed in the solution form ionic channels through the membrane; other proteins act as carriers to transport ions across the membrane (for an early review see Bangham, 1968).

An undoped bilayer membrane is called a black lipid membrane (BLM) because the membrane is so thin it appears black to reflected light; light incident on the membrane undergoes opposite phase reversal at the water/lipid interface and the lipid/water interface. Since the membrane is

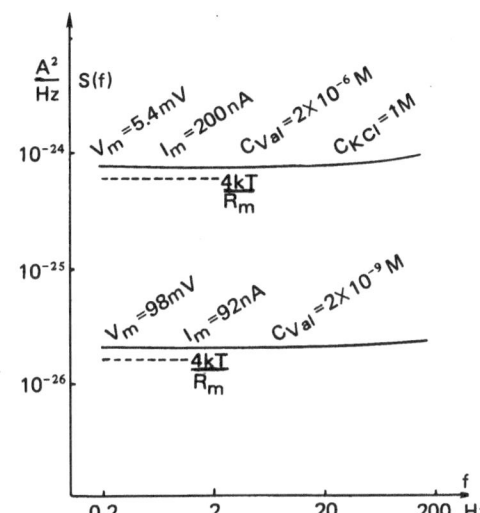

Figure 69.1. Current noise spectral densities (solid lines) of a BLM treated with two different concentrations of valinomycin. Bathing concentration: 1 M KCl; room temperature. $S(f)$ is normalized to a 1-cm² membrane. (Courtesy E. Wanke.)

much thinner than a wavelength of light, the reflected waves from these two surfaces cancel.

A doped BLM is a lipid bilayer in the presence of specific proteins that have altered the electrical conductance.* Typically, an undoped BLM has a specific resistance greater than 10^8 Ω cm².

PROBLEM. Calculate an effective resistivity for the lipid bilayer system by assuming a membrane thickness of 100 Å. A doped membrane may have a resistance of about 10^4 Ω cm².

If current is passed through an undoped BLM separating two salt solutions, $1/f$ noise is generated. One difficulty with doing this experiment is the large absolute resistance of the noise source. Even fairly large membrane areas will have resistances of over 100 MΩ.

Doped BLMs may show several types of excess noise; the type of excess noise depends on the nature of the molecules added to the bathing medium. Several examples are given below (Wanke, 1975; Wanke and Prestipino, 1976).

In the presence of valinomycin, white noise is observed in the range 0.2–20 Hz. Figure 69.1 shows current spectral densities from a BLM doped

---

* Excitability-inducing material (EIM) and other components incorporated into the membrane in the proper ionic environment can induce time- and voltage-dependent conductances that mimic excitability. For an introduction to this work see Mueller *et al.* (1962) and Mueller and Rudin (1963).

with two different concentrations of valinomycin. The solid lines are experimental curves; the dashed lines are theoretical curves. The membranes are bathed in 1 M KCl in both cases. In 2×10^{-6} M valinomycin, membrane resistance is about 27 kΩ; in 2×10^{-9} M valinomycin, membrane resistance is about a MΩ. In both cases, measured noise is only slightly above the theoretically expected Johnson noise, indicating that under these conditions, little excess noise is generated. Valinomycin is believed to increase membrane conductance by a carrier mechanism.

Kolb and Läuger (1978) treat carrier-mediated ion transport in model membranes in equilibrium in great detail. Kolb and Frehland (1979) treat carrier-mediated ion transport in systems in a nonequilibrium state.

In the presence of nystatin, $1/f$ noise is generated. This is shown in Figure 69.2 (upper curve) where the fluctuations were averaged for 10 min; shorter averaging times may result in Lorentzian-shaped spectra (lower curve). In this experiment, 20 μg/ml of nystatin was added to the bath after the BLM was formed. Membranes formed in a solution of 0.2 M NaCl and 5 μg/ml nystatin do not generate excess noise for currents up to 28 nA (E. Wanke, personal communication).

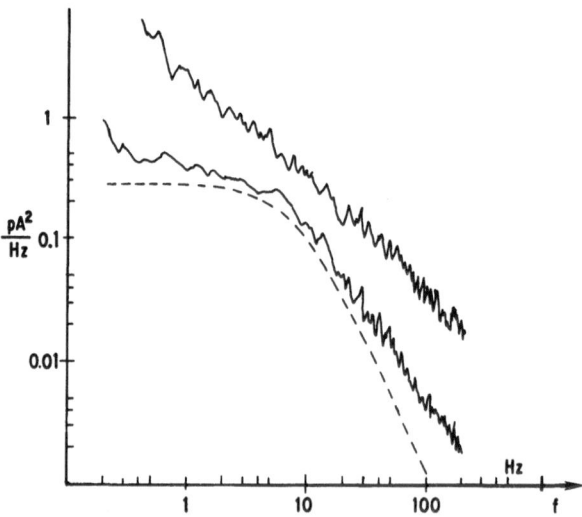

Figure 69.2. Current spectral density against frequency for a nystatin-doped membrane (20 μg/ml) in 0.2 M NaCl. *Upper curve.* Averaging time 10 min. $I = 2$ nA and $V = 40$ mV; membrane area: 4.4×10^{-3} cm²; room temperature. The lower curve is representative of spectral densities observed under the same conditions but for short averaging times (less than 30 sec). (Courtesy E. Wanke.)

PROBLEM. Calculate the *g* value (Section 66) of the nystatin-doped membrane shown in Figure 69.2, assuming the empirical law

$$S_I(f) = gI^2/f$$

for $1/f$ noise. Compare this *g* value with those found in carbon composition resistors and collodion membranes.

PROBLEM. Calculate the theoretical Johnson noise expected from the membrane of Figure 69.2.

Nystatin is believed to form a channel through the membrane. Direct evidence for this has been provided by Ermishkin *et al.* (1976); nystatin induces discrete changes in BLM conductance indicative of channel formation. These step changes in conductance may last over a minute, long compared with similar changes in the presence of other antibiotics (see below). Romine *et al.* (1977), have found that nystatin-doped BLMs have current spectral densities that are nearly white below 10 Hz, compatible with a carrier model rather than a pore. Such differences may be due to the formation of the membrane and the type of lipid used. It is clear, however, that both doped and undoped lipid bilayers may exhibit $1/f$ noise under certain conditions.

In the presence of monazomycin, noise with a Lorentzian spectral density is observed. Figure 69.3 shows current noise from a BLM doped with 2 μg/ml monazomycin and bathed in 25 mM KCl under three different experimental conditions. Membrane conductance is voltage dependent; the theoretical Johnson noise is indicated for each membrane voltage. Current noise is approximately Lorentzian for −56 and −60 mV but becomes more complex for higher membrane voltages. Monazomycin is believed to induce channels in the membrane (see also Moore and Neher, 1976).

PROBLEM. Approximate the Lorentzian cutoff frequency and the variance of the noise from the 60-mV spectrum in Figure 69.3. Assume a two-state model (Sections 72 and 73). Estimate the single-channel conductance and the density of monazomycin channels for the 60-mV case.

Lorentzian current spectral densities are also found for gramicidin-doped membranes. (Kolb *et al.*, 1975; Kolb and Bamberg, 1977; Kolb and Läuger, 1977). Gramicidin-A is believed to form channels in the membrane by the formation and disappearance of conducting dimers. For the structure

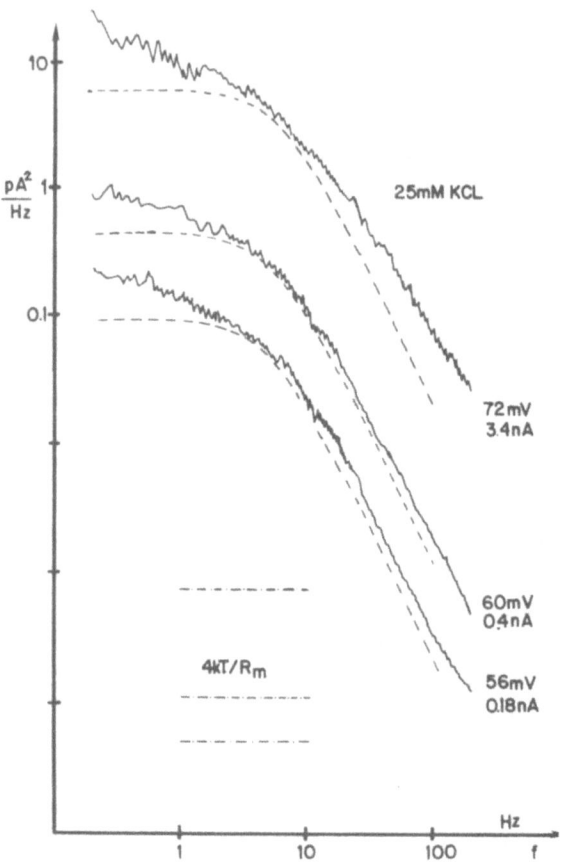

Figure 69.3. Current spectral density against frequency for a monazomycin-doped membrane (2 μg/ml) in 25 mM KCl. Three different transmembrane voltages are applied. The dashed curves are Lorentzians; expected Johnson noise is shown in the lower left-hand corner for the three cases. Membrane area: 4.4×10^{-3} cm^2; room temperature. (Courtesy E. Wanke.)

of gram-A and for details of the dimerization process see Urry (1971) and Urry *et al.* (1971). In Section C below a simple model of this process leads to a Lorentzian spectral density. If two gramicidin molecules are chemically bound before they are added to the membrane, the excess noise is $1/f$ noise similar to that shown for nystatin in Figure 69.2. The work on noise from BLMs doped with dimerized gramicidin has been done by Sauvé and Bamberg (1978). Bamberg has shown that the mean lifetime of channels formed by the dimerized gramicidin A is increased by a factor of 100 over the normal gramicidin-A channels. It is believed that the $1/f$ noise is generated in open channels, rather than in the kinetics of opening and closing, although this is difficult to prove.

The formation of single ionic channels was first demonstrated in BLMs using EIM (Bean *et al.*, 1969; Hladky and Haydon, 1970). These observations were influential in the construction of the two-state channel model for biological membranes (Section 72).

Suggested Reading. For additional information on channel formation in BLMs and the influence of this work on theories of channel formation in biological membranes with special reference to membrane noise see Ehrenstein *et al.* (1970), Ehrenstein (1971), Lattorre *et al.* (1972), Hladky and Haydon (1972), Ehrenstein *et al.* (1974), Zingsheim and Neher (1974), Neher and Zingsheim (1974), Lecar *et al.* (1975), Conti and Wanke (1975), Ehrenstein (1976), Ehrenstein and Lecar (1977), Neher and Stevens (1977), Kolb and Boheim (1978), and Boheim and Kolb (1978).

Another widely studied doping agent for BLMs is alamethicin. *Suggested reading* includes Gordon and Haydon (1972), Mauro *et al.* (1972), Eisenberg *et al.* (1973), and Kolb and Baheim (1976).

70. 1/f Noise and Concentration Gradients

In Sections 68 and 69 we considered $1/f$ noise from membranes bathed in the same salt solution on both sides. If a membrane separates different concentrations of the same salt solution, and the mobility of anion and cation differ, a diffusion potential exists across the membrane (Section 17). One may ask what effect this will have on $1/f$ noise.

The first experiments on $1/f$ noise from ionic systems were done on biological membranes. These experiments result from the collaboration of A. A. Verveen, a neurobiologist, and H. E. Derksen, an electrical engineer. Verveen had finished his doctoral thesis on "Fluctuation in Excitability" in 1961 using myelinated and unmyelinated nerve as an experimental model. It appeared that nerves were probabilistic devices and Verveen was looking for an explanation of this in the nerve membrane. Derksen's interests were in the application of communication theory to the nervous system. Their collaboration resulted in a series of articles, published jointly, and Derksen's doctoral thesis, "Axon Membrane Voltage Fluctuations," finished in 1965. The following are their publications from this period:

Verveen, A. A. (1961). Fluctuation in Excitability. Ph. D. thesis, Amsterdam: Netherlands Central Institute for Brain Research.

Verveen, A. A. and H. E. Derksen (1965). Fluctuations in membrane potential of axons and the problem of coding. *Kybernetick* 2:152–160.

Derksen, H. E. (1965). Axon membrane fluctuations. *Physiol. Pharmacol. Neerl.* 13:373–466. Doctoral thesis.

Derksen, H. E. and A. A. Verveen (1966). Fluctuations of resting neural membrane potential. *Science* 151:1388–1389.

Verveen, A. A., H. E. Derksen, and K. L. Schick (1967). Voltage fluctuations of neural membrane. *Nature* 216:588–589.

Verveen, A. A. and H. E. Derksen (1968). Fluctuations phenomena in nerve membrane. *Proc. IEEE* 56:906–916.

Verveen, A. A. and H. E. Derksen (1969). Amplitude distribution of axon membrane voltage noise. *Acta Physiol. Pharmacol. Neerl.* 15:353–379.

These papers established the transmembrane voltage in nerve as inherently noisy and introduced the use of spectral densities to characterize this noise.

Other workers began to study excess noise in ionic systems at about the same time. The experimental models ranged from simple holes in Mylar films to nerve axon membranes. Some patterns may be observed in these studies:

(i) A concentration gradient which results in membrane potential will shift the minimum of $1/f$ noise away from zero; DeFelice and Firth (1971), glass microelectrodes; Hooge and Gaal (1971), concentration cells; Poussart (1971), nerve axon. This result supports the work of Verveen and Derksen, who found a minimum in $1/f$ noise near the resting potential.

(ii) The spectral density is $1/f$ noise at the minimum. In carbon composition resistors, this is not true (Figure 65.1); the variance of $1/f$ noise goes through a minimum at zero current through the network and at the minimum the noise is Johnson noise. In ionic systems zero net current implies only a balance between ionic species (Section 13). In general, this balance is a nonequilibrium state of the system and Johnson noise is not expected.

(iii) Individual ionic currents (e.g., anionic and cationic currents) contribute to $1/f$ noise according to the driving forces they sense. As an example, consider two concentrations (n'' and n') of KCl separated by membrane more permeable to K than Cl. The side with the higher concentration will be negative with respect to the other side (Section 15). The driving force seen by K will be proportional to $V - V_K$; the driving force seen by Cl will be proportional to $V - V_{Cl}$. V is the transmembrane potential and V_K and V_{Cl} are the Nernst potentials for the two ions. For a tenfold difference in concentration (at room temperature),

$$V_K = -60 \text{ mV}; \qquad V_{Cl} = +60 \text{ mV}$$

Excess noise due to the K current will go through a minimum at $V = -60$ mV; excess noise due to the Cl current will go through a minimum at $+60$ mV. The variance of these noise sources add (DeFelice and Firth, 1971, glass microelectrodes).

EXERCISE. Consider a membrane permeable only to K. What noise is expected when the current through the membrane is zero? Consider a p–n junction in semiconductors passing a current. Do holes generate $1/f$ noise?

Suggested Reading. For a general discussion of membrane noise and its interpretation see Stevens (1972, 1975a, 1977), DeGoede and Verveen (1977), and Feher (1978).

71. Theories of Excess Noise

Theories of excess noise may be divided into two categories.

One may assume that excess noise is caused by an external perturbation to the system. In such theories, the passage of current through a resistor interacts with it; both are necessary to produce the observed excess noise. An example of this approach is provided by Handel (1975), who investigated the scattering of electrons by atoms in the lattice of a conductor in an attempt to derive a formula for $1/f$ noise. I have cast the experimental work on $1/f$ noise in terms of energy dissipation in the belief that the interaction between the charge carrier and the conductor is of fundamental importance.*

One may assume that excess noise is due simply to a modulation of the current by resistance fluctuations. Such theories are similar to how microphones work. If sound waves can be made to vary a resistor, current passing through the resistor at constant voltage will copy the sound wave. Some of the early studies on excess noise were made on carbon resistor microphones (Section 68). An example of the modulation approach to excess noise is generation–recombination noise in semiconductors [see, for

---

* Dorset and Fishman (1975) and Cole (personal communication, 1979) have suggested that $1/f$ noise occurs only when there is an anisotropic distribution of current in the conductor.

example, van der Ziel (1976), pp. 46–49]. The number of carriers in the semiconductor is assumed to fluctuate due to spontaneous processes of generation and recombination. There are an average number of carriers and a fluctuating, instantaneous number. These fluctuations are detected by passing a constant current through the sample (voltage fluctuations), or by applying a constant voltage across the system (current fluctuations). Since carrier fluctuations appear as resistance fluctuations, the voltage or current noise will have the spectral density of the carrier noise.

In modulation theories of $1/f$ noise, the $1/f$ fluctuations are going on all the time, even at equilibrium. The fluctuations are simply read out as a current of voltage. Interaction theories take the opposite view.

PROBLEM. Assume that a resistor fluctuates such that it has a mean value R but an instantaneous value $R(t)$ with the spectral density

$$S_R(f) = R^2/f$$

What is the open-circuit voltage spectral density at equilibrium? Does the Nyquist derivation of Johnson noise (Section 59) make any assumptions about the constancy of R?

PROBLEM. Assume the same resistor as above. What would be the result of an impedance measurement on such a resistor? (Imagine that the impedance is measured by injecting constant current sine waves of different frequencies and observing the voltage drop across the resistor.)

One should distinguish between a noise source and the impedance of a system. For example, in Johnson noise, the fundamental noise source is white (the thermal agitation of charge). If a resistor is placed in parallel with a noiseless capacitor, the open-circuit voltage spectral density is a Lorentzian (Section 35). This is simply an effect of filtering.

In Section C we provide a modulation theory of excess noise in which the relationship between the noise source, the system impedance, and the measured noise will be made explicit.

Suggested Reading. Suggested reading in the theory of $1/f$ noise generation includes: Richardson (1950), Schönfeld (1955), Halford (1968), Handel (1971), Offner (1971a, 1971b, 1792), Bird (1974a, 1974b), Lündström and McQueen (1974), Frehland and Läuger (1974), Neumcke (1975), Clay and Shlesinger (1976, 1977a,b), Weissman (1976, 1977), Voss and Clark (1976), Frehland (1976, 1977), Holden (1976a), Holden and Rubio (1976). and Green (1976, 1977).

Neumcke (1978) has written a good review of $1/f$ noise theories and experiments related to membranes. Also of interest is the Proceedings of the Symposium on $1/f$ Fluctuations held in Tokyo, July, 1977 (Fukuyo, 1977). These proceedings contain a theoretical review by van der Ziel and other articles on the theory and nature of $1/f$ noise. The Second International Symposium was held in Gainesville, Florida, 17–19 March 1979 (van der Ziel, 1979). Other symposia are the International Conferences on Noise; the last one was held in Bad Nauheim, West Germany, 13–16 March, 1978 (Wolf, 1978).

A selected reading list of experimental studies of $1/f$ noise and other types of noise found in electronic conductors but of relevance to membrane noise includes the following: Moullin (1938), Arguimbau (1948), Pierce (1956), van der Ziel (1954, 1959, 1970, 1976), Bell (1960), Bennett (1960), van Vliet and Farrett (1965), King (1966), and Hooge (1969a,b). A fascinating account of fluctuation theory from a general viewpoint but with special reference to electrical noise is found in MacDonald (1962).

C. LORENTZIAN NOISE

As with white noise and $1/f$ noise, the term "Lorentzian noise" implies a category of spectral density and not a process. The name comes from the work of H. A. Lorentz on the theory of radiation; the paragraph below is translated from this work (Lorentz, 1919, p. 62).

After a brief introduction to Planck's formula for blackbody radiation, Lorentz writes:

We shall not follow his [Planck's] calculations here, but simplify and shorten the derivation through the assumption that the resonator consists of an electron that is quasielastically bound to an equilibrium state and can move in but one direction along the X axis. From the theory of electrons, we may borrow the idea that the vibration which determines its radiation is subject to an apparent resistance proportional to its velocity. If we call the coefficient of resistance g, the electron mass m, the electron charge e, and the movement along the X axis x, then the quasielastic force is $-Fx$ and the equation of motion of the electron becomes

$$m\ddot{x} + g\dot{x} + Fx = ed_x$$

where d_x is the X component of the electric field. Suppose that the electric field d is a plane wave with amplitude a; if the frequency of this wave is n we may expect that the solution of this equation shall have a periodic vibration with the same frequency n. Suppose the solution of the equation is given by the real part of pe^{int}, and d_x by $a_x e^{int}$. Then it follows that

$$p(-mn^2 + ign + F)e^{int} = ea_x e^{int}$$

If a_x is real then p is complex; this expresses the fact that the phase of the stimulated vibration differs from the phase of the wave. If we wish to know the average energy of the electron, it must come from the amplitude, or modulus, of p. The average total energy is

$$\frac{1}{2} F \mid p \mid^2 = \frac{1}{2} Fe^2 \, \frac{a_x^2}{(F - mn^2)^2 + n^2 g^2}$$

where the square modulus of p is taken as the product of p with its complex conjugate.

After this passage, Lorentz calculates the average total energy of the electron over all frequencies contained in the stimulating wave. This is his introduction to the quantum hypothesis of Planck and the theory of black-body radiation.

Lorentz's method of solution of the equation of motion of the electron is equivalent to a solution by Fourier analysis. To solve the equation in our notation let $n = \omega = 2\pi f$ and let $d_x = -E_x$, the electric field (V/m) in the x direction. The Fourier transform of the equation of motion is

$$-m\omega^2 \bar{x}(f) + i\omega g \bar{x}(f) + F\bar{x} = -e\bar{E}_x(f)$$

where (Section 23)

$$\bar{x}(f) = \int_{-\infty}^{\infty} x(\varepsilon)e^{-i\omega t} \, dt, \text{ etc.}$$

and $x(t)$ is the solution of the original equation. Solving the transformed equation for $\bar{x}(f)$:

$$\bar{x}(f) = \frac{e}{m} \, \frac{\bar{E}_x}{(\omega^2 - \omega_0^2) - i\omega\Gamma}$$

where $\omega_0^2 = F/m$ and $\Gamma = g/m$. The notation now corresponds to that found in modern textbooks, e.g., Jackson (1975) Equation 13.19. (Jackson uses a different sign convention; his formula has $+ i\omega\Gamma$ in place of $-i\omega\Gamma$.)

PROBLEM.   Find the motion of the bound charge, $x(t)$, by taking the plus-i transform of $\bar{x}(f)$ assuming that $\bar{E}_x$ is constant. What is the input in the time domain?

We wish to calculate the energy transferred to the bound charge by the field. Let the velocity of the bound electron in the x direction be

$$v_x(t) = \dot{x}(t)$$

The force exerted by the field on the charge is

$$F_x(t) = -eE_x(t)$$

The rate of doing work (the power) in the x direction is

$$W = F_x v_x$$

which may be written

$$\frac{dE}{dt} = -eE_x v_x$$

E is the energy transferred from the field to the charge.

PROBLEM. Assume the electron can migrate in a conductor. Show that the equation $W = F_x v_x$ is equivalent to the familiar expression $W = IV$ for a constant linear electric field.

Let the total energy transferred from the field to the charge be ΔE:

$$\Delta E = \int_{-\infty}^{\infty} dE = -e \int_{-\infty}^{\infty} E_x v_x \, dt$$

This equation is equivalent to 13.22 in Jackson (1975) written for the x direction.

PROBLEM. Show that if $x(t)$ and $E_x(t)$ are real, then

$$\bar{x}(-f) = \bar{x}^*(f)$$

and

$$\bar{E}_x(-f) = \bar{E}_x^*(f)$$

The asterisk signifies the complex conjugate. Show that these expressions state that the Fourier transform of a real function has an even real part and an odd imaginary part.

We may now substitute the Fourier-transformed variables into the expression for ΔE to obtain an expression for the energy transferred in the frequency domain. Note that

$$E_x(t) = E_x{}^*(t) = \int_{-\infty}^{\infty} \bar{E}_x{}^*(f)e^{-i\omega t}\, df$$

and

$$\bar{v}_x(f) = i\omega \bar{x}(f)$$

Then

$$\varDelta E = -e \int_{-\infty}^{\infty} \left\{ \int_{-\infty}^{\infty} \bar{E}_x{}^*(f)e^{-i\omega t}\, df \right\} \left\{ \int_{-\infty}^{\infty} i\omega \bar{x}(f)e^{+i\omega t}\, df \right\} dt$$

$$= -e \int_{-\infty}^{\infty} \left\{ \int \int_{-\infty}^{\infty} \bar{E}_x{}^*(f)i\omega' \bar{x}(f')e^{-i\omega t}e^{+i\omega' t}\, df'\, df \right\} dt$$

or

$$\varDelta E = -e \int \int_{-\infty}^{\infty} \bar{E}_x{}^*(f)i\omega' \bar{x}(f') \left\{ \int_{-\infty}^{\infty} e^{-it(\omega-\omega')}\, dt \right\} df'\, df$$

ω' is used to indicate a dummy variable of integration. The quantity in curly brackets in the last equation is the delta function at $\omega = \omega'$ (Section 25). With this substitution

$$\varDelta E = -e \int_{-\infty}^{\infty} \bar{E}_x{}^*(f)i\omega \bar{x}(f)\, df$$

is the energy transfer in the frequency domain.

Since $\bar{E}_x{}^*$ and $\bar{x}$ have even real parts and odd imaginary parts, the cross terms between the real and imaginary parts integrate to zero over all frequencies leaving only

$$\varDelta E = -2e\, \mathrm{Re} \int_{0}^{\infty} \bar{E}_x{}^*(f)i\omega \bar{x}(f)\, df$$

We now insert the expression for $\bar{x}(f)$ that was obtained as a solution to the equation of motion of the bound charge:

$$\varDelta E = -2\,\frac{e^2}{m}\, \mathrm{Re} \int_{0}^{\infty} |\bar{E}_x(f)|^2 \left\{ \frac{i\omega}{(\omega^2 - \omega_0{}^2) - i\omega\varGamma} \right\} df$$

The real part of $\langle \ \rangle$ is obtained from

$$\mathrm{Re} \left\{ \frac{i\omega}{(\omega^2 - \omega_0{}^2) - i\omega\varGamma} \right\} \left\{ \frac{(\omega^2 - \omega_0{}^2) + i\omega\varGamma}{(\omega^2 - \omega_0{}^2) + i\omega\varGamma} \right\} = \frac{-\omega^2\varGamma}{(\omega^2 - \omega_0{}^2)^2 + \omega^2\varGamma^2}$$

Finally,

$$\varDelta E = 2\,\frac{e^2}{m} \int_{0}^{\infty} |\bar{E}_x{}^*(f)|^2 \left\{ \frac{\omega^2\varGamma}{(\omega^2 - \omega_0{}^2)^2 + \omega^2\varGamma^2} \right\} df$$

The quantity in curly brackets is called a Lorentzian function. If there is no damping (no quasielastic force), then $\omega_0 = 0$ and

$$\{\} = \frac{\omega^2 \Gamma}{\omega^4 + \omega^2 \Gamma^2} = \frac{1/\Gamma}{(\omega^2/\Gamma^2) + 1} = \frac{\theta}{1 + \omega^2 \theta^2}$$

if $\Gamma = 1/\theta$. This is the form of the Lorentzian function that appears in noise problems. (In physical acoustics this function is referred to as the Debye function.)

PROBLEM.   Assume the perturbing field is an impulse at $t = 0$:

$$E_x(t) = A_x\, \delta(t)$$

Plot ΔE as a function of ω ($\omega_0 \neq 0$). Repeat the problem, assuming the perturbing field is (a) constant for all t, $E_x(t) = E_x$; (b) a pure sinusoid, $E_x(t) = E_x \sin \omega' t$. Investigate ΔE as ω' approaches $\omega_0 \cdot \omega'$ is used to indicate a specific frequency of the stimulating wave.

The integrand in the equation for ΔE is a spectral density. This may be seen from the identity

$$\Delta E = \int \frac{d}{df}\, (\Delta E)\, df$$

PROBLEM.   If

$$\Delta E = \int S(f)\, df$$

what are the units of $S(f)$? By definition, ΔE is a variance (Section 34).

In 1973, Feher and Weissman used noise analysis to study the reaction kinetics of $BeSO_4$ in solution. This paper was one of the first papers to show the applicability of fluctuation analysis to kinetic problems in a model ionic system. The reaction

$$BeSO_4 \rightleftharpoons Be^{2+} + SO_4^{2-}$$

alters the conductance of the solution and by studying conductance fluctuations parameters of the reaction can be determined. The situation is similar in biological membranes and Feher and Weissman's article appeared just when conductance fluctuations were being observed at the neuromuscular junction (Anderson and Stevens, 1973). The following sections detail the relationship between the kinetics of channel formation in membranes and the measurement of conductance fluctuations. An analysis of the correspondence of the equations of excitability to kinetic schemes of conductance change was given by FitzHugh in 1965.

72. The Two-State Channel

As an example of a source of Lorentzian noise, consider a population of channels in a membrane; the channels flip between a closed state that does not conduct current and an open state that does. The channels modulate a steady current flowing through the membrane. The current is interrupted by the opening and closing channels and the statistics of the current fluctuations reflect the statistics of the channel fluctuations. The process of opening and closing is considered independent of the current flow; the current flow does not cause the channels to flip between the two states but is merely modulated by that process.

a. The Kinetic Equations

Let the total number of channels be N and the number of closed and open channels at any time be $N_c(t)$ and $N_o(t)$. Then at every instant of time,

$$N = N_c(t) + N_o(t)$$

No channel enters the open state without leaving the closed state, i.e., N is constant.

Let the rate constant for flipping from a closed to an open state be α; let the rate constant for flipping from an open to a closed state be β. Then the *average* behavior of the population of closed and open channels may be described by the kinetic scheme

$$N_c \underset{\beta}{\overset{\alpha}{\rightleftharpoons}} N_o$$

It is understood that N_c stands for the probable number of closed channels at time t and N_o the probable number of open channels at time t—not the actual number, but the probable number. Since the number of open and closed channels may only gain from one another, or lose from themselves, their time rate of change is governed by

$$\frac{dN_c}{dt} = \beta N_o - \alpha N_c$$

and

$$\frac{dN_o}{dt} = \alpha N_c - \beta N_o$$

These equations describe the probable behavior of N_o and N_c and are, in effect, definitions of α and β.

Since the instantaneous and the probable values of N_o and N_c are constrained by N, each of the equations above may be written for N_c or N_o alone. For example, the second equation becomes

$$\frac{dN_o}{dt} = \alpha(N - N_0) - \beta N_o$$

We will focus on the open channels because we are ultimately interested in the statistics of the current that flows through them.

The above equation may be solved for N_o. Rearranging terms gives

$$\frac{dN_o}{dt} + (\alpha + \beta)N_o = \alpha N$$

Notice that

$$\frac{d}{dt}(N_o e^{(\alpha+\beta)t}) = (\alpha + \beta)N_o e^{(\alpha+\beta)t} + \frac{dN_o}{dt} e^{(\alpha+\beta)t}$$

Therefore, the left-hand side of the differential equation may be written

$$\frac{d}{dt}(N_o e^{(\alpha+\beta)t})e^{-(\alpha+\beta)t}$$

or

$$d(N_o e^{t/\theta}) = (\alpha N e^{t/\theta})\, dt$$

where

$$\theta = \frac{1}{\alpha + \beta}$$

In this form, the integration of the equation is immediate:

$$\int_{t'}^{t''} d(N_o e^{t/\theta}) = \alpha N \int_{t'}^{t''} e^{t/\theta}\, dt$$

or

$$N_0(t'')e^{t''/\theta} - N_o(t')e^{t'/\theta} = \alpha N\theta(e^{t''/\theta} - e^{t'/\theta})$$

Multiply both sides of this equation by $e^{-t'/\theta}$ and let

$$\tau = t'' - t'$$

then

$$N_o(t' + \tau)e^{\tau/\theta} - N_o(t') = \alpha N\theta(e^{\tau/\theta} - 1)$$

or

$$N_o(t' + \tau) = N_o(t')e^{-\tau/\theta} + \alpha N\theta(1 - e^{-\tau/\theta})$$

PROBLEM. Show that as $\tau \to \infty$, $N_o(t' + \tau) \to \alpha\theta N$. We shall call this the average value of the probable number of open channels at long time, or $N(\infty)$.

b. *The Correlation Function*

The above equation may be written as

$$[N_o(t' + \tau) - \alpha\theta N] = [N_o(t') - \alpha\theta N]e^{-\tau/\theta}$$

Multiply both sides of this equation by $[N_o(t') - \alpha\theta N]$; then

$$[N_o(t' + \tau) - \alpha\theta N][N_o(t') - \alpha\theta N] = [N_o(t') - \alpha\theta N]^2 e^{-\tau/\theta}$$

The reason for these manipulations is that the average value of the left-hand side of the above equation is the correlation function of the process $N_o(t)$. The averaging must be done over all initial states of the process; using the symbols $\langle \ \rangle$ to mean this average,

$$\langle [N_o(t' + \tau) - \alpha\theta N][N_o(t') - \alpha\theta N] \rangle = C(\tau)$$

where $\alpha\theta N$ is the average value of the process (see Section 42).

It follows that

$$C(0) = \langle [N_o(t') - \alpha\theta N]^2 \rangle$$

and therefore

$$C(\tau) = C(0)e^{-\tau/\theta}$$

This is the correlation function for the process described by

$$N_c \underset{\beta}{\overset{\alpha}{\rightleftharpoons}} N_o \qquad \text{(written for a population)}$$

in which each channel in the population is flipping between an open and a closed state,

$$C \underset{\beta}{\overset{\alpha}{\rightleftharpoons}} O \qquad \text{(written for a single channel)}$$

Note that α and β are identical for both the single-channel kinetics and the population kinetics. We have not proved this equality, although it may be verified for simple membranes composed of a few channels. For example, consider a membrane with two channels, each with the same probability of being either open or closed. A series of snapshots of such a membrane should reveal an average state in which there is one open and one closed channel. By determining this average state (without keeping track of which channel is open or closed) one would find

$$\alpha = N_o/N = 1/2 \qquad \text{and} \qquad \beta = N_c/N = 1/2$$

These are exactly the α and β for each channel.

For large populations of channels where it is impossible to keep track of individual channels, it is assumed that all channels have the same average properties.

We stress again that the solution to the equation

$$\frac{dN_o}{dt} = \alpha N_c - \beta N_o$$

does not describe the actual number of open channels at time t but rather the expected number. To illustrate this point, let $t' = 0$ and $\tau = t'' = t$ in the previous development. Then

$$N_o(t) = N_o(0)e^{-t/\theta} + N_o(\infty)(1 - e^{-t/\theta})$$

Suppose all channels are closed at $t' = 0$, i.e., $N_o(0) = 0$. If α and β suddenly assume values that imply a steady state $N_o(0)$ other than zero, the population of open channels will approach the new steady state along the curve

$$N_o(t) = N_o(\infty)(1 - e^{-t/\theta})$$

This is illustrated in Figure 72.1 The random fluctuations around the average curve represent the actual number of open channels. Once the system evolves to the new steady state, no further average change occurs. The random opening and closing of channels, however, continues.

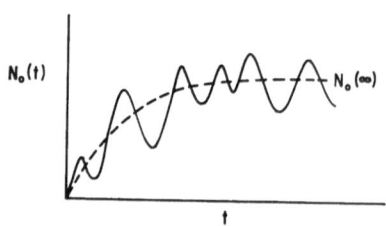

$N_0(t)$

$N_0(\infty)$

t

Figure 72.1. The average curve $N_0(t) = N_0(\infty)(1 - e^{-t/\theta})$ is shown as a dashed line. Around this curve, spontaneous fluctuations occur; these represent the actual number of channels open at any instant. Although the actual number of open channels is shown as a continuous curve, at high resolution this curve would be a random staircase.

PROBLEM.   Derive the solution of $dN_o/dt = \alpha N_o - \beta N_o$ in the frequency domain.

SOLUTION. $N = N_o + N_c$, $\theta = 1/(\alpha + \beta)$ and $p = \alpha\theta$. Then

$$\theta \frac{dN_o}{dt} + N_o = pN \tag{72.1}$$

Take the minus-*i* Fourier transform of both sides of this equation:

$$i\omega\theta \bar{N}_o + \bar{N}_o = pN\,\delta(f)$$

The solution in the frequency domain is

$$\bar{N}_o(f) = \frac{pN\,\delta(f)}{1 + i\omega\theta}$$

The solution in the time domain is the plus-*i* transform of $\bar{N}_o$:

$$N_o(t) = \int_{-\infty}^{\infty} \frac{pN\,\delta(f)}{1 + i\omega\theta}\,df = pN$$

The answer is

$$N_o(t) = pN = \frac{\alpha}{\alpha + \beta}\,N = \langle N_o \rangle$$

where $\langle N_o \rangle$ is the average number of channels in the open state; this frequency domain solution leads to the time domain solution at $t \to \infty$.

This solution may be understood by reexamining Equation (72.1). The right-hand side of Equation (72.1) is the driving force for the kinetic equation. The driving force is constant and the solution is the steady-state solution. The use of the term "steady state" is by analogy with problems in circuit theory where steady-state solutions are contrasted to transient solutions. It must be understood that the steady state also includes spontaneous fluctuations from the average value.

To obtain the analog to the transient solution, the driving force for the kinetic equation must be perturbed. Consider a step perturbation $H(t)$, where $H(t)$ is the Heaviside function (Section 26). Equation (72.1) becomes

$$\theta \frac{dN_o}{dt} + N_o = pNH(t)$$

The Fourier transform of both sides of this equation is now given by

$$\theta[i\omega \bar{N}_o - N_o(0)] + \bar{N}_o = pN \frac{1}{i\omega}$$

To arrive at this expression, the derivative theorem of the Fourier transform (Section 26) has been extended to include an initial value constant $\bar{N}_o(0)$; this procedure is equivalent to the Laplace transformation with transformation variable $s = i\omega$ (Goldman, 1949; Churchill, 1972). Solving for $\bar{N}_o$,

$$\bar{N}_o(f) = \frac{N_o(0)}{1 + i\omega\theta} + \frac{1}{i\omega} \frac{pN}{1 + i\omega\theta}$$

The plus-i transformation is (CF # I-438 and I-210):

$$N_o(t) = N_o(0)e^{-t/\theta} + pN(1 - e^{-t/\theta})$$

where $t > 0$. This is equivalent to our previous time domain solution and completes the problem.

$N_o(t)$ does not describe the actual number of open channels in real time—as the notation would normally imply—it may therefore be an advantage to replace t by τ when describing the average process. Since the correlation function for the two-state process is a relaxation proportional to $e^{-\tau/\theta}$, it may have been easier to write initially

$$\frac{dN_o}{d\tau} = \alpha N_c(\tau) - \beta N_o(\tau)$$

It remains to calculate $C(0)$. One may proceed directly from the expression

$$C(0) = \langle (N_o(t') - \alpha\theta N)^2 \rangle$$

where the prime (') notation indicates the initial state of the system. The average $\langle \rangle$ must then be over all possible initial states. Rather than follow this line we return to the statistics of single channels and develop an expression for the variance of a population of open channels. This variance is $C(0)$.

c. Bernoulli's Distribution

Consider a population of N channels, each channel in either an open or a closed state. We wish to calculate the probability that exactly N_o of them are open and N_c are closed. Let p be the probability that a channel is open and q be the probability that it is closed. If we were to take snapshots of the total population N, on average we should find that there are

$$\langle N_o \rangle = pN$$

open channels and

$$\langle N_c \rangle = qN$$

closed channels. p and q are the average fraction of open or closed channels. p and q are unitless quantities that vary between 0 and 1. Also since $N = N_o + N_c$,

$$p + q = 1$$

As a first example, consider two channels in parallel in a membrane. The possible combinations of open (O) and closed (C) states are

$$OO, \quad OC, \quad CO, \quad CC$$

The position of O and C indicates which of the two channels are open or closed. The probabilities associated with these states are

$$p^2, \quad pq, \quad qp, \quad q^2$$

For example, let $p = q = 1/2$ (equal change of a channel being either open or closed). The states OO and CC would each occur in only one-fourth of the snapshots. However, finding the system with one channel open and one closed would occur in one-half of the snapshots, since there are two ways this may happen. Thus,

$$\frac{1}{4}, \quad \left\{ \frac{1}{4}, \frac{1}{4} \right\}, \quad \frac{1}{4}$$

are the respective probabilities for $p = q = 1/2$. The two states in brackets have a similar effect; they are called like states. Notice the individual probabilities of the states sum to one as they must.

For three channels the possible combinations are

$$OOO, \quad OOC, \quad OCO, \quad OCC$$
$$COO, \quad COC, \quad CCO, \quad CCC$$

and the probabilities of occurrence of each state are

$$p^3, \quad p^2q, \quad p^2q, \quad pq^2$$
$$qp^2, \quad q^2p, \quad q^2p, \quad q^3$$

We now give two general rules: First, the number of ways to find exactly N_o open channels in a population of N is

$$\frac{N!}{N_o!N_c!} = \frac{N!}{N_o!(N - N_o)!}$$

For example, the number of ways we can find two open channels in a population of three channels is

$$\frac{3!}{2!(3 - 2)!} = 3$$

These are the OOC, OCO, and COO states above. Second, the probability that N_o channels are open in any given state is

$$p^{N_o}q^{N_c} = p^{N_o}q^{N-N_o}$$

This formula generates the eight cases listed above; for example, the probability of two open and one closed state is

$$p^2q^{3-2} = p^2q$$

Since there are three of these, the total probability of two open and one closed state in a population of three is $3p^2q$.

In general, the probability of obtaining exactly N_o open channels in a population of N is

$$B_N(N_o) = \frac{N!}{N_o!(N - N_o)!} p^{N_o}q^{N-N_o}$$

where p is the probability of a channel being open and q is the probability of a channel being closed. We may, of course, substitute $1 - p$ for q. The function $B_N(N_o)$ is called the Bernoulli distribution.

PROBLEM. Work out the case for four channels. Calculate the probability of obtaining each of the 16 states for two cases: $p = q = 1/2$ and $p = 1/3$, $q = 2/3$. Then calculate the probabilities for obtaining like-states, e.g., $OOCC$ is like $OCOC$. Which of the like states has the highest probability?

d. The Mean

We now calculate the average number of open channels in a Bernoulli distribution of open channels. In order to do this, we need the following identity: Let p and q be defined as above and let y be any number; then

$$(yp + q)^N = \sum_{N_o=0}^{N} B_N(N_o)y^{N_o}$$

To verify this equation for a simple case, let $N = 2$; then

$$(yp + q)^2 = y^2p^2 + q^2 + 2ypq$$

and

$$\sum_{N_o=0}^{2} B_2(N_o)y^{N_o} = \left\{\frac{2!p^0q^2}{0!2!}\right\}y^0 + \left\{\frac{2!p^1q^1}{1!1!}\right\}y^1 + \left\{\frac{2!p^2q^0}{2!0!}\right\}y^2$$
$$= q^2 + 2pqy + p^2y^2$$

These expressions are equal.

From the general equation, we may derive others by differentiating with respect to y. The first derivative with respect to y gives the equation

$$N(yp + q)^{N-1}p = \sum_{N_o=0}^{N} N_o B_N(N_o)y^{N_o-1}$$

The result is true for any y; letting $y = 1$ gives

$$N(p + q)^{N-1}p = \sum_{N_o=0}^{N} N_o B_N(N_o)$$

The left-hand side of this equation is pN, since $p + q = 1$. pN is defined as the average number of open channels, thus

$$pN = \sum_{N_o=0}^{N} N_o B_N(N_o) = \langle N_o \rangle$$

In general, the average of a quantity over a distribution is the product of the quantity and the distribution summed over all possibilities.

e. The Variance

The second derivative of our initial equation is

$$(N - 1)N(yp + q)^{N-2}p^2 = \sum_{N_o=0}^{N} N_o(N_o - 1)B_N(N_o)y^{N_o-2}$$

Let $y = 1$ and $p + q = 1$; then

$$(N - 1)Np^2 = \sum_{N_o=0}^{N} N_o(N_o - 1)B_N(N_o)$$

Expanding the right-hand side of this equation gives

$$\sum_{N_o=0}^{N} N_o(N_o - 1)B_N(N_o) = \sum_{N_o=0}^{N} N_o^2 B_N(N_o) - \sum_{N_o=0}^{N} N_o B_N(N_o)$$

$$= \langle N_o^2 \rangle - \langle N_o \rangle$$

Combining the previous two equations leads to

$$(N - 1)Np^2 = \langle N_o^2 \rangle - \langle N_o \rangle$$

or by rearrangement,

$$\langle N_o \rangle - Np^2 = \langle N_o^2 \rangle - N^2 p^2$$

Since $pN = \langle N_o \rangle$,

$$pN - Np^2 = \langle N_o^2 \rangle - \langle N_o \rangle^2$$

The right-hand side is the variance in N_o (Section 33), thus

$$\sigma^2 = pN(1 - p)$$

The initial problem was to calculate the variance defined as

$$\sigma^2 = \langle [N_o(t') - \alpha\theta N]^2 \rangle$$

where $\langle \ \rangle$ stands for an average over all initial states. Ultimately we calculated the variance defined as

$$\sigma^2 = \langle N_o^2 \rangle - \langle N_o \rangle^2$$

Let $t' = 0$. The first definition of σ^2 may be rewritten as

$$\langle [N_o(0) - N_o(\infty)]^2 \rangle = \langle N_o^2(0) \rangle + N_o^2(\infty) - 2N_o(\infty)\langle N_o(0) \rangle$$

If

$$\langle N_o(0) \rangle = N_o(\infty)$$

the two definitions of σ^2 are identical. This equality is a condition of the steady state, for in the steady state the average value of N_o is constant.

We note some alternative ways of writing the variance of a population of N independent two-state channels:

$$\sigma^2 = Npq$$
$$\sigma^2 = q\langle N_o\rangle$$
$$\sigma^2 = p\langle N_c\rangle$$
$$\sigma^2 = qN(1 - q)$$
$$\sigma^2 = pN(1 - p)$$

It is perhaps simplest to remember this rule: The variance of a random two-state process is the probability of being open times the probability of being closed. The expected variance in the number of open channels is the same as that in the number of closed channels since the two processes are locked to one another by N.

Previously we showed that the correlation function associated with the number of open channels in a population of N channels flipping between the open and closed states is

$$C(\tau) = C(0)e^{-\tau/\theta}$$

Since $C(0) = \sigma^2$, the final result is

$$C(\tau) = pN(1 - p)e^{-\tau/\theta}$$

We may use a subscript to identify this correlation function, thus $C_{N_o}(\tau)$: this is the correlation function of the steady-state fluctuation in the number of open channels. The expected number of open channels at long times is

$$N_o(\infty) = \alpha\theta N = pN$$

The expected number at long times is the average number; also

$$p = \alpha\theta = \frac{\alpha}{\alpha + \beta}$$

and

$$q = 1 - p = \frac{\beta}{\alpha + \beta}$$

Using these identities,

$$C(\tau) = \frac{N\alpha\beta}{(\alpha + \beta)^2} e^{-(\alpha+\beta)\tau}$$

PROBLEM. Show that the spectral density of this process is

$$S(f) = \frac{4N\alpha\beta(\alpha + \beta)}{(\alpha + \beta)^2 + \omega^2}$$

EXERCISE. Follow the derivation in Hill and Chen (1972) for the case $x = 1$, to conclude that

$$C(0) = Nn(\infty)[1 - n(\infty)]$$

$n(\infty)$ is identical to our p. See also Stevens (1972) in the same volume of *Biophysical Journal*. An example of this type of problem is given in T. L. Hill (1968), Chap. 7.

EXERCISE. The method developed in this paragraph uses a differential equation to describe the average process

$$N_c \underset{\beta}{\overset{\alpha}{\rightleftharpoons}} N_o$$

derived from N channels. Each channel is a random process represented by

$$C \underset{\beta}{\overset{\alpha}{\rightleftharpoons}} O$$

A more direct method for calculating the correlation function of a single channel is given in Lee (1960), Chap. 8. Lee assumes a Poisson distribution for the zero crossings of the rectangular wave associated with a single channel. This wave is called a random telegraph signal. The Poisson distribution is

$$P_t(n) = \frac{(\nu t)^n}{n!} e^{-\nu t}$$

where ν is the average rate of zero crossings and $P_t(n)$ is the probability of finding exactly n crossings in the time t. Lee assumes the probability of opening equals the probability of closing and calculates the correlation function by averaging over $P_t(n)$.

For a rectangular wave of amplitude E_m, he derives

$$C(\tau) = E_m{}^2 e^{-2\nu|\tau|}$$

Show the correspondence between this result and ours. Repeat Lee's derivation for the case $p \neq q$.

Exercise. Beginning with the Bernoulli distribution

$$B_N(N_o) = \frac{N!}{N_o!(N - N_o)!} \, p^{N_o}(1 - q)^{N - N_o}$$

show that for $N \gg N_o$ and N very large, then

$$B_N(N_o) \rightarrow P_N(N_o)$$

where

$$P_N(N_o) = \frac{(pN)^N}{N_o!} \, e^{-pN}$$

This is the Poisson distribution; it is the probability of finding exactly N_o open channels in a population of N under the stated conditions. p is the probability of a channel being open and $pN = \langle N_o \rangle$. This derivation is given in Goldman (1948), Chap. 7. Discuss the correspondence between $P_N(N_o)$ and $P_t(n)$ given in the previous exercise.

Problem. Consider N channels initially closed. Let the channels open in sequence according to

$$N^o \xrightarrow{\nu} N_1 \xrightarrow{\nu} N_2 \cdots$$

where ν is the rate constant for opening, N_1 represents the state "one open channel in N," N_2 represents "two open channels in N," and so on.

The rate equations for this process are

$$\frac{dN^o}{dt} = -\nu N^o$$

$$\frac{dN_1}{dt} = \nu N^o - \nu N_1$$

$$\frac{dN_2}{dt} = \nu N_1 - \nu N_2$$

etc. The solution to the first equation is

$$N^o(t) = N(0)e^{-\nu t}$$

where $N(0)$ is the number of channels at $t = 0$; $N(0) = N$. Substituting $N(t)$ into the second equation gives

$$\frac{dN}{dt} = \nu N e^{-\nu t} - \nu N_1$$

Show that the solution to this equation is

$$N_1(t) = N\nu t e^{-\nu t}$$

Show that the general solution for the nth equation is

$$N_n(t) = N \frac{(\nu t)^n}{n!} e^{-\nu t}$$

Defining

$$P_t(n) = \frac{N_n(t)}{N}$$

as the probability of obtaining exactly n open channels in the time t provides a derivation of the Poisson distribution. Compare this derivation with that obtained as a limiting case of Bernoulli's distribution.

Suggested Reading. For comprehensive texts on probability and the theory of random processes see Feller (1960), Papoulis (1965), Pfeiffer (1965), Gerault (1966), and Davenport (1970). An excellent collection of papers on noise theory and stochastic processes is Wax (1964). For the relationship of noise and probability to information theory and communication see Gabor (1950) and Goldman (1953).

73. The Relationship between Channel Noise and Current Noise

Consider the stochastic function $N_o(t)$, composed of the independent superposition of N channels flipping between two states. The variable t is real time and $N_o(t)$ is now the actual number of open channels at time t. Imagine some way to record the number of open and closed channels at every moment and plot $N_o(t)$ against t. The correlation function of $N_o(t)$ is

$$C_{N_0}(\tau) = Np(1 - p)e^{-\tau/\theta}$$

where p and θ have their previous meaning (Section 72).

Ordinarily there is no direct way to record the actual number of open and closed channels (for example, taking photographs of all N channels and counting those in the open and closed state in each successive photograph). An indirect method is usually used.

Suppose that a channel separates two solutions of KCl and is perfectly selective for the K ion. An equivalent circuit for such a channel is shown

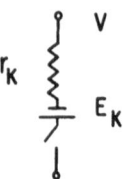

Figure 73.1. An equivalent circuit for a K channel.

in Figure 73.1. Let V be the voltage across the channel; the current through the channel is from Ohm's law:

$$i_K = \gamma_K(V - E_K)$$

γ_K is the single-channel conductance $(1/r_K)$ and E_K is the Nernst potential (Section 14). The switch in the equivalent circuit is a random telegraph switch that opens and closes the channel. γ, V, and E are assumed constant; $i(t)$ is the random function shown in Figure 73.2.

Consider N such channels in parallel in a membrane. The total current through these channels at any moment is given by the number of open channels times the current through each channel:

$$I_K(t) = N_o(t)i_K$$

The average current $\langle I_K \rangle$, usually written as I_K, is

$$I_K = pNi_K = pN\gamma_K(V - E_K)$$

$pN\gamma_K$ is called the K conductance of the membrane:

$$G_K = pN\gamma_K$$

The maximum conductance is $N\gamma_K$.

The correlation function of $I_K(t)$ is the product

$$C_{I_K}(\tau) = \langle I_K(t + \tau) - I_K\rangle\langle I_K(t) - I_K\rangle$$

(Sections 42–44).

The average current, I_K, is subtracted from the instantaneous current, $I_K(t)$, to calculate the correlation function of the fluctuating component.

Figure 73.2. The instantaneous current through a single channel.

Writing I_K in terms of the microscopic current i_K gives

$$C_{I_K}(\tau) = \langle N_o(t + \tau)i_K - pNi_K\rangle\langle N_o(t)i_K - pNi_K\rangle$$
$$= i_K{}^2\langle N_o(t + \tau) - pN\rangle\langle N_o(t) - pN\rangle$$

The time dependence of $I_K(t)$ is entirely through $N_o(t)$. The current is modulated (chopped) by the population of two-state channels; the statistics of the current fluctuations reflect the statistics of the channel fluctuation exactly.

In Section 72 we derived the correlation function for a population of N channels flipping between two states. Using these results and the definition of i_K implied by Figure 73.1, the correlation function of $I_K(t) - I_K$ is

$$C_{I_K}(\tau) = \gamma_K{}^2(V - E)^2 Np(1 - p)e^{-\tau/\theta} \qquad (73.1)$$

Since $I_K = pN\gamma_K(V - E_K)$ is the average current,

$$C_{I_K}(\tau) = I_K{}^2\frac{1 - p}{Np}e^{-\tau/\theta}$$

Recall that $\theta = 1/(\alpha + \beta)$ and $p = \alpha\theta$; $1/\alpha$ is the average channel closed time and $1/\beta$ is the average channel open time (Section 72). Another form of the equation is

$$C_{I_K}(\tau) = I_K{}^2\frac{\beta/\alpha}{N}e^{-(\alpha+\beta)\tau}$$

Consider an experiment in which both C_{I_K} and I_K are measured. At $\tau = 0$ the variance is

$$C_{I_K}(0) = \sigma_K{}^2 = I_K{}^2\frac{\beta/\alpha}{N}$$

or

$$\frac{\beta/\alpha}{N} = \left(\frac{\sigma_K}{I_K}\right)^2$$

From a plot of $C_{I_K}(\tau)$ vs. τ, θ may be determined; thus $(\alpha + \beta) = 1/\theta$ is also known.

PROBLEM. Show that $p(1 - p)$ has its maximum value if $p = 1/2$ ($\alpha = \beta$). From Equation (73.1), the maximum noise occurs for $\alpha = \beta$. $p = 1/2$ also implies half the maximum average current is flowing when the noise is at its maximum level.

Suppose α and β are varied by an external parameter, e.g., the voltage across the membrane; if this parameter is varied to maximize σ_K^2 at that point $\alpha = \beta$ and

$$\frac{1}{N} = \left(\frac{\sigma_K^{\max}}{I_K}\right)^2 = 4\left(\frac{\sigma_K^{\max}}{I_K^{\max}}\right)^2$$

Thus N may be determined. Once N is known, β/α and $\alpha + \beta$ are known for every other value of σ_K and I_K and α and β may be determined independently.

PROBLEM. Show that the average channel open time is

$$\frac{1}{\alpha} = \theta + N\left(\frac{\sigma_K}{I_K}\right)^2 \theta$$

and the average channel closed time is

$$\frac{1}{\beta} = \theta + \frac{1}{N}\left(\frac{I_K}{\sigma_K}\right)^2 \theta$$

PROBLEM. Devise an experiment to determine the single-channel conductance γ_K.

PROBLEM. Show that the spectral density of the current noise is

$$S_{I_K}(f) = \frac{4 I_K^2}{N}\frac{1-p}{p}\frac{\theta}{1+\omega^2\theta^2}$$

where $\omega = 2\pi f$.

PROBLEM. C_{I_K} and S_{I_K} were calculated above for the noise component of $I_K(t)$ only, i.e., for $I_K(t) - I_K$. Calculate the correlation function and spectral density for $I_K(t)$. Interpret the functions as $\tau \to \infty$ and $f \to 0$.

PROBLEM. Show that the variance of the membrane conductance, σ_g^2, is related to the variance of the membrane current, σ_I^2, by

$$\sigma_g^2 = \frac{\sigma_I^2}{(V - E)^2}$$

This relationship is true at every τ.

74. Two-State Channels in Series

In Sections 72 and 73 the model membrane consisted of a population of two-state channels in parallel. Let ↑ represent an open channel (O) and let ↓ represent a closed channel (C). A three-channel membrane may, at some instant, be in the state

$$↑ ↓ ↑$$

and at another instant

$$↓ ↑ ↑$$

and so on. Consider a population of such channels in the steady state. In the steady state α and β are constant and the average number of open channels is

$$N_o(\infty) = \frac{\alpha}{\alpha + \beta} N$$

Suppose α and β change abruptly; let the new values be α^* and β^*. The population of channels will approach a new steady state along the curve

$$N_o(t) = N_o(0)e^{-t/\theta^*} + N_o^*(\infty)(1 - e^{-t/\theta^*})$$

where t is measured from the moment α and β change to their new values α^* and β^* and where $N_o(t)$ represents the average number of open channels at time t. The initial average number of open channels, $N_o(0)$, is the previous steady-state value. Writing the expression in terms of the initial and final values of α and β,

$$N_o(t) = N\left\{\frac{\alpha}{\alpha + \beta} e^{-t(\alpha^*+\beta^*)} + \frac{\alpha^*}{\alpha^* + \beta^*} (1 - e^{-t(\alpha^*+\beta^*)})\right\}$$

or

$$N_o(t) = N\left\{\frac{\alpha\beta^* - \alpha^*\beta}{(\alpha + \beta)(\alpha^* + \beta^*)} e^{-t(\alpha^*+\beta^*)} + \frac{\alpha^*}{\alpha^* + \beta^*}\right\}$$

To calculate the properties of two-state channels in series we interpret the equation for $N_o(t)$ for a single channel.

The steady-state probability of a channel opening is

$$p = \frac{\langle N_o \rangle}{N}$$

Let

$$p(t) = \frac{N_o(t)}{N}$$

Then

$$p(t) = p(0)e^{-t/\theta} + p(\infty)(1 - e^{-t/\theta})$$

The * has been dropped from this equation; *it is understood that α and β refer to the new steady state* toward which $p(t)$ is headed. $p(t)$ is the probability of finding a given channel in the open state at time t after an abrupt change in α and β. Although α and β change instantaneously, it takes time for the channel to adjust to its new average rate of opening and closing and to come to a new steady state. This is illustrated in Figure 74.1. This adjustment period results from the probabilistic nature of opening and closing. To see this, suppose the new values of α and β shorten the average open time as illustrated in Figure 74.1; after α and β change (at $t = 0$) there may still be rather long open times; this will depend on the initial state of the channel. The probability of finding the channel in the open state is independent of the initial state only at times much longer than $1/\theta$.

With this background, we now calculate the noise due to two-state channels in series.

Consider a three-channel membrane. Each channel is constructed such that it requires two independent flips to the open state to open it. This may be represented by two arrows in series; at some instant, the three-channel membrane may be in the state

↑↓↑

↑↑↓

and at another instant,

↓↑↑

↑↑↑

and so on. An arrow is now a subunit of the channel, whereas previously it was the entire channel. The probability of finding a subunit open is $p(t)$; the probability of finding two subunits open at the same time is $p^2(t)$. This

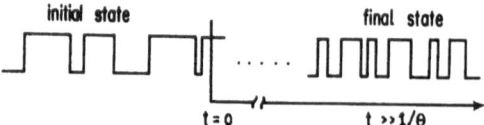

initial state   final state

t = 0   t ≫ 1/θ

Figure 74.1. Diagram illustrating the transition of a single channel to a new steady state.

may be verified by an example: let $p(t) = 1/2$; at this particular time there is an equal chance of finding a channel subunit open or closed. The possible channel states are

$$\uparrow \quad \uparrow \quad \downarrow \quad \downarrow$$
$$\uparrow, \quad \downarrow, \quad \uparrow, \quad \downarrow$$

Only one state in four is open and $p^2(t) = 1/4$.

PROBLEM. Consider a channel with three independent subunits. Verify by example that $p^3(t)$ is the probability of finding an open channel.

The probability of a channel having two independent two-state sub-units open at the same time is therefore

$$p^2(t) = [p(0)e^{-t/\theta} + p(\infty)(1 - e^{-t/\theta})]^2$$
$$= p^2(0)e^{-2t/\theta} + p^2(\infty)(1 - e^{-t/\theta})^2 + 2p(0)p(\infty)(e^{-t/\theta} - e^{-2t/\theta})$$

Solving for $p^2(t) - p^2(\infty)$ gives

$$p^2(t) - p^2(\infty) = e^{-2t/\theta}[p^2(0) + p^2(\infty) - 2p(0)p(\infty)] + e^{-t/\theta}[2p(0)p(\infty) - 2p^2(\infty)]$$

Multiplying both sides of the equation by $p^2(0) - p^2(\infty)$ gives

$$[p^2(t) - p^2(\infty)][p^2(0) - p^2(\infty)] = [p^2(0) - p^2(\infty)][p(0) - p(\infty)]^2 e^{-2t/\theta}$$
$$+ 2p(\infty)[p^2(0) - p^2(\infty)][p(0) - p(\infty)]e^{-t/\theta}$$

The average value of the left-hand side of this equation is the correlation function for the $p^2(t)$ process. As before, the average must be taken over all initial states; the correlation function is written

$$C_{p^2}(t) = \langle [p^2(t) - p^2(\infty)][p^2(0) - p^2(\infty)] \rangle$$

To preserve the distinction between real time t and correlation time τ, let τ replace t (see Section 72). Then

$$C_{p^2}(\tau) = \langle [p^2(0) - p^2(\infty)][p(0) - p(\infty)]^2 \rangle e^{-2\tau/\theta}$$
$$+ 2p(\infty)\langle [p^2(0) - p^2(\infty)][p(0) - p(\infty)] \rangle e^{-\tau/\theta} \qquad (74.1)$$

Note that $p(\infty) = p$ is the steady-state probability of a subunit being open; $p(\infty)$, $\exp(-2\tau/\theta)$, and $\exp(-\tau/\theta)$ represent average values and may be taken outside the $\langle \ \rangle$ brackets.

The above equation is written for a single channel. For a population of channels with two independent two-state subunits,

$$N_o(t) = p^2(t)N$$

and the correlation function for the process $N_o(t)$ is simply N times the correlation function for $p^2(t)$:

$$C_{N_o} = NC_{p^2}$$

To calculate the variance $\sigma_{p^2}^2$ consider

$$C_{p^2}(0) = \langle [p^2(0) - p^2(\infty)][p(0) - p(\infty)]^2 \rangle$$
$$+ 2p(\infty)\langle [p^2(0) - p^2(\infty)][p(0) - p(\infty)] \rangle$$

The average must be taken over all initial states; to avoid this calculation we derive the Bernoulli distribution for a population of channels with two subunits and calculate the variance directly, as we did in Section 72 for a population of channels with only one subunit.

75. Bernoulli's Distribution for Two Independent Two-State Subunits

In Section 72 we derived the probability of obtaining exactly N_o open channels in a population of N; this was called the Bernoulli distribution $B_N(N_o)$. If each channel is composed of two subunits, the probability of obtaining N_o open channels in N is

$$B_N{}^2(N_o) = \frac{N!}{N_o!(N - N_o)!} \, p^{2N_o}(1 - p^2)^{N - N_o}$$

[The 2 on $B_N{}^2(N_o)$ implies two subunits, not squaring.]

To verify this formula, consider two channels in parallel. The possible combinations of states are 16; assume the subunits are equally likely to be open or closed ($p = 1/2$):

OO	*OO*	*OC*	*OC*
OO	*OC*	*OO*	*OC*
OO	*OO*	*OC*	*OC*
CO	*CC*	*CO*	*CC*
CO	*CO*	*CC*	*CC*
OO	*OC*	*OO*	*OC*
CO	*CO*	*CC*	*CC*
CO	*CC*	*CO*	*CC*

The probabilities for these states are p^4, p^3q, etc. The distribution of states are one p^4 state, four p^3q states, six p^2q^2 states, four pq^3 states, and one q^4 state. The probability of obtaining exactly two open channels in a population of two is

$$B_2{}^2(2) = \frac{2!}{2!0!} \left(\frac{1}{2}\right)^4 \left(1 - \frac{1}{4}\right)^0 = \frac{1}{16}$$

corresponding to the single p^4 state. The probability of obtaining one open channel is

$$B_2{}^2(1) = \frac{2!}{1!1!} \left(\frac{1}{2}\right)^2 \left(1 - \frac{1}{4}\right)^1 = \frac{6}{16}$$

corresponding to the four p^3q states and two of the p^2q^2 states, and so on.

Following the method of Section 72, we now calculate the mean and the variance of the $N_o(t) = p^2(t)N$ process.

PROBLEM.   Show that

$$(yp^2 + r)^N = \sum_{N_o=0}^{N} B_N{}^2(N_o)y^{N_o}$$

where $r = 1 - p^2$. (Verify the equality for $N = 1$, $N = 2$, etc.)

Taking the first derivative of the above expression with respect to y and setting $y = 1$ results in

$$N(p^2 + r)^{N-1}p^2 = \sum_{N_o=0}^{N} N_o B_N{}^2(N_o)$$

The right-hand side is the average value of N_o over the Bernoulli distribution for two subunits; since $p^2 + r = 1$,

$$Np^2 = \langle N_o \rangle$$

as expected.

Taking the second derivative with respect to y and setting $y = 1$ gives

$$N(N - 1)(p^2 + r)^{N-2}p^4 = \sum_{N_o=0}^{N} N_o(N_o - 1)B_N{}^2(N_o)$$

Since $p^2 + r = 1$

$$N^2p^4 - Np^4 = \sum_{N_o=0}^{N} N_o{}^2 B_N{}^2(N_o) - \sum_{N_o=0}^{N} N_o B_N{}^2(N_o)$$

By our previous result for $\langle N_o \rangle$ this equation may be written

$$\langle N_o \rangle^2 - Np^4 = \langle N_o{}^2 \rangle - Np^2$$

or

$$Np^2 - Np^4 = \langle N_o{}^2 \rangle - \langle N_o \rangle^2$$

The right-hand side is the variance in N_o (Section 33), thus

$$\sigma_{N_o}^2 = p^2 N(1 - p^2)$$

This is the variance in $N_o(t)$ in a population of N channels each composed of two identical two-state subunits; each subunit has the probability p of being in the open state and both subunits must be open to have an open channel. The variance is proportional to the probability of the channel being open times the probability of the channel being closed; recall that the same rule applies for channels with one subunit (Section 72).

The variance is related to the correlation function by

$$\sigma_{N_o}^2 = C_{N_o}(0) = N C_{p^2}(0)$$

therefore

$$\sigma_{p^2}^2 = p^2(1 - p^2)$$

By definition

$$\sigma_{p^2}^2 = \langle [p^2(0) - p^2]^2 \rangle$$
$$= \langle p^4(0) + p^2 - 2p^2(0)p^2 \rangle$$
$$= \langle [p^2(0)]^2 \rangle - \langle p^2 \rangle^2$$

The last equality follows only if

$$\langle p^2(0) \rangle = p^2$$

Equating the two expressions for $\sigma_{p^2}^2$ gives

$$\langle p^4(0) \rangle - p^4 = p^2(1 - p^2)$$

or

$$\langle p^4(0) \rangle = p^2$$

It may seem strange that the averages of both $p^2(0)$ and $p^4(0)$ have the same value. This is a consequence of a channel that has only two subunits; the open-channel probability is never more than p^2 and the average value is never more than $\langle p^2 \rangle$.

PROBLEM.   Show that $\langle p^3(0)\rangle = p^2$. In general,

$$\langle p^n(0)\rangle = p^2, \qquad n \geq 2$$

since there are only two subunits and both must be open for the channel to be open.

Finally, since the process we are considering is in the steady state,

$$\langle p(0)\rangle = p(\infty) = p$$

These equations are used in the next paragraph to calculate an explicit expression for the correlation function of the two-subunit channel in terms of p.

PROBLEM.   Derive the Bernoulli distribution, $B_N{}^4(N_o)$, for a population of N channels each composed of four independent two-state subunits.

76. Correlation Functions for Two-State Channels in Series

From Equation (74.1), the correlation function of the open-channel probability is

$$C_{p^2}(\tau) = \langle p^4(0) - 2pp^3(0) - p^4 + 2p^3 p(0)\rangle e^{-2\tau/\theta}$$
$$+ 2p\langle p^3(0) - p^2 p(0) - pp^2(0) + p^3\rangle e^{-\tau/\theta}$$

Using the results (Section 75)

$$\langle p(0)\rangle = p$$
$$\langle p^n(0)\rangle = p^2, \qquad n \geq 2$$

and multiplying both sides of the equation by N gives

$$C_{N_o}(\tau) = N(p^2 + p^4 - 2p^3)e^{-2\tau/\theta} + 2pN(p^2 - p^3)e^{-\tau/\theta}$$

or

$$C_{N_o}(\tau) = Np^2[(1-p)^2 e^{-2\tau/\theta} + 2p(1-p)e^{-\tau/\theta}]$$

PROBLEM.   Show that the spectral density for N_o is

$$S_{N_o}(f) = 8N\theta p^2(1-p)\left(\frac{1-p}{4+\omega^2\theta^2} + \frac{p}{1+\omega^2\theta^2}\right)$$

PROBLEM. Show that the correlation function for a population of N channels, each channel composed of four two-state subunits in series, is

$$C_{N_o}(\tau) = Np^4[(1 - p)^4 e^{-4\tau/\theta} + 4p(1 - p)^3 e^{-3\tau/\theta}$$
$$+ 6p^2(1 - p)^2 e^{-2\tau/\theta} + 4p^3(1 - p)e^{-\tau/\theta}]$$

PROBLEM. Assume that $N_o(t)$ gates a K current (I_K) similar to that described in Section 73. Show that the current noise spectral density of a population of N_K K channels with four two-state subunits is

$$S_{I_K} = \frac{4I_K^2\theta}{N_K}\left[\left(\frac{1 - \nu}{p}\right)^4 \frac{4}{16 + \omega^2\theta^2} + \left(\frac{1 - p}{p}\right)^3 \frac{12}{9 + \omega^2\theta^2}\right.$$
$$\left. + \left(\frac{1 - p}{p}\right)^2 \frac{12}{4 + \omega^2\theta^2} + \frac{1 - p}{p} \frac{4}{1 + \omega^2\theta^2}\right]$$

where $p = \alpha/(\alpha + \beta)$ and $\theta = 1/(\alpha + \beta)$.

Let $p = 1/2$ and $\theta = 1$ msec. Plot each of the four terms in square brackets on log–log paper. Compare these four terms with their sum. (p is equivalent to the H–H n parameter for the K conductance.)

PROBLEM. Derive an expression for the integral spectrum (Sections 45–47) corresponding to S_{I_K} in the above problem. Plot each of the four terms of the integral spectrum, and their sum, for $p = 1/2$ and $\theta = 1$ msec.

77. The Relationship between Channel Models and Kinetic Schemes

In Section 72 a population of two-state channels,

$$C \underset{\beta}{\overset{\alpha}{\rightleftharpoons}} O$$

was described by the average kinetic equation

$$\frac{dN_o}{dt} = \alpha(N - N_o) - \beta N$$

Here we explore the general relationship between channel models and the kinetic equations that govern the average behavior of these channels.

First, consider a channel composed of two two-state channels (or subunits) in series (Section 74). The channel may be found in one of four states

$$C \qquad \left\{ \begin{matrix} C & O \\ O & C \end{matrix} \right\}, \qquad O$$
$$C' \qquad \qquad \qquad O$$

The first state is closed by two subunits; call the number of channels in this state N_2. The second and third states are closed by one or the other subunit; call the number of channels closed in this way N_1. The fourth state is the only open state and the number of open channels is N_0.

For any subunit, α and β describe the average rates of flipping between the open and closed states just as they did in the channel with only one subunit. The difference is that there are now three ways a channel can be closed. If both subunits are closed there are two ways that the channel may go to the middle state (in curly brackets); therefore the rate from the doubly closed channel to the middle state will be twice α. Once the channel is in one or the other of the middle states, there is only one way to get back to the doubly closed state, and the rate of doing so is β. This may be diagrammed by

$$\begin{matrix} C \\ C \end{matrix} \underset{\beta}{\overset{2\alpha}{\rightleftharpoons}} \left\{ \begin{matrix} \\ \end{matrix} \right\}$$

where the two states in the curly brackets are indistinguishable. A similar argument holds for the open state. The final kinetic scheme is

$$N_2 \underset{\beta}{\overset{2\alpha}{\rightleftharpoons}} N_1 \underset{2\beta}{\overset{\alpha}{\rightleftharpoons}} N_0$$

where N_1 is doubly degenerate.

The equations that govern this kinetic scheme are

$$\frac{dN_2}{dt} = \beta N_1 - 2\alpha N_2$$

$$\frac{dN_1}{dt} = 2\alpha N_2 + 2\beta N_0 - (\alpha + \beta)N_1$$

$$\frac{dN_0}{dt} = \alpha N_1 - 2\beta N_0$$

and

$$N = N_1 + N_2 + N_0$$

In principle, these equations can be solved for $N_0(t)$ in terms of α, β, and N. To determine the correlation function for $N_0(t)$, form the product

$$\langle [N_0(t) - N_0(\infty)][N_0(0) - N_0(\infty)] \rangle$$

where $\langle \ \rangle$ implies an average over all initial states. This procedure—relatively simple for a channel with one subunit (Section 72)—becomes rapidly more difficult as the number of subunits increases. In Sections 74–76 we avoided this calculation for a channel composed of two subunits by using the definition of variance to show that $\langle p^n(0) \rangle = p^2$ for $n \geq 2$. Similar results may be derived for channels composed of more than two subunits.

Consider the Hodgkin–Huxley model of Na conduction in nerve membrane. The conductance is governed by a gate with both an activation parameter m and an inactivation parameter h. Consider each channel to be controlled by three m subunits and one h subunit; each of the four subunits may flip between an open and a closed state:

$$C_m \underset{\beta_m}{\overset{\alpha_m}{\rightleftharpoons}} O_m$$

$$C_h \underset{\beta_h}{\overset{\alpha_h}{\rightleftharpoons}} O_h$$

Consider the m system first. There is only one way the m system can be triply closed. However, there are three ways the m system can be doubly closed, namely, if any one of the three closed m subunits flips to the open state. Thus

$$N_3 \overset{3\alpha_m}{\longrightarrow} N_2$$

Once in the N_2 state, there are only two ways a channel may proceed to the singly closed state; e.g., one of the N_3 states is

which can go to N_2 by either of the bottom two arrows flipping but not the top arrow.

PROBLEM.   Write down the complete kinetic scheme for the m system for N_3, N_2, N_1, and N_0. Add the fourth component to the channel (the h inactivation parameter) and write down the complete kinetic diagram for the m^2h channel assuming m and h are independent.

PROBLEM. Derive the correlation function for m^3h scheme of the H–H Na channel. Show that the current noise spectral density is given by

$$S_{\text{Na}}(f) = \frac{4I_{\text{Na}}^2}{N_{\text{Na}}} \left\{ \frac{1-h}{h} \frac{\theta_h}{1+\omega^2\theta^2} + \theta_m \sum_{i=1}^{3} \frac{1}{i} \binom{3}{i} \left(\frac{1-m}{m}\right)^i \left[\frac{1}{1+(\omega\theta_m/i)^2} \right. \right.$$
$$\left. \left. + \frac{1-h}{h} \frac{\theta_h}{1+i\theta_h} \frac{1}{1+[\omega\theta_h\theta_m/(\theta_m+i\theta_h)]^2} \right] \right\}$$

where

$$\binom{3}{i} = \frac{3!/i!}{(3-i)!}$$

$$\theta_h = \frac{1}{\alpha_h + \beta_h}, \qquad h = \alpha_h\theta_h$$

$$\theta_m = \frac{1}{\alpha_m + \beta_m}, \qquad m = \alpha_m\theta_m$$

and I_{Na} is the average current through the open Na channels. The total number of Na channels (open plus closed) is N_{Na}.

PROBLEM. Let $m = h = 1/2$ and let $\theta_h = 1$ msec and $\theta_m = 0.1$ msec. Expand the above expression for $S_{\text{Na}}(f)$; plot all seven terms on log–log paper and compare these with their sum. Show that a reasonable approximation to $S_{\text{Na}}(f)$ above 100 Hz is

$$S_{\text{Na}}(f) = \frac{4I_{\text{Na}}^2}{N_{\text{Na}}} \frac{\theta_m}{h} \sum_{i=1}^{3} \frac{1}{i} \binom{3}{i} \frac{[(1-m)/m]^i}{1+(\omega\theta/i)^2}$$

The inactivation subunits (h), whose kinetics are generally much slower than the activation subunits (m), affect primarily the amplitude of the high-frequency region of the Na current noise spectral density, not the roll-off.

D. CAMPBELL'S THEOREM

Previously we considered three specific sources of noise: white noise, $1/f$ noise, and Lorentzian noise. The spectral density of these sources was more or less related directly to a random physical process, e.g., the two-state channel in the case of Lorentzian noise. Although a specific physical process implies a unique spectral density, the reverse is not true. This section develops a general theorem (Campbell, 1909a,b; Rice, 1944) that relates average properties of noise signals to underlying events without regard to the specific mechanism of the source.

78. The Mean and the Variance

A compelling description of a source of random noise is the shot noise theory. Suppose N impulses $g(t)$ arrive randomly in the time interval T. The sum of these will result in a random noise signal $x(t)$. This is shown qualitatively in Figure 78.1. The average effect of one of the impulses is (Section 33)

$$\langle g(t) \rangle = \frac{1}{T} \int_0^T g(t) \, dt$$

For N arrivals, the average effect is

$$\langle x(t) \rangle = N \langle g(t) \rangle = \frac{N}{T} \int_0^T g(t) \, dt$$

if all impulses arrive independently.

Let N and T become large but assume the ratio N/T tends to the limit

$$\nu = \frac{N}{T}$$

ν is the average rate of arrivals of elementary events $g(t)$; the average value of $x(t)$ is then

$$\langle x(t) \rangle = \nu \int_0^\infty g(t) \, dt \tag{78.1}$$

This is the first part of Campbell's theorem.

PROBLEM. Let $g(t) = ae^{-t/\theta}$. Let $a = 10$ mV, $\theta = 1$ msec, and $T = 10$ msec. Plot $g(t)$ and $\langle g(t) \rangle$. Assume 1000 such impulses arrive in 1 sec. Using Equation 78.1, show that

$$\langle x(t) \rangle = \nu a \theta$$

Calculate $\langle x(t) \rangle$ and compare this value with $N \langle g(t) \rangle$. Show that as $T \to 0$, $\langle g(t) \rangle \to a$. Reconcile this with the formula $N \langle g(t) \rangle = \nu a \theta$.

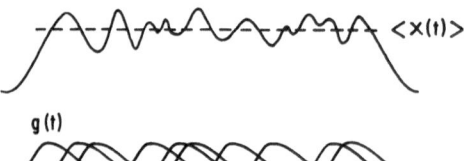

Figure 78.1. Diagram illustrating the sum of N impulses $g(t)$ arriving randomly in the time interval T. The solid line represents the sum effect, the dashed line represents the average effect.

It is required to calculate the variance of $x(t)$ from the random sum of elementary events $g(t)$. The variance is the mean square value minus the square of the mean. Suppress, for a moment, the mean value of $g(t)$; the mean square value is by definition

$$\langle g^2(t) \rangle = \frac{1}{T} \int_0^T g^2(t) \, dt$$

Setting the mean of $g(t)$ arbitrarily to zero implies that $\langle g^2(t) \rangle$ is the variance of $g(t)$. N independent impulses would result in a total variance in $x(t)$ of

$$\sigma_x^2 = N\sigma_g^2 = \nu \int_0^\infty g^2(t) \, dt \tag{78.2}$$

where N/T tends to ν in the limit of large N and T. This is the second part of Campbell's theorem.

We have calculated $\langle x(t) \rangle$ and σ_x^2 by considering $\langle g(t) \rangle$ and $\langle g^2(t) \rangle$ separately and multiplying by N. Actually, the full formula for σ_g^2 includes both, namely,

$$\sigma_g^2 = \langle g^2(t) \rangle - \langle g(t) \rangle^2$$

$$= \frac{1}{T} \int_0^T g^2(t) \, dt - \frac{1}{T^2} \left(\int_0^T g(t) \, dt \right)^2$$

In the limit $T \to \infty$, but $N/T \to \nu$, the second term approaches zero and

$$\sigma_x^2 = \frac{N}{T} \left[\int_0^T g^2(t) \, dt - \frac{1}{T} \left(\int_0^T g(t) \, dt \right)^2 \right]$$

$$\simeq \nu \int_0^\infty g^2(t) \, dt$$

which is equivalent to Equation (78.2). The validity of this approximation is demonstrated below for a special case.

Let $g(t) = ae^{-t/\theta}$. Then,

$$\int_0^T g^2(t) \, dt = a^2 \int_0^T e^{-2t/\theta} \, dt = \frac{a^2\theta}{2} (1 - e^{-2T/\theta})$$

and

$$\left(\int_0^T g(t) \, dt \right)^2 = a^2 \left(\int_0^T e^{-t/\theta} \, dt \right)^2 = a^2\theta^2(1 - e^{-T/\theta})^2$$

For large T,

$$\sigma_g^2 \simeq \frac{1}{T}\left(\frac{a^2\theta}{2} - \frac{a^2\theta^2}{T}\right)$$

Dropping terms in T^2,

$$\sigma_g^2 \simeq \frac{a^2\theta}{2T}$$

and finally

$$\sigma_x^2 = N\sigma_g^2 \simeq \tfrac{1}{2}a^2\theta\nu$$

This result is also obtained directly from Equation (78.2).

PROBLEM. Show that as $T \to 0$, $\sigma_g^2 \to a^2$ if the limit is taken as

$$\sigma_g^2 = \lim_{T\to\infty}[\langle g^2\rangle - \langle g\rangle^2]$$

rather than in two separate steps as above. Reconcile this with the formula $N\sigma^2 = a^2\theta^2\nu/2$.

The mean and the variance of $x(t)$ contain no temporal information about $x(t)$. It is possible, for example, to construct very different wave forms that have the same value of $\langle x\rangle$ and $\langle x^2\rangle$. To illustrate this point, imagine a sine wave and a random noise signal. Obviously these may be set up in such a way that the variance of both the sine wave and the noise signal are the same; a measurement of $\langle x^2\rangle$ would not distinguish them. One way to tell them apart, however, would be to measure their spectral densities. Although the spectral density does include temporal information it still does not provide a unique determination of $x(t)$, for the spectral density is also an average property of a signal (see Section 34).

In order to calculate the spectral density of $x(t)$ from $g(t)$ we need Rayleigh's theorem (Section 34). The theorem states that

$$\int_{-\infty}^{\infty}|g(t)|^2\,dt = \int_{-\infty}^{\infty}|\bar{g}(f)|^2\,df$$

where $\bar{g}(f)$ is the Fourier transform of $g(t)$. For an arbitrary, real signal $g(t)$, which is nonzero only for $0 \le t \le T$, Rayleigh's theorem reduces to

$$\int_0^T g^2(t)\,dt = 2\int_0^{\infty}|\bar{g}(f)|^2\,df$$

$| \bar{g}(f) |$ is an even function. Multiply both sides of this equation by N/T. Then from Equation (78.2),

$$\sigma_x^2 = \int_0^\infty \frac{2N}{T} | \bar{g}(f) |^2 \, df$$

By definition (Section 34)

$$\sigma_x^2 = \int_0^\infty S_x(f) \, df$$

and therefore

$$S_x(f) = 2v | \bar{g}(f) |^2 \tag{78.3}$$

in the limit $N/T \to v$ for large N and T. This equation predicts the spectral density of the wave form $x(t)$ that results from a random sum of elementary events $g(t)$.

PROBLEM.   Show that an equivalent expression to Equation (78.3) is

$$C_x(\tau) = v \int_{-\infty}^\infty g(t)g(t + \tau) \, dt$$

This equation predicts the correlation function of the wave form $g(t)$ that results from a random sum of elementary events (excluding the mean value).

EXERCISE.   Derive Campbell's theorem by assuming Poisson distribution of the arrival times of the impulses $g(t)$. This problem is solved by Rice (1954, pp. 145–150). The probability that exactly N impulses arrive in the interval T is

$$P(N) = \frac{(vT)^N}{N!} \, e^{-vT}$$

The signal $x_N(t)$ due to N impulses is

$$x_N(t) = \sum_{n=1}^N g(t - t_n)$$

The average value of $x_N(t)$, at some particular t and over a large number of

328

Chap. 5 • Noise Sources

intervals, each having exactly N arrivals, is

$$\langle x_N(t)\rangle = \sum_{n=1}^{N} \frac{1}{T} \int_0^T g(t - t_n)\, dt_n$$

This is an ensemble average. The average value of $x(t)$ over all intervals, not only those having N arrivals, is

$$\langle x(t)\rangle = \sum_{N=0}^{\infty} P(N)\langle x_N(t)\rangle$$

Show that this leads to Equation (78.1). Similarly, show that

$$x_N^2(t) = \sum_{n=1}^{N} \sum_{m=1}^{N} g(t - t_n)g(t - t_m)$$

and

$$x^2(t) = \sum_{n=0}^{\infty} P(N)\langle x_N^2(t)\rangle$$

lead to

$$\langle x^2(t)\rangle = v \int_{-\infty}^{\infty} g^2(t)\, dt + \langle x(t)\rangle^2$$

This is the general expression for $\langle x^2\rangle$; previously we considered the special case

$$g(t) = 0 \qquad \text{for } t < 0$$

With this condition, the integrations that appear in Campbell's theorem reduce to $0 \le t \le \infty$ and the above expression for $\langle x^2\rangle$ is equivalent to Equation (78.2) since $\sigma_x^2 = \langle x^2\rangle - \langle x\rangle^2$.

EXERCISE. Compare Campbell's theorem with Lee's analysis of Poisson waves of pulses of various shape and pattern of arrival (Lee, 1960, pp. 336–344; note especially Equations 62 and 70).

Suggested Reading. Suggested reading in the theory and nature of shot noise (noise generated by the arrival of impulses) includes: Campbell (1909a,b), Schottky (1918, 1926), Johnson (1925), Rice (1944), Blaquiérer (1966), Halford (1968), Heiden (1969), Schick (1974), and Schick and Verveen (1974). Two outstanding examples of the application of Campbell's theorem to membranes are found in Hagins (1965, Section 110) and Katz and Miledi (1970, Section 90).

79. Noise Spectra from Campbell's Theorem

Summarizing Section 78, we find that if a large number of impulses arrive independently with a mean rate v, the random sum of these elementary events has the following properties:

$$\langle x(t) \rangle = v \int_{-\infty}^{\infty} g(t)\, dt$$

$$\sigma_x^2 = v \int_{-\infty}^{\infty} g^2(t)\, dt \tag{79.1}$$

$$S_x(f) = 2v\, |\, \bar{g}(f)\, |^2$$

The integration in the first two equations reduces to

$$\int_{-\infty}^{\infty} \rightarrow \int_{0}^{\infty}$$

if $g(t) = 0$ for $t < 0$. If $g(t)$ is known, the three properties summarized in Equations (79.1) can be calculated. The reverse is not true; therefore, theories of noise derived from these formulas are *ad hoc* theories.

a. White Noise

Let $i(t) = q\, \delta(t)$ be an elementary current caused by an abrupt movement $\delta(t)$ of a charge q. Suppose the $i(t)$ occur randomly; the sum effect of a large number of these elementary currents is the fluctuating current $I(t)$. From Equations (79.1),

$$\langle I(t) \rangle = v \int_{-\infty}^{\infty} q\, \delta(t)\, dt = qv \equiv I$$

Since v is the mean rate of arrival, qv is by definition the average current I. The variance of $I(t)$ is

$$\sigma_I^2 = v \int_{-\infty}^{\infty} q^2\, \delta^2(t)\, dt$$

$$= vq^2 \int_{-\infty}^{\infty} \delta(t)\, \delta(t)\, dt = vq^2\, \delta(0)$$

Since $\delta(0)$ is infinite, the variance of $I(t)$ is also infinite. Finally,

$$S_I(f) = 2v \left|\, \int_{-\infty}^{\infty} q\, \delta(t) e^{-i\omega t}\, dt\, \right|^2 = 2vq^2$$

or

$$S_I(f) = 2qI$$

This is the formula for shot noise first derived by Schottky (1918, pp. 541–567) in 1918. Evidently, the variance defined as

$$\sigma_I^2 = \int_0^\infty S_I(f)\, df$$

is again infinite; this is a consequence of the infinitely small width of the delta function.

PROBLEM. Assume $i(t)$ is a rectangular impulse of finite width Δt. Calculate $\langle I \rangle$, σ_I^2 and $S_I(f)$.

PROBLEM. Find a form for the elementary current necessary to give Nyquist's formula for Johnson noise (Section 59). Interpret the fundamental bandwidth Δf of the quantum theory of Johnson noise (Section 61) in terms of the movement of electrons in the conductor and the equation $i(t) = q\,\delta(t)$ used above.

b. 1/f Noise

Let $i(t) = b/t^{1/2}$ be the elementary current. Note that

$$b^2 \ominus A^2 \sec$$

Let $i(t) = 0$ for $t < 0$. No interpretation of $i(t)$ is given; simply assume that a large number of such currents occur randomly, the sum effect being $I(t)$. From Equation (79.1) it is evident that $\langle I \rangle$ and σ_I^2 are infinite. However,

$$S_I(f) = 2\nu\,|\,\bar{\imath}(f)\,|^2$$

is defined. From CF # I-522.1, the Fourier transform of $b/t^{1/2}$ is

$$\int_{-\infty}^\infty \frac{b}{t^{1/2}}\, e^{-i\omega t}\, dt = \frac{b\pi^{1/2}}{(i\omega)^{1/2}}, \qquad t > 0$$

$\omega = 2\pi f$. Therefore

$$S_I(f) = \frac{\nu b^2}{f}$$

PROBLEM. Show that the Fourier transform of $|t|^{-1/2}$ is $|f|^{-1/2}$. These are called identical pairs since the two Fourier mates are the same function of their parametric variables t and f. CF # I-522.1 (used above) is a special case of this identical pair for $t > 0$.

PROBLEM. Restrict $i(t) = b/t^{1/2}$ to a pulse of finite duration beginning at $t = t_1$ and ending at $t = t_2$. Outside this range $i(t) = 0$; within this range $i(t)$ falls off as $t^{1/2}$. Calculate $\langle I(t) \rangle$, σ_I^2 and $S_I(f)$.

c. Lorentzian Noise

Let $i(t) = i_0 e^{-t/\theta}$ be the elementary current. Let $i(t) = 0$ for $t < 0$. These currents turn on infinitely fast to their initial value, i_0, and decay exponentially with time constant θ. From Equations (79.1),

$$\langle I(t) \rangle = \nu \int_0^\infty i_0 e^{-t/\theta} \, dt = \nu \theta i_0 = I$$

The variance of $I(t)$ is

$$\sigma_I^2 = \nu \int_0^\infty i_0^2 e^{-2t/\theta} \, dt = \frac{1}{2} \nu \theta i_0^2$$

The Fourier transform of $e^{-t/\theta}$ for $t > 0$ is given in CF # I-438:

$$\int_{-\infty}^\infty e^{-t/\theta} e^{-i\omega t} \, dt = \frac{1}{i\omega + 1/\theta}, \qquad t > 0$$

Therefore

$$S_I(f) = 2\nu \left| \frac{i_0}{i\omega + 1/\theta} \right|^2 = \frac{2\nu\theta^2 i_0^2}{1 + \omega^2\theta^2}$$

Alternative forms are

$$S_I(f) = \frac{2I\theta i_0}{1 + \omega^2\theta^2}$$

and

$$S_I(f) = \frac{4\theta\sigma_I^2}{1 + \omega^2\theta^2}$$

PROBLEM. What is another form for $i(t)$ that would give Lorentzian spectral density?

PROBLEM. Let $I = 1$ nA and $\sigma_I = 1$ pA. Calculate the amplitude of the elementary current assuming $i(t) = i_0 \exp(-t/\theta)$. If $\theta = 1$ msec, what is their average rate of arrival?

PROBLEM. Let

$$i(t) = cte^{-t/\theta}$$

where $c \ominus$ A/sec. Calculate $\langle I \rangle$, σ_I^2 and $S_I(f)$ from Equations (79.1). Plot $i(t)$ and $S_I(f)$ for $c = 1$ nA/sec and $\theta = 1$ msec.

Spectra calculated with Campbell's theorem are single-sided spectra (Section 43). The above formula, $S_I = 2I\theta i_0/(1 + \omega^2\theta^2)$, is one-half the value of a similar formula for the one-sided spectral density of a population of two-state channels (Section 72). This factor of one-half is a consequence of the difference between a random sum of exponentials and a random sum of the random switch function (Figure 73.2).

6

Membrane Impedance

80. *Equivalent Circuits of Kinetic Equations*

The kinetic equation that describes the average number of open channels in a population of N two-state channels is (Section 72)

$$\theta \frac{dN_o}{dt} + N_o = pN$$

The solution, $N_0(t)$, describes how the number of open channels changes in response to a change in α and β. In Section 72 it was shown that a step change in α and β at $t = 0$ gives

$$N_0(t) = N_0(0)e^{-t/\theta} + N_0(\infty)(1 - e^{-t/\theta})$$

where $\theta = 1/(\alpha + \beta)$ is the new value of θ immediately after the step. α and β are assumed to change instantaneously.

$N_0(t)$ is composed of two terms. $N_0(t)$ is the average number of open channels before the step; this initial state decays exponentially and is eventually forgotten. At the same time, the channels are climbing to their new steady-state average, $N_0(\infty)$, along the curve $1 - \exp(-t/\theta)$. Consider two cases: First, assume the new steady-state average is greater than the old average, i.e.,

$$N_0(\infty) > N_0(0)$$

The net effect will be an increase in N_0. This is shown qualitatively in Figure 80.1a. Second, assume that

$$N_0(\infty) < N_0(0)$$

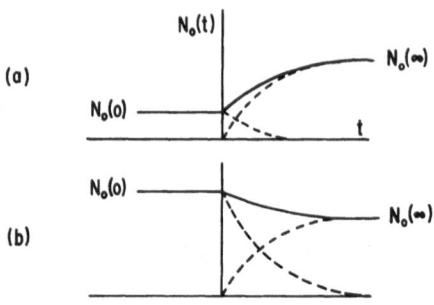

Figure 80.1. Diagram illustrating the average change in $N_0(t)$ for (a) $N_0(\infty) > N_0(0)$ and (b) $N_0(\infty) < N_0(0)$.

The net effect is a decrease in N_0 as shown in Figure 80.1b. N_0 controls the current through the membrane, and since $N_0(t)$ responds sluggishly to a sudden change in α and β, the response is called inductive; the fundamental property of an inductor is to resist sudden changes in current.*

The analogy between channel kinetics and inductance is made explicit by considering the current flowing through a resistor r in series with an inductor L. The current is

$$\bar{I}(f) = \frac{1}{r + i\omega L} \, \bar{V}(f)$$

If $V(t)$ is initially zero and steps suddenly to V at $t = 0$, then (Section 26)

$$\bar{V}(f) = V/i\omega$$

and

$$\bar{I}(f) = \frac{V}{i\omega} \frac{1/r}{1 + i\omega\theta}$$

where $\theta = L/r$. In the time domain (Section 26),

$$I(t) = \frac{V}{r} \, (1 - e^{-t/\theta})$$

PROBLEM. Show that a pure inductor is inadequate as an equivalent circuit of the kinetic equation. Give an intuitive description of the meaning of r and of L.

---

* In Section 24 the resistance of an inductor to a sudden change of current was a consequence of Faraday's law. The apparent inductance of a population of channels comes from their probabilistic open/close kinetics, as discussed in Section 72.

PROBLEM. Show that if the initial voltage is V_0, then

$$I(t) = \frac{V_0}{r} e^{-t/\theta} + \frac{V}{r} (1 - e^{-t/\theta})$$

Compare the two cases, $V_0 > V$ and $V_0 < V$.

The time course of the current through a resistor and inductor in series is analogous to the population kinetics of open channels; in the analogy

$$\frac{L}{r} = \frac{1}{\alpha + \beta}$$

One should not, however, equate $I(t)$ in the above problem with the full current through the membrane, for the total membrane current is composed of two parts, the kinetic part described above and an instantaneous part due to the state of the open channels at the moment the voltage step is applied.

Assume that $N_0(t)$ controls a specific ionic current as described in Section 73. The current through the membrane is

$$I(t) = \gamma[V(t) - E]N_0(t)$$

If $V(t)$ is a step from some initial value to the value V, then

$$I(t) = \gamma(V - E)[N_0(0)e^{-t/\theta} + N_0(\infty)(1 - e^{-t/\theta})]$$

In this formula, V is the voltage to which we have stepped. $N_0(\infty)$ and θ are evaluated at V but $N_0(0)$ is evaluated at the initial voltage.

At $t = 0$, the initial current is

$$I(0) = \gamma(V - E)N_0(0)$$

where

$$N_0(0) = \left(\frac{\alpha}{\alpha + \beta}\right)_{\text{initial}} N$$

Assuming the system has been at the initial potential for some time, $N_0(0)$ is a constant and therefore $I(0)$ depends linearly on V. When the voltage is stepped to V, an instantaneous current flows through $N_0(0)$; at the same moment, α and β sense the new voltage and N_0 begins to change along one of the curves shown in Figure 80.1, depending on whether the new steady-state value, $N_0(\infty)$, is greater or less than the old steady-state value, $N_0(0)$.

This point is made clear by reexamining Figure 23.1. The instantaneous change is along the dashed line; the slope of this line is the open-channel conductance $\gamma N_0(0)$. Figure 23.1 shows the voltage response to a current. Above we were considering the current response to a voltage. ΔV in the upper left quadrant of Figure 23.1 would give a fast rise along the dashed line followed by a gradual rise to the solid curve; this response is called inductive since it describes a current change. The response actually shown in Figure 23.1 is called capacitative for the opposite reason.

The exact form of the inductive response depends on the shape of the steady-state I–V curve and for a given I–V curve on the quadrant in which the step is applied. In general, the same ΔV results in a different response for each initial voltage.

The steady-state current as a function of voltage is

$$I(\infty) = \gamma(V - E)N_0(\infty)$$

The open-channel current, $\gamma(V - E)$, is ohmic. Although a more complex, nonlinear channel may be more appropriate for some membranes (Adrian, 1969; Clay and Shlesinger, 1977b; see also Section 107) in the present example, the nonlinear part of the I–V curve derives entirely from the voltage dependence of α and β.

Thus

$$N_0(\infty) = \frac{\alpha(V)}{\alpha(V) + \beta(V)} N$$

determines the curvature of a diagram like Figure 23.1. The steady-state conductance, $\gamma N_0(\infty)$, is the slope of the solid curve at a particular voltage. An outward rectifier (Figure 23.1) is explained in this model by an increase in the average number of open channels with depolarization. It follows that an inward rectifier (with linear instantaneous conductance) is explained by channels that close with depolarization.

The rL equivalent of the kinetic equation is inadequate as an equivalent circuit of the complete ionic conductance. The rL equivalent describes only the gradual change in $I(t)$ and not the instantaneous change. Evidently, another resistor in parallel with the rL branch is required. In the next paragraph we derive the total impedance at any voltage by considering the current change resulting from a small voltage perturbation and deriving the ratio $\delta V/\delta I$ in the frequency domain.

81. The Small-Signal Impedance of a Population of Ionic Channels

As a starting point, consider the two equations that govern the flow of a particular ionic current through a population of N two-state channels. The equations are

$$I = \gamma(V - E)N_0$$

and

$$\frac{dN_o}{dt} = \alpha(N - N_0) - \beta N_0$$

It is required to calculate the change in I due to a small change in V: Taking the variation of both equations gives

$$\delta I = \gamma N_0 \, \delta V + \gamma(V - E) \, \delta N_0$$

and

$$\frac{d}{dt}(\delta N_o) = (N - N_o) \, \delta\alpha - \alpha \, \delta N_o - N_o \, \delta\beta - \beta \, \delta N_o$$

PROBLEM. Show that if $y = y(x)$, then $\delta(dy/dx) = (d/dx)(\delta y)$.

Since E is constant, $\delta(V - E) = \delta V$ has been substituted into the first equation for δI. δI consists of two components. The first represents the change in current through channels open at the moment of change. The second is due to the change in the open channels resulting from the change in voltage. The second component changes with time as described by the equation for $(d/dt)(\delta N_0)$.

To eliminate δN_0 between the two equations, $\delta\alpha$ and $\delta\beta$ must be evaluated. Assume that

$$\alpha = \alpha(V), \qquad \beta = \beta(V)$$

are functions only of voltage. Then

$$\delta\alpha = \frac{d\alpha}{dV} \, \delta V \qquad \delta\beta = \frac{d\beta}{dV} \, \delta V$$

represent changes in the rate constants due to changes in voltage. The derivatives are taken at a particular V, thus $\delta\alpha$ and $\delta\beta$ change instantaneously with δV by an amount depending on their slope at V.

Substituting $\delta\alpha$ and $\delta\beta$ and rearranging terms gives

$$\frac{d}{dt}(\delta N_o) = [(N - N_o)\alpha' - N_o\beta']\,\delta V - (\alpha + \beta)\,\delta N_o$$

where α' and β' are the slopes of α and β at V; the factors of δV and δN_0 are constants at the same voltage. Taking the Fourier transform of both sides of the equation and rearranging terms gives

$$(\alpha + \beta + i\omega)\,\overline{\delta N_o} = [(N - N_o)\alpha' - N_o\beta']\,\overline{\delta V}$$

The Fourier transform of δI is

$$\overline{\delta I} = \gamma N_o\,\overline{\delta V} + \gamma(V - E)\,\overline{\delta N_o}$$

The factors of $\overline{\delta N_o}$ and $\overline{\delta V}$ are constant at V. Eliminating $\overline{\delta N_o}$,

$$\frac{\overline{\delta I}}{\overline{\delta V}} = \gamma N_o + \gamma(V - E)\frac{(N - N_o)\alpha' - N_o\beta'}{\alpha + \beta + i\omega}$$

By definition, this is the system admittance in the frequency domain. Consider the admittance of the circuit in Figure 81.1. By inspection,

$$\frac{1}{Z} = \frac{1}{R} + \frac{1}{r + i\omega L}$$

Rearranging the previous equation gives

$$\frac{\overline{\delta I}}{\overline{\delta V}} = \gamma N_o + \frac{\gamma(V - E)}{\alpha + \beta}\frac{(N - N_o)\alpha' - N_o\beta'}{1 + i\omega\theta}$$

where $\theta = L/r$. A term by term comparison between these equations results

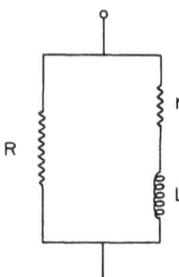

Figure 81.1. Equivalent circuit for a population of N channels each flipping between an open and a closed state. The complete circuit has a battery E in series with the resistor R.

in the following definitions of the components of the equivalent circuit:

$$R = \frac{1}{\gamma N_o}, \qquad r = \frac{\alpha + \beta}{\gamma(V - E)[(N - N_o)\alpha' - N_o\beta']}$$

$$L = r\theta, \qquad \theta = \frac{1}{\alpha + \beta}$$

When $V = E$, the equivalent circuit reduces to

$$R = \frac{1}{\gamma N_o(E)} \qquad (V = E)$$

for any frequency; $N_o(E)$ is the average number of open channels at the equilibrium potential for this ion. When $\omega \to \infty$, the equivalent circuit reduces to

$$R = \frac{1}{\gamma N_o(V)} \qquad (\omega \to \infty)$$

for any voltage; $N_o(V)$ is the average number of open channels at V and $1/R$ is the slope of the straight drawn between E and a point on the steady-state *I–V* curve at V (see Figure 23.1). This is the chord resistance. When $\omega = 0$, the equivalent circuit reduces to

$$\frac{rR}{r + R} = \frac{1}{\gamma N_o + \theta\gamma(V - E)[(N - N_o)\alpha' - N_o\beta']} \qquad (\omega = 0)$$

for any voltage. This is the parallel combination of r and R and is the slope resistance of the steady-state *I–V* curve at V. The slope resistance is the steady-state ratio $\Delta I / \Delta V$ in Figure 23.1 for ΔI and ΔV small.

The circuit elements in Figure 81.1 are linear components with constant value at every V. This circuit only applies to small perturbations. The same circuit holds for the entire *I–V* curve but has different values of R, r, and L at each point. It is unimportant whether the impedance at V is measured as a voltage response to an imposed current or a current response to an imposed voltage. The advantage of voltage control is that an excitable membrane may be held at voltages beyond threshold (Section 30).

PROBLEM. R and θ are positive definite. Find the conditions under which r (and L) are negative.

In Section 23, nonlinear *I–V* curves and apparent reactance were described for simple ionic systems in which the ionic profile changed to be simple, linear devices and·the nonlinear curves and impedance are due to the voltage dependence of the random open–close kinetics of the channels. R represents the channels open at the moment a change in V occurs (the old steady-state number). r is an additional resistance deriving from the change in N_0 caused by the change in V; it appears in parallel with R. L/r describes the rate at which N_0 changes. Since the circuit elements in Figure 81.1 represent average properties of open channels, the circuit itself is only the average impedance about which we may expect to find fluctuations.

The equivalent circuit in Figure 81.1 contains no battery; it represents the membrane impedance with all internal emf's shorted. The complete circuit does, however, contain a bias voltage $V = E$ at $I = 0$. When $V = e$, $r \to \infty$ and the circuit reduces to R; the open-circuit voltage at this point is E; therefore, the battery is in series with R.

This circuit (Figure 81.1 with E in series with R) is the macroscopic equivalent of many ionic-specific channels in parallel (see Section 73). For voltages other than E, the initial current that flows in response to a change in voltage is

$$I(0) = \gamma(V - E)N_o(0)$$

$I(0)$ senses the resistance $R = 1/\gamma N_o$ on a line that goes through E at $I = 0$. Therefore, at every V, the equivalent circuit contains a battery E in series with R. The R branch of the circuit represents the open channels, while the rL branch represents channel kinetics.

PROBLEM.   We have derived the impedance of a population of channels. What is the impedance of a single channel? Derive explicit expressions for a channel's resistance, inductance, and time constant in terms of $p = \alpha/(\alpha + \beta)$. Such expressions also represent average properties. Reconcile this model with the picture of instantaneous conductance changes that occur as the channel flips between the open and the closed states (the random telegraph signal; see Section 72 and Figure 73.2).

82. The Small-Signal Impedance of Channels in a Membrane

Consider a population of two-state, ion-specific channels embedded in a lipid matrix. Let the lipid be a perfect dielectric with no leakage. The equivalent circuit is then that shown in Figure 81.1 in parallel with a capacitor C. It is required to calculate the impedance of this circuit.

The RrLC Impedance. From Figure 81.1 (with C in parallel),

$$\frac{1}{Z} = \frac{1}{R} + \frac{1}{r + i\omega L} + i\omega C$$

By rearrangement

$$\frac{1}{Z} = \frac{(1 + i\omega\theta_1)(1 + i\omega\theta_2) + R/r}{R(1 + i\omega\theta_2)}$$

where

$$\theta_1 = RC, \qquad \theta_2 = L/r$$

Therefore

$$Z = R\,\frac{1 + i\omega\theta_2}{(1 - \omega^2\theta_1\theta_2) + R/r + i\omega(\theta_1 + \theta_2)} \tag{82.1}$$

The modulus square of Z is

$$|Z|^2 = R^2\,\frac{1 + \omega^2\theta_2{}^2}{[(1 - \omega^2\theta_1\theta_2) + R/r]^2 + \omega^2(\theta_1 + \theta_2)^2} \tag{82.2}$$

PROBLEM. Show that if $r = 0$, $|Z|^2$ has a resonance at $\omega = 1/(LC)^{1/2}$.

Z and $|Z|^2$ describe the membrane impedance at a particular V for small changes in V. To illustrate the general features of $|Z|^2$, consider the special case

$$R \simeq r$$

$$\theta_1 \simeq \theta_2 = \theta$$

Then

$$|Z|^2 \simeq R^2\,\frac{1 + \omega^2\theta^2}{4 + \omega^4\theta^4}$$

For this case, $|Z|^2$ has the value $(R/2)^2$ at dc, and goes to zero as $\omega \to \infty$. Between these frequencies $|Z|^2$ goes through a maximum at the resonance frequency. To find the resonance frequency, let $\omega^2\theta^2 = u$, and set

$$\frac{d}{du}\,\frac{1 + u}{4 + u^2} = \frac{1}{4 + u^2} - \frac{2u(1 + u)}{(4 + u^2)^2} = 0$$

This reduces to

$$4 + u^2 = 2u(1 + u)$$

or

$$u^2 + 2u - 4 = 0$$

The solutions of this equation are

$$u = -1 \pm 5^{1/2}$$

Taking the positive solution,

$$\omega^2_{\max} = \frac{1}{\theta^2} (5^{1/2} - 1)$$

and

$$|Z|^2_{\max} = \frac{R^2}{2(\sqrt{5} - 1)}$$

is the value of $|Z|^2$ at the resonance.

PROBLEM. Show that $u = 5^{1/2} - 1$ (the condition for zero slope) is the value of u for which $(1 + u)/(4 + u^2)$ is maximum. This may be done by taking the second derivative, or by simply plotting the function against u for $u > 0$.

PROBLEM. Show that $\theta_1 = \theta_2$ implies that the capacitance C is equal to

$$C = \frac{\gamma \alpha N}{(\alpha + \beta)^2}$$

where N is the total number of channels in the membrane. Show that this expression for C has the correct units. Assume

$$\alpha^{-1} = \beta^{-1} = 1 \text{ msec}$$

$$\gamma = 10 \text{ pS}$$

Show that a capacitance of 1 μF/cm^2 implies 4 channels/μ^2.

PROBLEM. Plot $|Z|^2 = R^2(1 + \omega^2\theta^2)/(4 + \omega^4\theta^4)$ for $R = 10$ kΩ cm^2 and $\theta = 1$ msec. Write Z as the complex number $Z = a + ib$ for the conditions $\theta_1 = \theta_2$ and $R = r$. Plot the phase

$$\phi = \tan^{-1} \frac{b}{a}$$

as a function of frequency and compare with $|Z|^2$.

PROBLEM. We have seen that for $\theta_1 \simeq \theta_2$ and $R/r \simeq 1$, the modulus square of the membrane impedance has a resonance. The position and the existence of this resonance depend on V. Investigate how the resonance shifts for R/r slightly greater than one and slightly less than one. Although R, θ_1, and θ_2 are always positive, r can be negative in certain voltage ranges (Section 81). Investigate $|Z|^2$ for $R/r = -2$, $\theta_1 = \theta_2$.

PROBLEM. Investigate Equation (82.2) for arbitrary values of the parameters and find the general condition for the existence of a resonance.

PROBLEM. Derive formulas for Johnson noise expected from a circuit described by Equation (82.1). Hint: Use the general expressions (Section 59)

$$S_V(f) = 4kTReZ$$

$$S_I(f) = 4kTRe(1/Z)$$

Compare your solutions to an equivalent circuit solution in which noise generators with spectral densities $4kTR$ and $4kTr$ are placed in series with R and r. Is it legitimate to include r in this way even though r can be negative?

83. Transient Response to the RrLC Circuit

Consider a step change in current in the *RrLC* circuit. What is the voltage response measured across the circuit? The impedance [Equation (82.1)] can be written

$$Z = \frac{R(1 + i\omega\theta_2)}{R/r + (1 + i\omega\theta_1)(1 + i\omega\theta_2)} \tag{83.1}$$

For the moment, let $p = i\omega$ (not to be confused with the same symbol used earlier for probability). Consider the denominator

$$R/r + (1 + p\theta_1)(1 + p\theta_2) = R/r + 1 + p(\theta_1 + \theta_2) + p^2\theta_1\theta_2$$

The roots of this equation are obtained by setting the right-hand side equal to zero and solving for p. There are two roots

$$p = \frac{-(\theta_1 + \theta_2) \pm [(\theta_1 + \theta_2)^2 - 4\theta_1\theta_2(1 + R/r)]^{1/2}}{2\theta_1\theta_2}$$

Consider the negative values of these roots, in particular, define

$$p_1 = \frac{+(\theta_1 + \theta_2) - [(\theta_1 + \theta_2)^2 - 4\theta_1\theta_2(1 + R/r)]^{1/2}}{2\theta_1\theta_2}$$

and

$$p_2 = \frac{+(\theta_1 + \theta_2) + [(\theta_1 + \theta_2)^2 - 4\theta_1\theta_2(1 + R/r)]^{1/2}}{2\theta_1\theta_2}$$

An equivalent expression for the impedance of the *RrLC* circuit is

$$Z = \frac{R(1 + i\omega\theta_2)}{\theta_1\theta_2(i\omega + p_1)(i\omega + p_2)}$$

PROBLEM.   Verify that

$$\theta_1\theta_2(i\omega + p_1)(i\omega + p_2) = R/r + (1 + i\omega\theta_1)(1 + i\omega\theta_2)$$

by direct expansion. Note that $R/\theta_1\theta_2 = r/LC$.

The general expression for the voltage across Z is

$$\bar{V}(f) = \bar{I}(f)Z$$

If $I(t)$ is a step current, zero for $t < 0$ and I for $t \geq 0$, then (Section 26)

$$\bar{I}(f) = \frac{I}{i\omega} \qquad \text{(step current)}$$

and

$$\bar{V}(f) = I\frac{Z}{i\omega}$$

For the *RrLC* circuit,

$$\bar{V}(f) = IR\frac{1/i\omega + \theta_2}{\theta_1\theta_2(i\omega + p_1)(i\omega + p_2)}$$

The solution in the time domain is given by the plus-i Fourier transformation. It is convenient to write

$$\frac{1}{(i\omega + p_1)(i\omega + p_2)} = \frac{1}{p_2 - p_1}\left(\frac{1}{i\omega + p_1} - \frac{1}{i\omega + p_2}\right)$$

PROBLEM.   Verify the above equality by expansion. This is a special case of the Heaviside expansion theorem discussed in Section 52.

The reason for this manipulation is to simplify the plus-i transformation of $\bar{V}(f)$. From CF #438

$$\text{plus-}i \text{ transform}\left(\frac{1}{i\omega + p_1}\right) = e^{-p_1 t}, \qquad t > 0$$

and from CF #I-210

$$\text{plus-}i \text{ transform}\left(\frac{1}{i\omega} \frac{1}{i\omega + p_1}\right) = \frac{1}{p_1}(1 - e^{-p_1 t}), \qquad t > 0$$

Similar expressions hold for terms in p_2. Therefore

$$V(t) = \frac{IR}{\theta_1\theta_2(p_2 - p_1)}\left[\frac{1}{p_1}(1 - e^{-p_1 t}) - \frac{1}{p_2}(1 - e^{-p_2 t}) + \theta_2(e^{-p_1 t} - e^{-p_2 t})\right]$$

$V(t)$ describes the time course of the voltage response to a step change in current from 0 to I.

PROBLEM. Show that an equivalent expression for $V(t)$ is

$$V(t) = I\frac{rR}{r + R}\left(1 + \frac{p_1 e^{-p_2 t} - p_2 e^{-p_1 t}}{p_2 - p_1}\right) + \frac{I}{C}\frac{e^{-p_1 t} - e^{-p_2 t}}{p_2 - p_1} \qquad (83.2)$$

$rR/(r + R)$ is the steady-state (or slope) resistance measured at a particular voltage V. If Z represents a cell, $rR/(r + R)$ is the cell's input resistance measured by injecting I. $V(t)$ is the sum of a time-independent term proportional to input resistance, plus two time-dependent terms: one proportional to the cell's input resistance and the other proportional to the inverse of the cell's input capacitance.

PROBLEM. Show that $p_2 - p_1$ can be written in the form

$$p_2 - p_1 = \frac{1}{LC}[(rC - L/R)^2 - 4LC]^{1/2}$$

Consider the response $V(t)$ for the following cases:

1. $\sqrt{} = 0$ ($\sqrt{} \equiv [(rc - L/R)^2 - 4LC]^{1/2}$). This condition implies that $p_2 = p_1$, or

$$(\theta_1 + \theta_2)^2 = 4\theta_1\theta_2(1 + R/r)$$

The value of $p_2 = p_1 = p_0$ is

$$p_0 = \frac{\theta_1 + \theta_2}{2\theta_1\theta_2}$$

$1/p_0$ is the harmonic mean of θ_1 and θ_2. Since

$$\lim_{p_2 \to p_1} \frac{p_1 e^{-p_2 t} - p_2 e^{-p_1 t}}{p_2 - p_1} = -(1 + p_0 t)e^{-p_0 t}$$

and

$$\lim_{p_2 \to p_1} \frac{e^{-p_1 t} - e^{-p_2 t}}{p_2 - p_1} = t e^{-p_0 t}$$

then Equation (83.2) becomes

$$V(t) = I \frac{rR}{r + R} [1 - (1 + p_0 t)e^{-p_0 t}] + \frac{I}{C} (t e^{-p_0 t})$$

PROBLEM. Let $rR/(r + R) = 10$ kΩ cm^2, $p_0^{-1} = 1$ msec, and $C = 1$ μF/cm^2. Plot the step response of a cell of area 10^{-3} cm^2 for an injected current of 1 nA.

PROBLEM. Show that $p_2 = p_1 = p_0$ implies

$$p_0 = \frac{1}{2} \left((\alpha + \beta) + \frac{\gamma \alpha N}{C(\alpha + \beta)} \right)$$

PROBLEM. Show that $p_2 = p_1 = p_0$ implies

$$(rC - L/R)^2 = 4LC$$

2. $\sqrt{}$ *Real.* This condition implies that p_1 and p_2 are real, and

$$(\theta_1 + \theta_2)^2 > 4\theta_1 \theta_2 (1 + R/r)$$

In this case

$$p_1 = \frac{(\theta_1 + \theta_2) - \sqrt{}}{2\theta_1 \theta_2} > 0$$

and

$$p_2 = \frac{(\theta_1 + \theta_2) + \sqrt{}}{2\theta_1 \theta_2} > p_1$$

Therefore $p_2 - p_1 > 0$. Since p_2 and p_1 are positive, the exponential terms in Equation (83.2) decay with time.

PROBLEM. Let input resistance, capacitance, I, and cell area be the same as in the last problem. Plot $V(t)$ from Equation (82.2) for the $\sqrt{}$ real and $p_1^{-1} = 1$ msec and $p_2^{-1} = 2$ msec.

3. $\sqrt{\ }$ *Imaginary.* This condition implies that p_2 and p_1 are imaginary, and

$$(\theta_1 + \theta_2)^2 < 4\theta_1\theta_2(1 + R/r)$$

In this case, p_1 and p_2 are of the form

$$p_1 = a - ib$$

and

$$p_2 = a + ib$$

where

$$a = \frac{\theta_1 + \theta_2}{2\theta_1\theta_2}$$

and

$$b = \frac{[(\theta_1 + \theta_2)^2 - 4\theta_1\theta_2(1 + R/r)]^{1/2}}{2\theta_1\theta_2}$$

To calculate $V(t)$ from Equation (82.2) consider

$$\frac{p_1 e^{-p_2 t} - p_2 e^{-p_1 t}}{p_2 - p_1} = \frac{(a - ib)e^{-(a+ib)t} - (a + ib)e^{-(a-ib)t}}{2ib}$$

$$= \frac{e^{-at}}{2ib} [a(e^{-ibt} - e^{ibt}) - ib(e^{-ibt} + e^{ibt})]$$

Similarly,

$$\frac{e^{-p_1 t} - e^{-p_2 t}}{p_2 - p_1} = \frac{e^{-at}}{2ib} (e^{ibt} - e^{-ibt})$$

Since

$$e^{\pm i\theta} = \cos\theta \pm i\sin\theta$$

then

$$e^{-ibt} - e^{ibt} = -2i\sin bt$$

and

$$e^{-ibt} + e^{ibt} = 2\cos bt$$

Therefore

$$V(t) = I\frac{rR}{r+R}\left[1 - e^{-at}\left(\frac{a}{b}\sin bt + \cos bt\right)\right] + \frac{I}{C}\left[e^{-at}\left(\frac{1}{b}\sin bt\right)\right]$$

PROBLEM. Plot $V(t)$ from Equation (82.2) for the same conditions as the last two problems, letting

$$a^{-1} = 2 \text{ msec}, \qquad b^{-1} = 1 \text{ msec}$$

PROBLEM.   Show that as $b \to 0$, the above expression for $V(t)$ reduces to case one.

The voltage response of the *RrLC* circuit to a step change in current depends on the values of the circuit elements. In our model, the circuit elements R, r, and L depend on the transmembrane voltage; therefore, the above cases are generated in this model by varying V.

The first two conditions on $\sqrt{}$ lead to simple exponential relaxations. The third condition leads to exponentials times sine and cosine functions, or what is described above as a damped oscillation. Recall that

$$\theta_1 \simeq \theta_2, \qquad R \simeq r$$

implies that $|Z|^2$ has a resonance (Section 82). This also implies $\sqrt{}$ imaginary, since

$$(\theta_1 + \theta_2)^2 - 4\theta_1\theta_2(1 + R/r) \simeq 4\theta^2 - 8\theta^2 < 0$$

In general, a resonance in $|Z|^2$ will give voltage responses that contain damped oscillations.

PROBLEM.   Derive the voltage response to a step change in current for the *RrLC* circuit letting $\theta_1 \simeq \theta_2 = \theta$ and $R/r = +1$ (assuming r is positive) and $R/r = -1$ (assuming r is negative).

In the above discussion, R is due entirely to the conduction through open channels. If an unspecified leak resistance exists in the membrane, it can be modeled by a resistor R_L in parallel with the circuit in Figure 81.1. In this case

$$R \to \frac{RR_L}{R + R_L}$$

in the above formulas, but all derivations remain the same.

PROBLEM.   Let the current injected in the *RrLC* circuit be steady, i.e., $I(t) = I$. Show from the formula

$$\bar{V}(f) = Z\bar{I}(f)$$

that $V = IrR/(r + R)$.

PROBLEM.   Derive the steady state $V(t)$ across the $RrLC$ circuit for $I(t) = I \sin \omega t$.

PROBLEM.   Derive the change in current through the $RrLC$ circuit in response to a step change in voltage. (This corresponds to a voltage clamp.)

SOLUTION.   Let $V(t) = 0$ for $t < 0$ and $V(t) = V$ for $t \geq 0$. Then $\bar{V}(f) = V/i\omega$ and for the $RrLC$ circuit,

$$\bar{I}(f) = \frac{1}{Z}\,\bar{V}(f) = \frac{\theta_1\theta_2}{R}\,\frac{(i\omega + p_1)(i\omega + p_2)}{1 + i\omega\theta_2}\,\frac{V}{i\omega}$$

or

$$\bar{I}(f) = VC\left(\frac{i\omega}{1/\theta_2 + i\omega} + \frac{p_1 + p_2}{1/\theta_2 + i\omega} + \frac{p_1 p_2}{i\omega(1/\theta_2 + i\omega)}\right)$$

Now $p_1 + p_2 = (\theta_1 + \theta_2)/\theta_1\theta_2$ and $p_1 p_2 = (1 + R/r)/\theta_1\theta_2$. From CF # 208

plus-i transformation$\left(\dfrac{i\omega}{1/\theta_2 + i\omega}\right) = \dfrac{d}{dt}\,(e^{-t/\theta_2}) = -\dfrac{1}{\theta_2}\,e^{-t/\theta_2}, \qquad t > 0$

Using our previous results for the second and third terms in $\bar{I}(f)$

$$I(t) = VC\left[\left(-\frac{1}{\theta_2} + \frac{\theta_1 + \theta_2}{\theta_1\theta_2}\right)e^{-t/\theta_2} + \frac{1 + R/r}{\theta_1}\,(1 - e^{-t/\theta_2})\right]$$

or

$$I(t) = \frac{V}{R} + \frac{V}{r}\,(1 - e^{-t/\theta_2})$$

There are no damped oscillations. Compare this solution with that obtained for Figure 81.1 (i.e., no parallel C). Derive $\bar{I}(f)$ and take the plus-i transformation to obtain $I(t)$. Repeat for $I(t) = I$ and $I(t) = I \sin \omega_0 t$.

84. *Voltage Noise from Channels Embedded in a Membrane*

Consider a population of two-state, ion-specific channels embedded in an ideal lipid matrix. In Section 73 it was shown that for the channels alone, the spectral density of the current noise is

$$S_I(f) = \frac{4I^2}{N}\,\frac{1 - p}{p}\,\frac{\theta}{1 + \omega^2\theta^2}$$

I is the mean current flowing through the channels, N is the total number of channels, p is the probability of a channel being open $[\alpha/(\alpha + \beta)]$, and θ is $1/(\alpha + \beta)$. Assume that the properties of the channels are unaltered in the lipid matrix. Since I is the closed-circuit current flowing through the membrane, the open-circuit voltage is

$$\bar{V}(f) = Z\bar{I}(f)$$

and therefore (Section 35)

$$S_V(f) = |Z|^2 S_I(f)$$

where Z is the source impedance. The equivalent impedance for the population of two-state channels undergoing small perturbations is given in Section 81. As a first approximation, we calculate S_V from S_I (above) and Z from Equation (82.2) (Z includes the parallel capacitor that represents the lipid matrix).

Thus

$$S_V(f) = R^2 \, \frac{(1 + \omega^2\theta_2{}^2)(4I^2/pN)(1 - p)}{[(1 - \omega^2\theta_1\theta_2) + R/r]^2 + \omega^2(\theta_1 + \theta_2)^2} \, \frac{\theta}{1 + \omega^2\theta^2}$$

Note that θ_2 in the impedance factor,

$$\theta_2 = L/r = 1/(\alpha + \beta)$$

and θ in the noise factor,

$$\theta = 1/(\alpha + \beta)$$

are equal. Also, since $I = \gamma N_o(V - E) = (V - E)/R$

$$I^2R^2 = (V - E)^2$$

Therefore

$$S_V(f) = \frac{4(V - E)^2}{N} \, \frac{1 - p}{p} \, \frac{\theta_2}{[(1 - \omega^2\theta_1\theta_2) + R/r]^2 + \omega^2(\theta_1 + \theta_2)^2}$$

PROBLEM. Show that an alternative expression for $S_V(f)$ is

$$S_V(f) = \frac{4(V - E)^2}{N} \, \frac{1 - p}{p} \, \frac{1}{\theta_1{}^2\theta_2(\omega^2 + p_1{}^2)(\omega^2 + p_2{}^2)}$$

where p_1 and p_2 are defined in Section 83.

PROBLEM. Show that $S_V(f)$ derived above has no resonance (even under conditions where $|Z|^2$ does).

In this model both $S_I(f)$ and $S_V(f)$ have their largest value at $f = 0$. The chief difference is that S_V senses the membrane impedance and rolls off faster than S_I. For example, if $\theta_1 = \theta_2 = \theta$ and $R = r$, then

$$S_V(f) = \frac{4(V-E)^2}{N} \frac{1-p}{p} \frac{\theta}{4 + \omega^4\theta^4}$$

Compare this formula with the first one in this paragraph.

To calculate the correlation function of the voltage noise, consider the general relationship between $C(\tau)$ and $S(f)$ described in Section 43:

$$C(\tau) = \int_0^\infty S(f) \cos \omega\tau \, df$$

For the case considered above

$$C_V(\tau) = \frac{4(V-E)^2}{N} \frac{1-p}{p} \frac{1}{\theta_1^2\theta_2} \int_0^\infty \frac{\cos \omega\tau \, df}{(\omega^2 + p_1^2)(\omega^2 + p_2^2)}$$

From BH # 175-1,

$$\int_0^\infty \frac{\cos \omega\tau \, df}{(\omega^2 + p_1^2)(\omega^2 + p_2^2)^2} = \frac{1}{4} \frac{p_2 e^{-p_1\tau} - p_1 e^{-p_2\tau}}{p_1 p_2(p_2^2 - p_1^2)}$$

Therefore

$$C_V(\tau) = \frac{(V-E)^2}{N} \frac{1-p}{p} \frac{1}{\theta_1^2\theta_2} \frac{p_2 e^{-p_1\tau} - p_1 e^{-p_2\tau}}{p_1 p_2(p_2^2 - p_1^2)}$$

PROBLEM. Consider $p_2 = p_1 = p_0$; show that

$$\lim_{p_2 \to p_1} \frac{p_2 e^{-p_1\tau} - p_1 e^{-p_2\tau}}{p_1 p_2(p_2^2 - p_1^2)} = \frac{1 + p_0\tau}{2p_0^3} e^{-p_0\tau}$$

In this case, $C_V(\tau)$ relaxes as a single exponential function.

PROBLEM. Show that the variance of the voltage noise is

$$\sigma_V^2 = \frac{(V-E)^2}{N} \frac{1-p}{p} \frac{\theta_2}{\left(\dfrac{\theta_1}{2\theta_2} + \dfrac{\theta_2}{2\theta_1} - \dfrac{R}{r}\right)(\theta_1 + \theta_2)}$$

and that $C_V(\tau)$ crosses zero when

$$\tau_0 = \frac{\ln(p_2/p_1)}{p_2 - p_1}$$

τ_0 is the logarithmic average of p_2 and p_1 (Table 19.1). What are the conditions that make τ_0 real?

PROBLEM.   Derive the expression

$$S_V(0) = \frac{4I^2}{N} \frac{1-p}{p} \frac{\theta_2}{\left(\dfrac{r+R}{rR}\right)^2}$$

and show that the case $p = 0$ is defined.

85. *The Equivalent Noise Source for Channel Noise*

Consider the *RrLC* circuit as a resistor R in parallel with Z', where

$$\frac{1}{Z'} = \frac{1}{r + i\omega L} + i\omega C$$

The previous expression for the *RrLC* impedance Z is related to Z' by

$$\frac{1}{Z} = \frac{1}{R} + \frac{1}{Z'}$$

From the first equation

$$Z' = \frac{r + i\omega L}{1 + (r + i\omega L)(i\omega C)}$$

The *RrLC* impedance illustrated in Figure 85.1 has a voltage noise source in the R branch. The dc bias potential E is omitted because we are only interested in an expression for the equivalent noise source of the membrane.

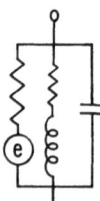

Figure 85.1. The equivalent *RrCL* circuit for channel noise.

From Figure 85.1 the open-circuit voltage is (Section 31)

$$\bar{V}(f) = \frac{\bar{e}Z'}{R + Z'}$$

Z' is considered noiseless. Therefore, the voltage noise spectral density is (Section 35)

$$S_V = S_e \left| \frac{Z'}{R + Z'} \right|^2$$

From Section 84,

$$S_V = \frac{4I^2}{N} \frac{1-p}{p} \frac{\theta}{1 + \omega^2\theta^2} |Z|^2$$

Since

$$|Z|^2 = \left| \frac{RZ'}{R + Z'} \right|^2$$

and $I^2R^2 = (V - E)^2$, the equivalent noise source is

$$S_e(f) = \frac{4(V - E)^2}{N} \frac{1-p}{p} \frac{\theta}{1 + \omega^2\theta^2} \qquad \text{(channel noise)}$$

A voltage noise source with a spectral density $S_e(f)$ placed in the *RrLC* circuit as shown in Figure 85.1 gives $S_V(f)$ and $S_I(f)$ derived previously, Since a voltage source in series with R can be replaced by a current source i in parallel with R, where $i = eE/R$,

$$S_i(f) = S_e/R^2 = S_I(f)$$

This formula justifies placing the noise source entirely in the R branch and not in the *rLC* network; the closed-circuit current derives from channels randomly opening and closing and R represents the parallel combination of N channels (Figure 73.1).

To calculate Johnson noise from the *RrLC* circuit, consider

$$S_V = 4kTReZ \qquad \text{(Johnson noise)}$$

This formula gives the same answer obtained by assuming a noise source e_R in series with R and a noise source e_r in series with r, where

$$S_{e_R} = 4kTR$$

$$S_{e_r} = 4kTr$$

To compare channel noise with Johnson noise at $f = 0$, set

$$S_V(0) = \frac{4(V - E)^2}{N} \frac{1 - p}{p} \frac{\theta_2}{(1 + R/r)^2}$$

equal to

$$S_V(0) = 4kT \frac{rR}{r + R}$$

Since $(V - E) = IR$ and $\theta_2 = \theta$

$$\frac{W\theta}{N} \frac{1 - p}{p} \frac{r}{r + R} = kT$$

is the condition that channel noise equal Johnson noise at $f = 0$. $W = I^2 R$ is the energy dissipation in the channels. If $\alpha \simeq \beta$ and $R \simeq r$,

$$\frac{W\theta}{2N} \simeq kT$$

is the condition that the two sources of noise are about equal at $f = 0$.

PROBLEM. Show that an equivalent condition to the one immediately above is

$$(V - E)^2 \gamma \theta \simeq 4kT$$

To calculate $1/f$ noise from the $RrLC$ circuit, assume a voltage noise source in series with R, where

$$S_{1/f} = g(V - E)^2/f$$

and g is a constant of the open channel (Section 68). To compare $1/f$ noise and channel noise at 1 Hz assume that for channel noise,

$$S_V(1 \text{ Hz}) \simeq S_V(0)$$

At 1 Hz

$$\frac{4\theta}{N} \frac{1 - p}{p} \left(\frac{r}{r + R}\right)^2 = g/(1 \text{ Hz})$$

is the condition that channel noise equal $1/f$ noise. If $\alpha \simeq \beta$ and $R \simeq r$,

$$N \simeq \theta/g$$

In this formula, θ is a pure number in seconds. For example, if $\theta = 10^{-3}$

sec and $g = 10^{-8}$, then

$$N \simeq 10^5$$

is the condition that channel and $1/f$ noise are approximately equal at 1 Hz.

Channel noise and $1/f$ noise go to zero when $V = E$ since both are generated by a mean current passing through channels. Johnson noise is finite at $V = E$ since it represents noise at equilibrium.

PROBLEM. Above we compared spectral densities from channel noise and Johnson noise at $f = 0$. Derive the condition that makes the variance of these noise sources equal.

86. *Current Noise Parameters Derived from Voltage Noise and Impedance*

Current noise provides direct information about changes in membrane conductance; by measuring the closed-circuit current, changes that occur in R are directly deflected in I without distortion by the rLC part of the network. Voltage measurements see changes in R across the rLC circuit; therefore, voltage fluctuations are indirectly related to the fluctuations that occur in R.

$S_I(f)$ is therefore a simpler test of channel models than $S_V(f)$. For technical reasons it may be easier to measure voltage noise and impedance and to calculate S_I rather than measure S_I directly. We have already seen that these three quantities are related by

$$S_I(f) = \frac{S_V(f)}{|Z|^2}$$

Here we derive the relationship between $S_I(0)$ and the voltage noise correlation function and input resistance.

The general relationship between C and S is

$$S_V(f) = 4 \int_0^\infty C_V(\tau) \cos \omega\tau \, d\tau$$

At $f = 0$, $\cos \omega\tau = 1$ and, for the $RrLC$ circuit,

$$Z(0) = \frac{rR}{r + R}$$

If the *RrLC* circuit represents a cell, $Z(0)$ is the cell's input resistance. Combining these last two results gives

$$\left(\frac{rR}{r+R}\right)^2 S_I(0) = 4 \int_0^\infty C_V(\tau) \, d\tau \qquad (86.1)$$

Knowing the input resistance and the voltage noise correlation function, $S_I(0)$ can be calculated. The input resistance is the slope of the steady-state *I-V* curve at a particular V. Although $C_V(\tau)$ sees the full membrane impedance, the filtering effect of Z is taken into account by the area under $C_V(\tau)$; generally, the correlation function crosses zero and has negative values; the area is the algebraic sum of the positive and negative portions.

PROBLEM.   Consider a parallel *RC* circuit with a white voltage noise source in series with *R*. Calculate the open circuit voltage noise $S_V(f)$ and from this calculate $C_V(\tau)$ and $S_I(f)$. Use these results to demonstrate that Equation (86.1) is valid. Repeat the calculation assuming that the noise from the parallel *RC* circuit is filtered by a high-pass *CR* network with the same time constant.

PROBLEM.   Show that

$$C_V(\tau) = C_I(\tau) * z(\tau)$$

where * stands for convolution (Section 41) and $z(\tau)$ is the plus-*i* Fourier transformation of $|Z|^2$.

PROBLEM.   Show that

$$C_V(0) = 4 \int_0^\infty \int_0^\infty C_I(u) \, |Z|^2 \cos \omega u \, du \, df$$

PROBLEM.   Derive a relationship between the variance of the voltage noise and the variance of the current noise for the model we have been considering.

87. Small-Signal Impedance of the HH-Axon Membrane

We have considered previously the simplest case of the impedance in which each channel is composed of a single subunit. Consider a channel composed of four subunits in series (see Section 74). This leads to the expressions

$$I = \gamma N_o (V - E)$$

where

$$N_o = p^4 N$$

instead of $N_o = pN$ as previously defined. $p = \alpha/(\alpha + \beta)$ is now the probability of a single subunit being open; all four must be open for the channel to conduct. The kinetic equation for p is the same:

$$\frac{dp}{dt} = \alpha(1 - p) - \beta p$$

The current through the channels is

$$I = \gamma N p^4 (V - E)$$

Let $\gamma N = \bar{g}_K$ (the maximum K conductance) and $p \rightarrow n$ (the HH variable for K conductance), then

$$I_K = \bar{g}_K n^4 (V - E)$$

is the standard HH equation for potassium current in the squid giant axon.

PROBLEM. Assuming p is a fourth-order variable, show that the same circuit (Figure 81.1) is obtained for the small signal impedance, but now

$$R = \frac{1}{\gamma N p^4}, \qquad r = \frac{\alpha + \beta}{4\gamma N p^3 (V - E)[\alpha' - p(\alpha + \beta)']}$$

$$L = r\theta, \qquad \theta = 1/(\alpha + \beta)$$

Similar expressions may be derived for the $m^3 h$ Na current in the giant axon. The final equivalent circuit for the small-signal impedance of the HH model is shown in Figure 87.1. The standard notation has been introduced to distinguish the K and the Na systems, and a leakage pathway and the membrane capacitance have been placed in parallel with the ion-specific pathways.* The resistors R_K, R_{Na}, and R_L in Figure 87.1 contain batteries (not shown) with the values at 6.3 °C:

$$E_K = -72 \text{ mV}, \qquad E_{Na} = +55 \text{ mV}, \qquad E_L = -10 \text{ mV}$$

In the HH model, r_n (and therefore L_n) is negative for $V < E_K$ (i.e., V hyperpolarized with respect to E_K); also, r_h is positive for $V < E_{Na}$. r_m, however,

---

* Details of the derivation can be found in A. Mauro *et al.* (1970) and A. L. Hodgkin and A. F. Huxley (1952).

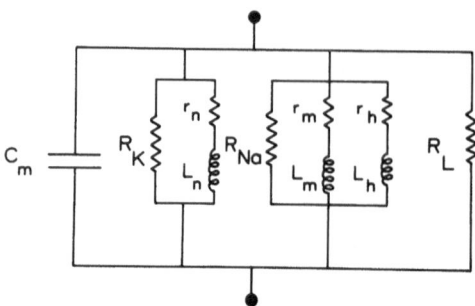

Figure 87.1. The theoretical small-signal impedance of the squid giant axon membrane.

is negative for $V < E_{\text{Na}}$. Since $V = -60$ mV at rest, all circuit elements in Figure 86.1 are positive at rest except r_m and L_m.

Figure 87.1 is the membrane impedance at a particular voltage V for small perturbations about V. Figure 87.2 shows the theoretical curves for the standard HH axon in normal sea water (NSW). The modulus of the impedance, $|Z|$, is plotted against frequency on a log–log scale for different

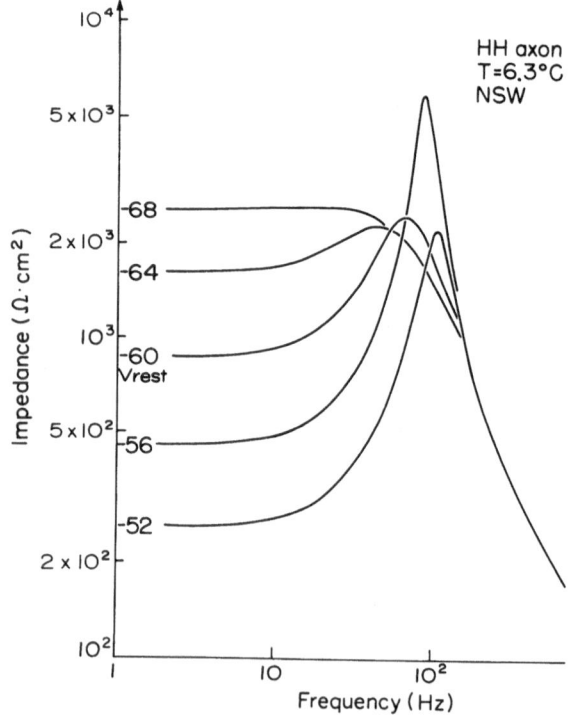

Figure 87.2. The modulus of the small signal impedance of the squid giant axon for membrane potentials near rest.

values of the mean membrane potential. When the membrane is hyper-polarized, the circuit is a simple *RC* network. Near rest, and at voltages depolarized to rest, the circuit is a resonance network with a peak near 100 Hz.

Figure 87.3 examines this behavior in more detail in three-dimensional perspective. $|Z|$ is plotted against f on linear axes for f between 2 and 400 Hz and mean membrane potential between -95 and -25 mV. The peak nearest the front of the diagram corresponds to the resonance curves in Figure 87.2. The impedance at the height of the peak (near -54.5 mV and 93 Hz) is infinite. This represents a balance between the inward and outward currents in the membrane; at some voltage and frequency these

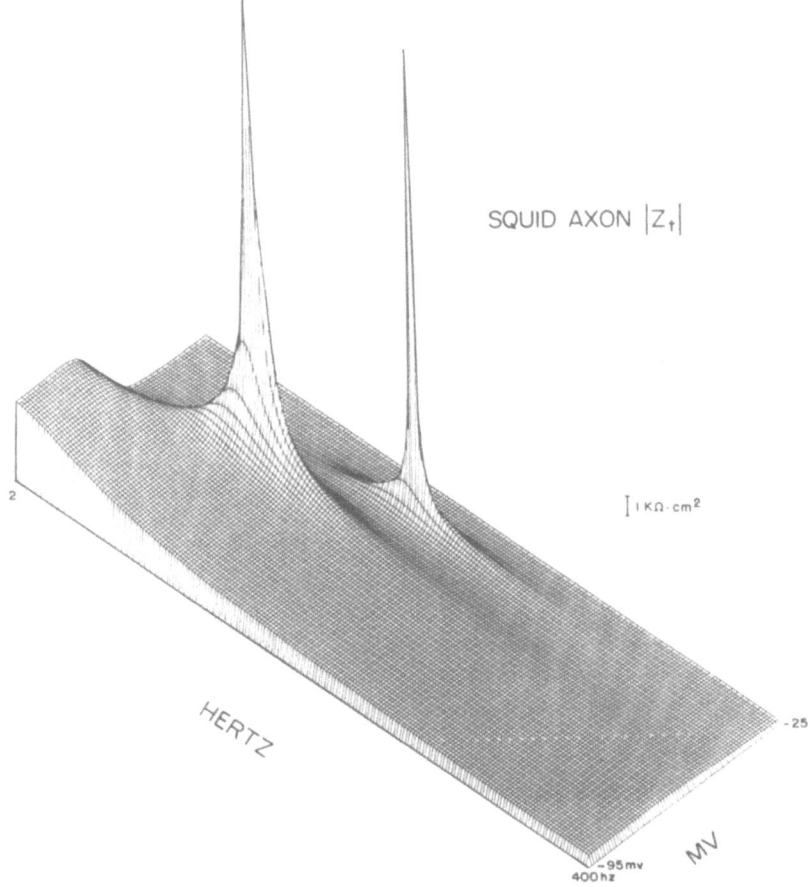

Figure 87.3. The modulus square of the small-signal impedance of the squid giant axon membrane in three-dimensional perspective. (Courtesy D. E. Clapham.)

currents sum exactly to zero and give an apparent infinite resistance, zero net current for finite driving potential.

PROBLEM. Show that in the reduced model of Section 80, where only one ionic species is present, $|Z|$ is finite at every voltage and frequency.

Figure 87.3 shows a second resonance near -38 mV and 170 Hz. This resonance is narrower than the first in both the voltage and the frequency range; it is due to an exact balance of inward and outward currents for a particular set of conditions.

Figure 87.4 is Figure 87.3 rotated to show the reverse side of the

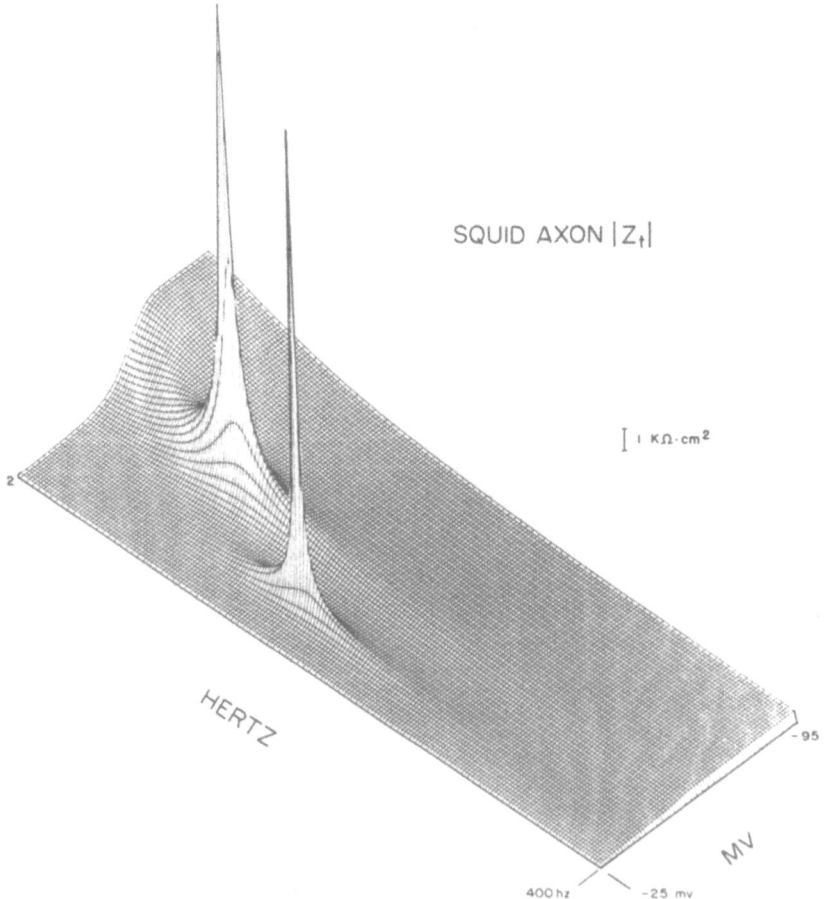

SQUID AXON $|Z_t|$

Figure 87.4. The same data as Figure 87.3 rotated to show the reverse side. (Courtesy D. E. Clapham.)

resonance peaks. The extreme left curve at 2 Hz (into the plane of the paper) is approximately the steady-state input resistance (0 Hz) of the axon membrane. This is the parallel combination of all R's and r's in Figure 87.1 and is the slope of the steady-state I–V curve.

In spite of the complex nature of the axon's small-signal impedance, near rest it may be approximated by the single $RrLC$ circuit discussed earlier (Figure 81.1 in parallel with C); in this approximation r and L are determined solely by the K system, but R is now the parallel combination of the R_K, R_{Na}, r_m, and R_L. Conti (1970) has given the values

$$R = 4.35 \text{ k}\Omega \text{ cm}^2, \qquad r = 1.7 \text{ k}\Omega \text{ cm}^2$$
$$L = 6.43 \text{ H cm}^2, \qquad C = 1 \text{ }\mu\text{F/cm}^2$$

(See also A. Mauro *et al.*, 1970, Figure 19 b.) Therefore

$$RC = 4.34 \text{ msec}, \qquad L/r = 3.78 \text{ msec}$$

These are reasonably close to the theoretical cases considered in Section 80.

PROBLEM.   Plot $|Z|$ vs. f (log–log). Z represents the $RrLC$ circuit with the values of the circuit elements provided above. Calculate the Johnson noise expected from this impedance. Plot $S_V(f)$ vs. f and derive the expected variance and the expected peak-to-peak Johnson noise.

88. The Heaviside Line and the RrLC Cable

In Section 29 we derived the passive properties of a cable that represented an unmyelinated nerve axon. The membrane was modeled by a parallel $r_m c_m$ membrane. The actual membrane requires a more complicated equivalent circuit to describe its subthreshold behavior. To introduce the problem of passive transmission along a cable with an $RrLC$ membrane (recognizing that the $RrLC$ circuit derives in part from the active properties of the membrane) we first consider the Heaviside transmission line.

First we reexamine the cable of Figure 29.1. The membrane is represented by r_m and c_m in parallel, the longitudinal cytoplasmic resistance is represented by r'', and the longitudinal external resistance is represented by r'. $V(x, t)$ is the transmembrane potential at a distance x along the line at time t. $V(x, t)$ is described by the equations

$$V = V'' - V'$$
$$dV'' = -i''(r'' \, dx)$$
$$dV' = -i'(r' \, dx)$$

Since

$$i_m = -\frac{di''}{dx} = +\frac{di'}{dx}$$

then (see Section 29)

$$\frac{d^2V}{dx} = \frac{r' + r''}{r_m} V + (r' + r'')c_m \frac{dV}{dt}$$

This is the cable equation for V. Dividing by $(r' + r'')c_m$ and rearranging terms gives

$$\frac{1}{(r' + r'')c_m} \frac{d^2V}{dx} = \frac{V}{\tau_m} + \frac{dV}{dt} = e^{-t/\tau_m} \frac{d}{dt}(Ve^{t/\tau_m})$$

where $\tau_m = r_m c_m$. Let $W(x, t) = V(x, t)e^{t/\tau_m}$ and $D^{-1} = (r' + r'')c_m$. The cable equation can now be written as a diffusion equation

$$\frac{d^2W}{dx^2} = \frac{1}{D} \frac{dW}{dt} \qquad (88.1)$$

To solve this equation, define the Fourier transform with respect to the spatial dimension x as

$$\overline{W} = \int_{-\infty}^{\infty} We^{-iux}\, dx$$

The Fourier transform of Equation (88.1) is

$$-u^2\overline{W}(u, t) = \frac{1}{D} \frac{d}{dt} \overline{W}(u, t)$$

Assume the initial condition

$$W(x, 0) = V(x, 0) = V_0(x)$$

The solution to the last two equations is

$$\overline{W}(u, t) = \overline{V}_0(u)e^{-u^2Dt}$$

PROBLEM. Show that u is in units of cm^{-1} and D is in cm^2/sec.

To obtain $W(x, t)$, take the inverse transform of the last equation. By the convolution theorem (Section 41)

$$W(x, t) = \overline{\overline{V_0(u)}} * \overline{\exp(-u^2\, Dt)}$$

Since e^{-au^2} is even,

$$\int_{-\infty}^{\infty} e^{-au^2} e^{iau}\, du = 2 \int_{0}^{\infty} e^{-au^2} \cos au\, du$$

From BMP # 1.4-11,

$$\int_{0}^{\infty} e^{-au^2} \cos ux\, du = \frac{1}{2}\, (\pi/a)^{1/2} e^{-x^2/4a}$$

Therefore

$$W(x,\, t) = V_0(x) * (\pi/Dt)^{1/2} e^{-x^2/4Dt}$$

or

$$V(x,\, t) = e^{-t/\tau_m} (\pi/Dt)^{1/2} \int_{-\infty}^{\infty} e^{-y^2/4Dt} V_0(x - y)\, dy$$

This is the general result for the transmembrane potential at x and t for arbitrary input $V_0(x)$ at $t = 0$.

For example, let $V_0(x)$ be a delta function at $x = 0$. Then

$$V_0(x) = v_0\, \delta(x)$$

where v_0 has the units of V cm. Then

$$V(x,\, t) = v_0 (\pi/Dt)^{1/2} e^{-t/\tau_m} e^{-x^2/4Dt}$$

PROBLEM. Solve Equation 88.1 for a step change input at $t = 0$ and $x = 0$, i.e., $V(x,\, t) = 0$ for $x < 0$, $t < 0$ and $V(x,\, t) = V_0$ for $x \geq 0$, $t \geq 0$.

For the cable in Figure 29.1, input signals travel along the line, but they lose their original shape and decay rapidly with time.

Suggested Reading. For a survey of solutions of diffusion and wave equations see Hochstadt (1973, Chap. 5). The book deals with the general theory of integral transformations, including nonlinear integral transformations; Laplace, Fourier, and Hilbert transformations are seen as special cases and their application to physical problems is presented with clarity. Highly recommended. A useful but somewhat diffuse text that compliments Hochstadt is by Davis (1960). It includes the solution to many problems involving integral equations.

The Time-to-Maximum Effect. Rewrite the solution for the delta functions input as

$$V(x, t) = \mathscr{K}\, \frac{1}{t^{1/2}}\, \exp\left[-(x^2/4Dt + t/\tau_m)\right]$$

Therefore

$$\frac{1}{\mathscr{K}}\frac{dV}{dt} = \frac{1}{t^{1/2}}\left[-\left(-\frac{x^2}{4Dt^2} + \frac{1}{\tau_m}\right)\right]\exp\left[-(x^2/4Dt + t/\tau_m)\right] + \frac{(-1/2)}{t^{3/2}}$$

$$= \exp\left[-(x^2/4Dt + t/\tau_m)\right]\frac{1}{t^{1/2}}\left(\frac{x^2}{4Dt^2} - \frac{1}{\tau_m} - \frac{1}{2t}\right)$$

The condition $dV/dt = 0$ leads to

$$\frac{x^2\tau_m - 4Dt^2 - 2Dt\tau_m}{4Dt^2\tau_m} = 0$$

which implies

$$t^2 + \frac{\tau_m}{2}\, t - \frac{x^2\tau_m}{4D} = 0$$

At any x, the time to maximum effect is

$$t_{\max} = \tfrac{1}{4}\left[-\tau_m \pm (\tau_m{}^2 + 4x^2\tau_m/D)^{1/2}\right]$$

For $r_m \to \infty$ (no leakage),

$$t_{\max} = \frac{x^2}{2D} = \frac{(r' + r'')c_m}{2}\, x^2$$

which is the familiar solution.

PROBLEM. Let $r'' = 0.5\,\text{M}\Omega$ cm, $r_m = 0.16\,\text{M}\Omega$ cm, $c_m = 6.28\,\text{nF/cm}$, and $r' = 0$. (These values are from Section 28 for a typical axon.) Plot $V(x, t)$ [for delta function input $v_0\,\delta(x)$, $v_0 = 1$ V cm] (a) for $x = 0$, (b) $x = 1$ mm, (c) $x = 1$ cm, (d) $x = 1$ m.

Consider a cable with an inductor l' in place of r' in Figure 29.1. $l' \ominus$ H/cm. In this case

$$dV' = -\frac{di'}{dt}\,(l'\,dx)$$

and therefore

$$\frac{dV}{dx} = l'\frac{di'}{dt} - r''i''$$

The equation relating i_m to i'' and i' still applies, therefore

$$\frac{d^2V}{dx^2} = l'\frac{di_m}{dt} + r''i_m$$

$$= \left(l'\frac{d}{dt} + r''\right)\left(\frac{V}{r_m} + c_m\frac{dV}{dt}\right)$$

$$= \frac{r''}{r_m}V + \left(r''c_m + \frac{l'}{r_m}\right)\frac{dV}{dt} + l'c_m\frac{d^2V}{dt^2}$$

This equation is known as the telegraphist's equation.

Consider the special case $r'' = 0$ and $r_m = \infty$. Then the line is a simple *LC* line and the telegraphist's equation becomes

$$\frac{d^2V}{dx^2} = l'c_m\frac{d^2V}{dt^2}$$

The equation is now in the form of a wave equation.

Rearrange the telegraphist's equation to the form

$$\frac{d^2V}{dx^2} = l'c_m\left[\frac{d^2V}{dt^2} + \left(\frac{r''}{l'} + \frac{1}{r_mc_m}\right)\frac{dV}{dt} + \frac{r''}{l'r_mc_m}V\right]$$

$$= l'c_m\left(\frac{d}{dt} + \frac{r''}{l'}\right)\left(\frac{d}{dt} + \frac{1}{r_mc_m}\right)V$$

l'/r'' and r_mc_m have the units of time. Let $r_mc_m = \tau_m$ and let $l'/r'' = \tau_a$, then

$$\frac{d^2V}{dx^2} = l'c_m\left(\frac{d}{dt} + \frac{1}{\tau_a}\right)\left(\frac{d}{dt} + \frac{1}{\tau_m}\right)V$$

Suppose that the circuit elements are such that

$$\tau_a = \tau_m = \tau$$

Then

$$\frac{d^2V}{dx^2} = l'c_m\frac{d^2}{dt^2}(Ve^{t/\tau})e^{-t/\tau}$$

which is easily checked by expansion. $V = V(x, t)$ describes the transmembrane voltage of the transmission line. Define

$$W = V(x, t)e^{t/\tau}$$

The telegraphist's equation for the special case $\tau_m = \tau_a = \tau$ becomes the wave equation

$$\frac{d^2W}{dx^2} = \frac{1}{\theta^2}\frac{d^2W}{dt^2}$$ (88.2)

where

$$\theta^2 = \frac{1}{l'c_m}$$

PROBLEM.   Solve the wave equation assuming the initial conditions

$$W(x, 0) = V(x, 0) = V_0(x)$$

$$\left.\frac{dW}{dx}\right|_{t=0} = 0$$

SOLUTION.   It may be shown by the method of Fourier transformation that the solution is

$$W(x, t) = \tfrac{1}{2}[V_0(x + \theta t) + V_0(x - \theta t)]$$

and therefore,

$$V(x, t) = \tfrac{1}{2}[V_0(x + \theta t) + V_0(x - \theta t)]e^{-t/\tau}$$

where $\tau = \tau_a = \tau_m$. In this case, $\theta = 1/(l'c_m)^{1/2}$. The solution $V(x, t)$ predicts a waveform that retains its initial shape, travels with uniform velocity θ, and decays in amplitude exponentially. For $r_m \to \infty$ and $r'' \to 0$, then

$$V(x, t) = \tfrac{1}{2}[V_0(x + \theta t) + V_0(x - \theta t)]$$

in which case the initial signal is propagated without distortion or attenuation at one-half the initial height. (The two terms in the solution correspond to waves traveling left and right from the input.)

The telegraphist's equation describes the Heaviside transmision line. If the Heaviside line is balanced such that the time constant of the loss along the line equals the time constant of the axial properties, the input signal travels without distortion. Such lines were actually used in the early days of telegraph and telephone.* The Heaviside line was used by Nyquist in 1928 in his derivation of the formula for Johnson noise (Section 59).

---

* Some of the controversy accompanying the initial use of the Heaviside line in communication networks is presented in L. J. DeFelice (1978).

PROBLEM. Show that the telegraphist's equation can be written in the form

$$\frac{d^2W}{dx^2} - l'c_m \frac{d^2W}{dt^2} = -hW$$

$$h = \frac{l'c_m}{4}\left(\frac{1}{\tau_a} - \frac{1}{\tau_m}\right)^2$$

Since h involves the square of the difference between τ_m and τ_a, small differences can be tolerated giving virtually distortion-free solutions. This formulation leads to solutions of the form

$$W(x, t) = \tfrac{1}{2}\,|\,f(x + \theta t) + f(x - \theta t)\,| + O(h)$$

where h is defined above, $\theta = 1/(l'c_m)^{1/2}$, and

$$W(x, t) = V(x, t)\exp\left[\frac{1}{2}\left(\frac{1}{\tau_a} + \frac{1}{\tau_m}\right)t\right]$$

The size of h plays the critical role. Low leakage (τ_m large) and large inductance (τ_a large) lead to low values of h.

The RrLC Line. Consider the cable of Figure 29.1 in which the *RrLC* circuit (Figure 81.1 with C in parallel) replaces the parallel $r_m c_m$ circuit. Let $r' = 0$. Such a cable is the equivalent circuit of a cylindrical axon for small perturbations near rest.

To introduce the *RrLC* circuit into the cable equations, it is convenient to define these parameters: Let the *RrLC* circuit refer to a unit membrane area; then

$$R \ominus \Omega \text{ cm}^2, \qquad C \ominus \text{F/cm}^2$$

$$r \ominus \Omega \text{ cm}^2, \qquad L \ominus \text{H cm}^2$$

To introduce the *RrLC* circuit into the cable equations, it is convenient to define

$$r_m = \frac{R}{2\pi a} \ominus \Omega \text{ cm}$$

$$r_n = \frac{r}{2\pi a} \ominus \Omega \text{ cm}$$

$$l_n = \frac{L}{2\pi a} \ominus \text{H cm}$$

$$c_m = 2\pi a C \ominus \text{F/cm}$$

where a is the radius of a cylindrical axon. (See Section 29 for details. Do not confuse r_m for the HH variable for the Na system used earlier.)

Since it has been assumed that $r' = 0$,

$$dV' = 0$$

and

$$\frac{dV}{dx} = -r''i''$$

Therefore

$$\frac{d^2V}{dx^2} = r''i_m$$

For the *RrLC* circuit,

$$i_m = \frac{V}{r_m} + \frac{V - V_L}{r_n} + C_m \frac{dV}{dt}$$

where V is the drop across the *RrLC* membrane and V_L is the drop across L. Now

$$V_L = l_n \left(\frac{di}{dt} \right)_L$$

where $(di/dt)_L$ is the time rate of change of current through the inductor. It is evident that

$$\left(\frac{di}{dt} \right)_L = \frac{di_m}{dt} - \left(\frac{di}{dt} \right)_R - \left(\frac{di}{dt} \right)_C$$

$$= \frac{d}{dt} \left(\frac{1}{r''} \frac{d^2V}{dx^2} \right) - \frac{d}{dt} \frac{V}{r_m} - \frac{d}{dt} C_m \frac{dV}{dt}$$

Substituting these equations into the equation for i_m gives

$$i_m = V \left(\frac{1}{r_m} + \frac{1}{r_n} \right) - \frac{l_n}{r_n} \frac{d}{dt} \left(\frac{1}{r''} \frac{d^2V}{dx^2} - \frac{V}{r_m} - C_m \frac{dV}{dt} \right) + C_m \frac{dV}{dt}$$

Therefore

$$\frac{d^2V}{dx^2} = \frac{r''(r_n + r_m)}{r_m r_n} V + \left(\frac{r''l_n}{r_n r_m} + r''C_m \right) \frac{dV}{dt} + \frac{r''l_n C_m}{r_n} \frac{d^2V}{dt^2} - \frac{l_n}{r_n} \frac{d^3V}{d^2x\,dt}$$

Rearranging this equation gives

$$\frac{r_n}{r''} \frac{d^2V}{dx^2} = l_n C_m \left[\frac{d^2V}{dt^2} + \left(\frac{1}{r_m C_m} + \frac{r_n}{l_n} \right) \frac{dV}{dt} + \frac{r_n + r_m}{r_m l_n C_m} V \right] - \frac{l_n}{r''} \frac{d^3V}{d^2x\,dt}$$

Define*

$$\tau_n = \frac{l_n}{r_n} = \frac{L}{r}$$

$$\tau_m = r_m c_m = RC$$

Substituting these definitions into the last equation gives

$$\frac{r_n}{r''}\frac{d^2 V}{dx^2} = l_n c_m\left[\frac{d^2 V}{dt^2} + \left(\frac{1}{\tau_n} + \frac{1}{\tau_m}\right)\frac{dV}{dt} + \frac{(1 + r_m/r_n)}{\tau_n \tau_m}V\right] - \frac{l_n}{r''}\frac{d^3 V}{d^2 x\, dt}$$

$$= l_n c_m\left[\frac{d^2 V}{dt^2} + \left(\frac{1}{\tau_n} + \frac{1}{\tau_m}\right)\frac{dV}{dt} + \frac{1}{\tau_n \tau_m}V\right] + V - \frac{l_n}{r''}\frac{d^3 V}{d^2 x\, dt}$$

$$= l_n c_m\left[\left(\frac{d}{dt} + \frac{1}{\tau_n}\right)\left(\frac{d}{dt} + \frac{1}{\tau_m}\right)V\right] + V - \frac{l_n}{r''}\frac{d^3 V}{d^2 x\, dt}$$

This is the final expression for the transmembrane potential $V(x_1 t)$ at any distance along the cylindrical axon at any moment if the membrane is represented by an *RrLC* circuit and the external longitudinal resistance is negligible.

Suppose that $\tau_n = \tau_m = \tau,$[†] and

$$\frac{l_n}{r''}\frac{d^3 V}{dx^2\, dt} = V \tag{88.1}$$

Then

$$\frac{r_n}{r''}\frac{d^2 V}{dx^2} = l_n c_m\left(\frac{d}{dt} + \frac{1}{\tau}\right)^2 V$$

$$= l_n c_m\frac{d^2}{dt^2}\left(Ve^{t/\tau}\right)e^{-t/\tau}$$

The last step can be verified by expansion.

Let $\int W(x, t) = V(x, t)e^{t/\tau}$. Then the differential equation for the *RrLC* line assumes the form of a wave equation in W, namely,

$$\frac{r_n}{r''}\frac{d^2 W}{dx^2} = l_n c_m\frac{d^2 W}{dt^2}$$

---

* τ_m and τ_n are identically θ_1 and θ_2 of Section 81. Different symbols are used here to identify cable properties; these time constants are independent of the dimensions of the cylindrical cable.

† In the giant axon at rest, $\tau_n = 3.78$ msec and $\tau_m = 4.34$ msec (Conti, 1970).

Let

$$\theta = \left(\frac{r_n/r''}{l_n c_m} \right)^{1/2}$$

then

$$\frac{d^2 W}{dx^2} = \frac{1}{\theta^2} \frac{d^2 W}{dt^2}$$

PROBLEM.   Show that θ is in units of velocity. Derive the shape of the voltage wave form that would propagate as a nondistorted wave by solving Equation (89.1). Hint: Take the Fourier transform of Equation (88.1) with respect to the spatial variable, solve for $\bar{V}(u, t)$ for appropriate boundary conditions, and invert.

Mauro *et al.* (1972b) derived the cable equation for the *RrLC* line and showed that the propagated subthreshold response to a current step in the squid giant axon has the qualitative features expected for such a line. See Detwiler *et al.* (1978) for an interesting application of subthreshold propagation for slow, time-variant conductance changes in the turtle retina.

7

Experimental Results

89. *Miniature End-Plate Potentials*

Although it is impossible to single out one paper as being the first to study membrane noise, subsequent events in the field have isolated the 1950 note of Fatt and Katz on "biological noise" at the neuromuscular junction as a starting point. In their full paper in 1952 they write the following:

> In the course of some earlier work, while recording from the surface of isolated muscle fibers, we occasionally noticed a spontaneous discharge of small monophasic action potentials. The potentials varied somewhat in size, but had a very consistent time course, rising rapidly in 1–2 msec, and declining more slowly, to one-half in about 3–4 msec. They were localized at one region of the fibre, and in their shape and spatial spread resembled the end-plate potential... Not much attention was paid to the phenomenon at the time, and it was suspected to be due to local injury....

Miniature end-plate potentials, it was shown, are a natural feature of the innervated muscle fiber. The study of mepp's, and the study of the molecular events underlying them, have formed the most complete story of membrane noise available at present. (For an excellent summary see Stevens, 1975.)

An example of the effect is given in Figure 89.1. The response of the postsynaptic membrane to a nerve impulse is shown (A) at the end-plate and (B) 2 mm away. In each case, the subthreshold spontaneous activity is also shown. The randomly occurring mepp's occur at the end plate; here the end-plate potential (the foot of the action potential) is large and the muscle spike originates. At a distance of 2 mm from the end plate the mepp's are not evident and the muscle spike occurs after a delay due to the conduction

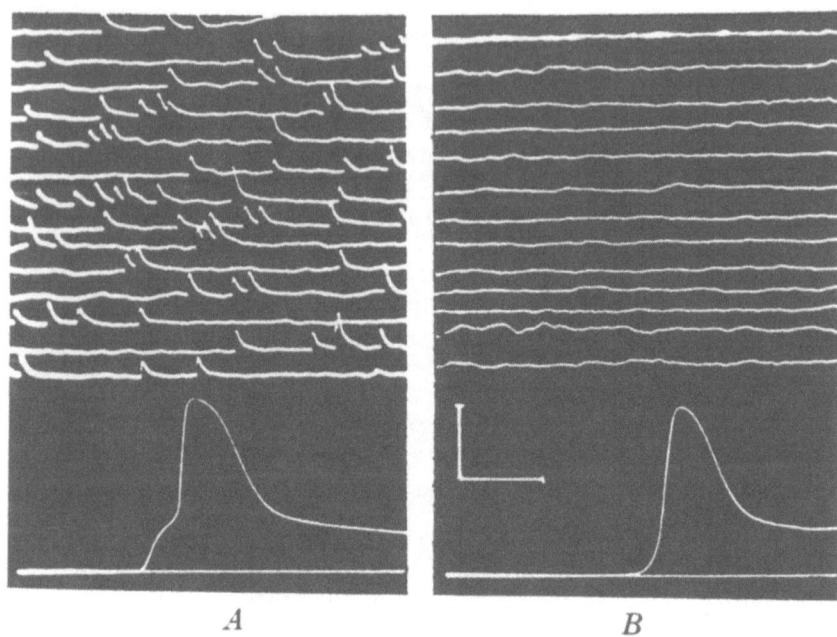

$$A \qquad\qquad\qquad B$$

Figure 89.1. Localization of spontaneous activity from Fatt and Katz (1952). A is an intercellular recording at the end-plate and B is an intracellular recording 2 mm away in in the same muscle fiber. The lower trace for which the scale is 5 mV and 2 msec, is the response the postsynaptic membrane to a nerve stimulus applied at the beginning of the sweep. The upper traces for which the scale is 3.6 mV and 47 msec, show the spontaneous activity in the postsynaptic membrane. Each small transient is a miniature end-plate potential (mepp).

time. Fatt and Katz asked whether the mepp's could be molecular leakage of acetylcholine (ACh) from nerve endings. Since there was thought to be a slow continuous escape of the ACh molecules that were being synthesized at the cholinergic nerve endings, it seemed possible that a "random collision of individual molecules of ACh with the end-plate" could be the cause of the mepp's. This was ruled out on several grounds: (1) From perfusion experiments it was known that about 10^6 molecules of ACh were released per impulse. As approximate as this estimate was (probably too low), a single mepp would probably involve thousands of molecules since it is roughly one-thousandth the size of an end-plate potential. (2) Reapplication of ACh to the external bathing solution should greatly increase the frequency of the mepp's, contrary to observation; ACh sufficient to cause a few milli-volts depolarization in the postsynaptic membrane did not change the frequency of the spontaneous discharges. The only effect was a small reduction in the amplitudes of the mepp's.

PROBLEM. Explain the reduction in amplitude of mepp's in the presence of ACh in the external bath.

Fatt and Katz (1952) conclude that "...if individual molecular collision between ACh and the end-plate builds up a steady depolarization then the molecular units of this depolarization must be much smaller than the recorded miniature end-plate potential."

Let us consider what was meant by the random nature of the spontaneous discharge. The sequence of events appeared to be irregular under normal conditions, that is, the occurrence of one event independent of previous activity. For such processes the interval between successive discharges is distributed exponentially. The number of times (n) one would expect to find an interval of length t between successive events is given by the formula (Section 72)

$$n = \frac{N \, \Delta t}{T} \, e^{-t/T} \tag{89.1}$$

where N is the total number of observations, and should be large. Δt is the time window in which n is counted, and should be small compared to the average time interval T. Figure 89.2 shows the result of this test of randomness for spontaneous mepp's from the frog neuromuscular junction. What causes this random discharge is unknown.

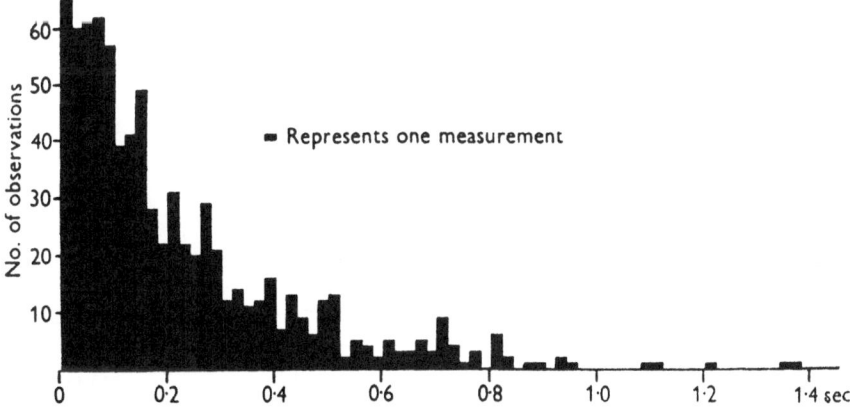

Figure 89.2. Distribution of time intervals between spontaneous mepp discharges. From Fatt and Katz (1952). $N = 800$, $\Delta t = 20$ msec, $T = 0.221$ sec. [See Equation (89.1).]

PROBLEM.   Plot $n = [(N\,\Delta t)/T]\exp(-t/T)$ using the values in the caption of Figure 89.2.

Fatt and Katz also recorded the mepp's with an extracellular electrode. The externally recorded event is negative going and is much faster than the intracellularly recorded event. Figure 89.3 compares the two recordings.

PROBLEM.   Explain with a circuit equivalent why the external mepp is more brief than the internally recorded event and why its polarity is reversed. If one recorded with an internal and external electrode simultaneously, would there be a one-to-one correlation between the mepp's?

Speculating on the cause of the spontaneous excitation at the nerve terminals, Fatt and Katz suggested (1950) membrane voltage noise and later derived (1952) the formula for the Johnson noise expected from a nerve axon. To calculate Johnson noise, some model of the membrane is necessary. Fatt and Katz assumed a distributed RC cable model like that described in Section 29. Using the notation of that section, the impedance seen at the end of an infinite cable is

$$Z = (r''z_m)^{1/2} = [r_i''r_m/(1 + i\omega\theta)]^{1/2}$$

r'' is the internal resistor; it is assumed that $r'' \gg r'$ in Figure 29.1. The formula for the mean square voltage noise expected from this impedance is (Section 60)

$$\sigma_V{}^2 = 4kT \int_0^\infty \mathrm{Re}\, Z\, df$$

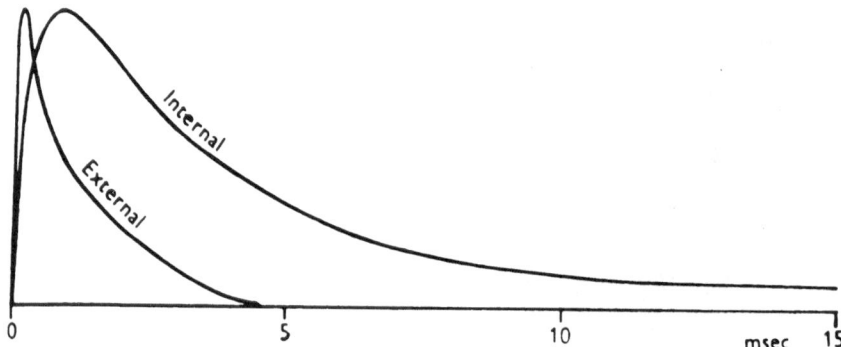

Figure 89.3. Internal and external mepps compared. From Fatt and Katz (1952). The external event has been scaled to the same amplitude as the internal event, and has been turned up for the comparison. The average size of an internal mepp is about 1 mV.

PROBLEM. Show that for $Z = [r''r_m/(1 + i\omega\theta)]^{1/2}$,

$$\text{Re } Z = A(r''r_m/2)^{1/2}$$

where

$$A = \left(\frac{1}{1 + \omega^2\theta^2} + \frac{1}{(1 + \omega^2\theta^2)^{1/2}} \right)^{1/2}$$

and $\omega = 2\pi f$, $\theta = r_m c_m$.

The cable parameters are related to the material properties of the membrane and the axon cytoplasm by the equations (Section 28)

$$r'' = \frac{\varrho''}{\pi a^2}, \qquad r_m = \frac{R_m}{2\pi a}$$

where ϱ'' is the axoplasmic resistivity (Ω cm), R_m is the areal resistivity of the membrane (Ω cm^2), and a is the axon radius. It is unnecessary to give the formula for c_m, since c_m enters only through θ. The expression for the noise variance is

$$\sigma_V{}^2 = \frac{8kT}{\pi(2a)^{3/2}} \left(\frac{\varrho''R_m}{2} \right)^{1/2} \int_0^\infty A \, df$$

Only the effective bandwidth $\int_0^\infty A \, df$ remains to determine the expected noise from a long unmyelinated axon. Since the integral $\int_0^\infty A \, df$ diverges, a simple cutoff f_{max} is used to calculate the bandwidth.

PROBLEM. Show that

$$\int_0^{f_{max}} A \, df = \frac{1}{2\pi\theta} [(1 + \beta_{max}^2)^{1/2} - 1]^{1/2}$$

where $\beta = \omega\theta$. For $\beta_{max} \gg 1$ (i.e., $f_{max} \gg 1/2\pi\theta$) the final expression for the noise is

$$\sigma_V{}^2 = \frac{8kT}{\pi(2a)^{3/2}} \left(\frac{\varrho''R_m}{2} \right)^{1/2} \frac{\beta_{max}^{1/2}}{2\pi\theta}$$

and since $\beta_{max} = 2\pi\theta f_{max}$ and $\theta = R_m C_m$, then

$$\sigma_V{}^2 = \frac{4kT}{(2\pi a)^{3/2}} \left(\frac{\varrho''f_{max}}{C_m} \right)^{1/2}$$

PROBLEM. Assume $a = 10$ μm, $\varrho'' = 200$ Ω cm, $C_m = 1$ μF/cm, and $f_{max} = 10,000$ Hz. Calculate $\sigma_V{}^2$. How small will the radius have to be for the peak–peak voltage fluctuations to reach 2 mV? Compare this radius with typical dimensions for nerve terminals. The final expression for $\sigma_V{}^2$ does not contain R_m. Can this still be described as noise due to the thermal agitation of ions within the membrane? The divergence of the integral $\int_0^\infty A\, df$ is not due to the membrane resistance, but to the axoplasmic resistance. This derivation should be compared to the result for a simple RC circuit with Johnson noise in R, namely,

$$\sigma_V{}^2 = \frac{kT}{C}$$

(see Section 58).

90. Acetylcholine Noise

A remarkable sequel to the Fatt and Katz (1952) work on miniature end-plate potentials was the discovery of acetylcholine (ACh) noise by Katz and Miledi (1970, 1971, 1972). This discovery introduced an entirely new electrophysical approach to the study of the drug–receptor interaction: "When a steady dose of ACh is applied to an end-plate, the resulting depolarization is accompanied by a significant increase in voltage noise" (Katz and Miledi, 1972). This is shown in Figure 90.1.

Once observed, the effect was large enough to make the first measurements of enlarged oscilloscope records. The peak–peak noise can be related to the rms (Section 33) and the control can be subtracted from the ACh-induced noise.

Although there can be large variations between preparations, from Katz and Miledi (1972),

> ... the basic phenomenon was seen in every experiment provided enough amplification was used. In fact, once one decides to look for it, "ACh-noise" is easy enough to detect. It is, of course, also easy to overlook for by comparison with a discrete phenomenon such as miniature e.p.p.s. [Figure 90.1], the appearance of noise is unimpressive.

When the membrane was depolarized by injecting a current, there was no increase of noise comparable to the ACh noise. The ACh noise was unaffected by shifting the position of the extracelluler pipette containing the drug. And it could also be observed by replacing the bath solution with one containing ACh.

Katz and Miledi initially analyzed records like that shown in Figure 90.1 by hand. The envelopes of the noise were traced and measured at

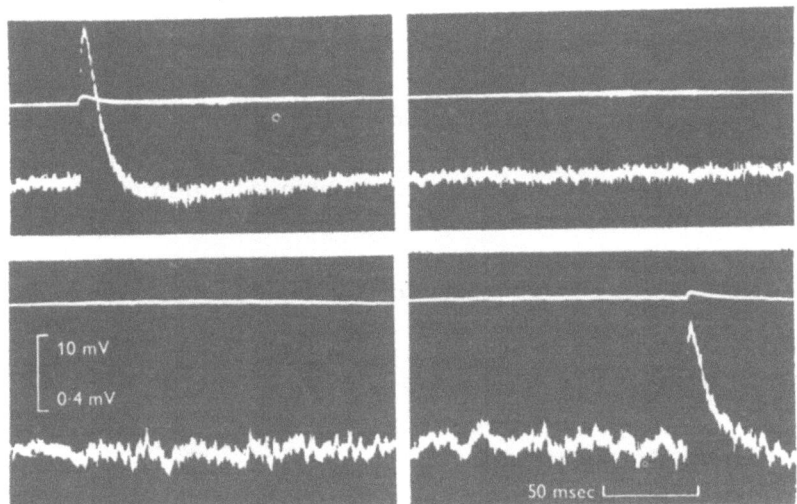

Figure 90.1. Intracellular recording from an end plate in frog sartorius at 21°C. From Katz and Miledi (1972). For each frame, the upper trace scale is 10 mV and the lower trace 0.4 mV. The top row shows controls with no ACh added; the bottom row shows the membrane noise in the presence of ACh, which is diffusing from a nearby micropipette at a steady rate. Note the accompanying depolarization. Two miniature end-plate potentials are also seen.

intervals of 0.75 msec for 200-msec records. The base-line noise gave a value of 13.1 μV for the rms of the fluctuations. During application of ACh, the base-line noise rose to 32 μV rms. The membrane potential changed from −93 to −84.5 mV. Automated measurements from about 100 experiments gave mean values of 19.4 and 47.7 μV comparing control noise with ACh noise at 10.4-mV depolarization, for a net change of 43.5 μV. The ACh-noise values are much more broadly distributed about their mean values than the control noise.

PROBLEM.   Calculate that the net rms value of extra ACh noise in going from 13.1- to 32-μV rms is 29.2 μV. Assume that the elementary events are uniform, exponentially shaped, and that they arrive randomly. Derive the formula $\sigma^2 = \Delta Va/2$ (Section 79) where σ^2 is the variance of the noise, ΔV is the mean change in potential, and a is the size of the exponential event. Calculate the size of the elementary event that gave rise to the ACh noise seen in Figure 90.1 Would it be visible on the scale used? Compare the expression for σ^2 for an exponentially shaped pulse with the same expression for a rectangular pulse.

It is usually assumed in the application of Campbell's theorem that the amplitude of the elementary event is constant. If the amplitude is allowed to vary in a random manner about some mean value then the expression for the amplitude must be replaced. Let the variance in a be given by

$$\sigma_a^2 = \overline{(a - \bar{a})^2} = \overline{a^2} - \bar{a}^2$$

then

$$\overline{a^2} = \bar{a}^2 + \sigma_a^2$$

must replace a^2 in all formulas.

A much more serious correction must be made when measuring voltage noise.

The added effect of the molecular bombardment that brings the mean membrane potential to a new level changes the conductance of the membrane across which the voltage noise is measured. (This is the same problem encountered in the nonlinear summation that occurs for mepp's and generator potentials.)

Assume a model for the conductance change at the end plate in which the ACh-sensitive resistance r is in parallel with the cell input resistance R. Each pathway has a potential source. Let E be the resting potential of the cell ($r \rightarrow \infty$) and e be the reversal potential of the ACh-induced ionic pathway. The voltage across the circuit is (Section 31)

$$V = \frac{Er + eR}{R + r}$$

Let $V_0 = e - E$. Then

$$V = \frac{Er + (V_o + E)R}{R + r} = E + \frac{V_0 R}{R + r}$$

and

$$V - E = \frac{V_0 R}{R + r} = \frac{V_0 R g}{R g + 1}$$

where we have set $r = 1/g$. $V - E$ is the actual depolarization from rest that occurs and it is a function of g. Let $V - E = \Delta V$. Then (Katz and

Miledi, 1970, Equation 8)

$$\frac{\Delta V}{V_0} = \frac{Rg}{Rg + 1}$$

Therefore, to have ΔV proportional to g, ΔV must be multiplied by the factor $Rg + 1$.

PROBLEM. Plot ΔV against g for $0 \leq g \leq 1/R$. Observe that V_0 is the maximum change in voltage that can occur.

PROBLEM. Show that $1 + gR = V_0/(V_0 - \Delta V)$.

Katz and Miledi assumed that g is directly proportional to the concentration of ACh, i.e., $g = g_0[ACh]$, in the range of concentrations they used. The conductance change is a linear sum of many elementary changes in parallel. It is evident that as the net conductance changes the depolarization each event produces will not be the same.

Consider the way in which a change in voltage (ΔV) varies with a small change in conductance. From the expression above for ΔV

$$\delta(\Delta V) = \frac{d(\Delta V)}{dg} \delta g = \frac{V_0 R}{(Rg + 1)^2} \delta g$$

Consider ΔV a small fluctuation in voltage due to the presence of ACh. Then $\delta(\Delta V)$ is the change in the fluctuation. In order that the noise variance be proportional to δg, it must be multiplied by $(Rg + 1)^4$.

Therefore, to account for the overall change in conductance, the mean change ΔV must be multiplied by $(Rg + 1)$ but the variance in ΔV by $(Rg + 1)^4$. If the variance is proportional to the mean, the net correction factor is $(Rg + 1)^3$. For example, exponentially distributed random depolarizing pulses of shape $ae^{-t/\theta}$ give (Section 79)

$$\sigma_V^2 = \frac{a}{2} \Delta V$$

To account for the change in conductance, the relationship between the variance and the mean becomes

$$\sigma_V^2 (Rg + 1)^4 = \frac{a}{2} \Delta V (Rg + 1)$$

or

$$\sigma_V^2 = \frac{a}{2} \frac{\Delta V}{(Rg + 1)^3} = \frac{a \Delta V}{2} \left(\frac{V_0 - \Delta V}{V_0} \right)^3$$

PROBLEM. Plot σ_V^2 against g. Show that a maximum occurs when $V = V_0/4$.

Recall that $Rg + 1 = V_0/(V_0 - \Delta V)$. Therefore, to obtain consistency when calculating a at different levels of depolarization, the value obtained from the formula

$$a = \frac{2\sigma_V^2}{\Delta V}$$

should be multiplied by $[V_0/(V_0 - V)]^3$. This procedure was adopted by Katz and Miledi, who found that, when corrected, the size of the elementary event was constant over the voltage range $0 \leq \Delta V \leq 23$ mV for an individual experiment. The corrected value of a, over a wide range of temperatures (1.3–25°C) and fiber sizes, was 0.42 μV, with variations extending from 0.064 to 1.54 μV. A decrease in temperature decreased the value of a from 0.78 μV at 22°C to 0.32 μV at 2°C.

PROBLEM. How does a change in R affect the value of a?

The most important contribution of the new technique may well be in drug kinetics. In the 1972 paper Katz and Miledi tested the effects of curare on ACh noise. If curare is a blocking agent that competes with ACh for the receptor molecules in the postsynaptic membrane, then curare should reduce the frequency of the occurrence but not the size of an elementary event. Thus a should remain the same but ΔV should be less in the presence of curare for a given dose of ACh. This was observed. Doses of curare which reduce the mepps to 1/10 their value had little effect on the noise or on the calculated value of a.

So far we have considered only the size of the elementary events, calculated from the voltage noise under the assumption of a specific shape of the elementary event. Can the shape of the elementary event be deduced from the noise? As we have previously pointed out consistency but not uniqueness may be shown; the assumption of an exponential shape predicts the spectrum, however, the same spectrum can be obtained from other assumptions (Sections 78 and 79). For random events of shape $ae^{-t/\theta}$, the spectrum is

$$S_V = \frac{2a\,\Delta V\,\theta}{1 + \omega^2\theta^2}$$

Measured spectra have this shape (Figure 90.2).

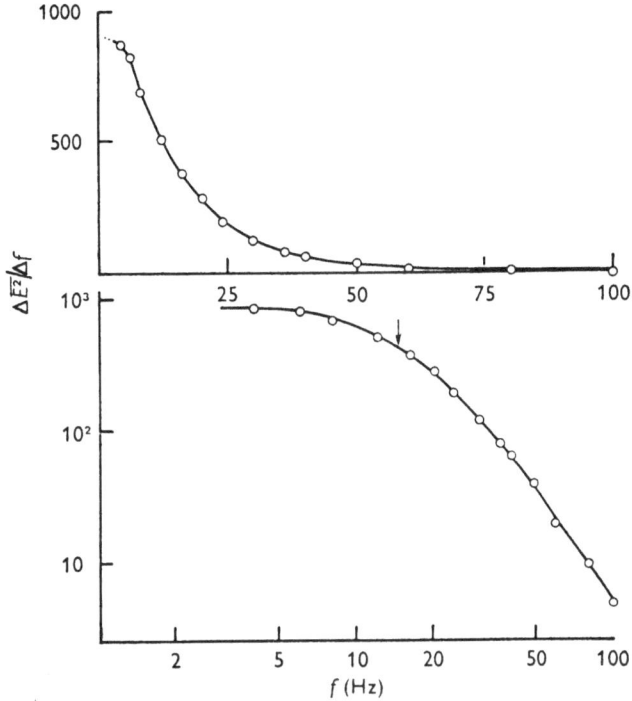

Figure 90.2. Voltage noise spectral density of intracellularly recorded ACh noise. From Katz and Miledi (1972). Temp. 25°C. A linear plot (top) is compared to a double logarithmic plot (bottom). The shape of S_V did not vary with the dose of ACh. The average half-power frequency was 14 Hz for T between 1.5 and 25°C. 14 Hz implies $\theta = 11.5$ msec. The vertical scale is relative.

PROBLEM.   The half-power frequency of the ACh noise (11.5 msec) is close to the RC time constant of the membrane (10 msec). What does the elementary event $ae^{-t/\tau}$ represent? Draw an equivalent circuit of a single ACh channel in parallel with an RC membrane with rest potential E. Let the bias voltage of the ACh channel be zero. The channel is suddenly turned on at $t = 0$ and turned off at $t = T$. What is the voltage change across RC? What is the expected spectral density of a random sum of such voltage changes?

PROBLEM.   Assume that the shape of the elementary conductance change due to the ACh–receptor interaction has the form

$$g(t) = \gamma \frac{t}{\theta} e^{-t/\theta}$$

Plot $g(t)$ and calculate the spectral density of the conductance fluctuations due to a random sum of these events. Calculate the spectral density of the voltage fluctuations that would be observed from a muscle fiber membrane modeled as a parallel RC circuit.

After 2 to 6 weeks of chronic denervation the half-power frequency of the voltage spectral density increased considerably. In other words, the ACh fluctuations were much slower than in normal fibers. This was interpreted as a prolongation of the ionic conductance channel formed by the ACh–receptor interaction.

It was evident from previous work that intracellularly recorded mepp's are limited in their time course by the RC time constant of the membrane (Figure 89.3). In the same way, intracellularly recorded noise is expected to be slower than extracellularly recorded noise. To find the true time course of the current fluctuations, and hence the conductance fluctuations, Katz and Miledi used a focal noise recording technique in which a voltage electrode is pressed into the surface of the membrane. The recorded voltage represents current flowing across the shunt between the surface electrode and the membrane. Because the shunt is not well characterized, the absolute scale of the current noise measured in this way is unknown. However, by this method the time course of the current fluctuations is undistorted by the membrane capacitor. The extracellular ACh noise had half-power frequencies in the range 100–180 Hz, about 10 times that for internally recorded noise. Thus the relaxation time of the elementary event underlying the voltage fluctuations is actually on the order of 1 msec. Katz and Miledi (1972) state that for

> ... membrane potential noise, the assumption of an approximately exponential shape is reasonable; to make a similar assumption for the underlying elementary current pulse seems more arbitrary and, indeed, unlikely. The opening and closing of an individual ionic channel is more probably a sudden on off event, but it might be argued that if the duration of individual "on" states varies in a random fashion, the final result might approximate to that for an exponential shape.

This quotation negates the oversimplification that Katz and Miledi in 1972 viewed the elementary conductance events as exponential. Clearly one cannot tell since the noise spectra are ambiguous. The random switch model of conductance channels, for which there is now direct evidence, was considered in these early experiments as likely to be the correct model.

PROBLEM. Draw an equivalent circuit which illustrates the focal extra-cellular noise analysis technique. By a circuit analysis show that the current measured extracellularly is undistorted by membrane capacitance.

A comparative study of ACh and carbachol noise recorded extracel-lularly revealed that the carbachol noise is faster. This is shown in Figure 90.3. The conclusion drawn from the spectral densities of Carb versus ACh noise is that the average conductance change induced by carbachol is considerably more brief than that produced by ACh. The carbachol channel stays open 0.3–0.4 msec compared to the 1-msec channel activated by ACh. This makes carbachol less effective.

PROBLEM. What would be the effect of carbachol on intracellularly recorded voltage noise? In particular what would be the effect on the value calculated for the size of the elementary voltage event?

To summarize, the elementary ACh event is on the order of 0.3 μV, about one-thousandth the size of a mepp. This alone is sufficient to obtain a rough estimate of the elementary conductance event. Since a mepp cor-responds to a 0.1 μS (microsiemens) change, the noise is assumed to derive from many random events each approximately 0.1 nS in size. The elemen-

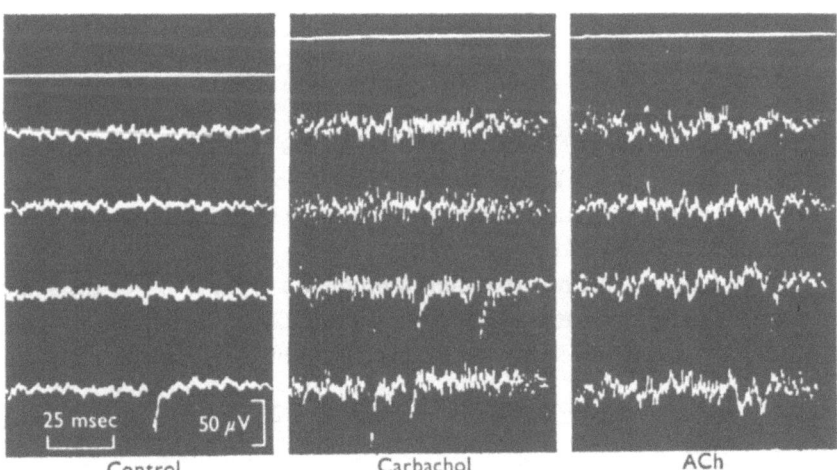

Figure 90.3. Sample records of focal extracellular noise during carbachol (Carb) and ACh applications. From Katz and Miledi (1972). Membrane potential was measured intracellularly (top traces). The depolarization in both cases was 5 mV compared to the control. External mepp's are also seen (transient downward deflections).

tary current can be calculated at a particular driving force, e.g., 75 mV implies that 4.7×10^4 ions pass in 1 msec; 1 msec is the approximate duration of an elementary event. These calculations are

$$i = 75 \text{ mV} \times 0.1 \text{ nS} = 7.5 \text{ pA}$$

$$q = 7.5 \text{ pA} \times 1 \text{ msec} = 7.5 \times 10^{-15} \text{ C}$$

$$\text{number} = \frac{7.5 \times 10^{-15} \text{ C}}{1.6 \times 10^{-19} \text{ C/ion}} = 4.7 \times 10^4 \text{ ions}$$

Direct measurements of flux rate from vesiculated particles containing the receptor had given vastly different values. Kasai and Changeaux (1971) calculated a flux rate of 5×10^3 Na ions/minute and an elementary conductance of 10^{-15} S. Also, the noise study suggested that about 1000 ionic gates are opened during the normal release of the packaged transmitter substance. Earlier estimates of the number of ACh molecules per quantal unit, made by direct chemical assay, were on the order of 50,000.

The technique of measuring noise by extracellular microelectrodes pressing down on the cell surface is an extremely powerful one. Schuetze *et al.* (1978) have perfected the technique to record ACh noise from chick myotubules during innervation. The procedure is simple and was introduced by Mauro and Schuetze into the Winter Course in Neurobiology at Woods Hole in 1979.

91. Other Types of Chemically Induced Noise

Katz and Miledi (1973) extended their studies to decamethonum (C10) acetylthiocholine (ATCh) and suberylcholine (SubCh) and compared their effects to acetylcholine and carbachol. All of these substances are depolarizing agents at the frog end plate.

More C10 than ACh is required to produce a given ΔV. For the same ΔV, much smaller fluctuations are associated with C10 noise than ACh noise. At 21°C, the mean values for the elementary event for frog end plate are

$$a_{C10} = 0.054 \ \mu V, \qquad a_{ACh} = 0.28 \ \mu V$$

or a mean ratio of about five to one. At 4°C the corresponding values are

$$a_{C10} = 0.17 \ \mu V, \qquad a_{ACh} = 0.54 \ \mu V$$

or a mean ratio of about three to one. Spectral analysis of extracellular

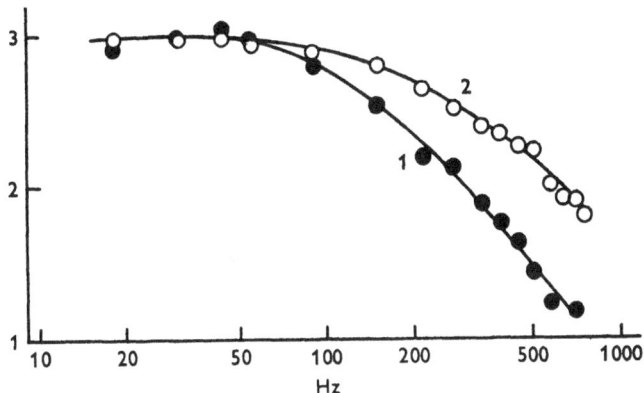

Figure 91.1. Intracellular voltage noise from rat diaphragm end plate. From Katz and Miledi (1973). Curve 1 is ACh-induced noise. Curve 2 is C10-induced noise. Temperature = 24°C. The two curves are normalized to a common maximum.

noise showed that C10 was similar to Carb, both agonists producing more brief conductance channels than ACh. However, the C10 is even less effective that Carb in that the size of the elementary even, as well as the mean duration, is smaller in C10 than in ACh. This conclusion was also established in rat diaphragm. In rat diaphragm, the intracellularly recorded mepp's have a much quicker exponential decay than in the frog, similar to that of extracellularly recorded mepp's from the frog end plate. Thus spectral analysis of the voltage noise gives higher cutoff frequencies. An example is shown in Figure 91.1.

In rat diaphragm, the mean values of the elementary voltage event are

$$a_{C10} = 0.5 \ \mu V, \qquad a_{ACh} = 0.71 \ \mu V$$

In frog end plate, ATCh produces much shorter duration conductance channels (0.12 msec) than ACh, and SubCh produces longer duration conductance channels (0.64 msec) than ACh. These comparisons were also made for noise recorded extracellularly.

Katz and Miledi considered the drug–receptor reaction as a reversible two-stage process:

$$A + R \underset{k_{-1}}{\overset{k_1}{\rightleftarrows}} AR \underset{k_{-2}}{\overset{k_2}{\rightleftarrows}} AR'$$

where A is the drug concentration, R the receptor concentration, AR an intermediate compound, and AR' the active compound which produces the elementary conductance change. The assumption they made in relating this kinetic scheme to the noise analysis is that the measured value of θ (from extracellular noise, or noise corrected for the time constant of the membrane if recorded intracellularly) is equal to $1/k_{-2}$:

$$\theta = 1/k_{-2}$$

The smaller θ, the less stable the active compound AR' and the less potent the agonist.

PROBLEM. How would a greater stability of the inactive compound AR influence the spectral density of the extracellular noise?

PROBLEM. Show that the assumption $\theta = 1/k_{-2}$ is equivalent to assuming $p \ll 1$ in the two-state model of Section 72. Compare this assumption to Anderson and Stevens, Section 92.

Suggested Reading. For additional information on ACh noise, reaction kinetics of chemically induced noise and the effect of local anesthetics see Sachs and Lecar (1973, 1977), Colquhoun (1975), Peper *et al.* (1975), Dreyer and Peper (1975), Dreyer *et al.* (1975, 1976a,b, 1977), Ruff (1976, 1977), Colquhoun and Hawkes (1977), Colquhoun *et al.* (1977), Neher and Steinbach (1978), and Noma and Trautwein (1978). For studies on human derived tissue see Cull-Candy, Miledi and Trautmann (1978, 1979a,b). For GABA and glycine noise in cultured neurons see Baker and McBurney (1978) and McBurney and Baker (1978).

92. ACh Noise under Voltage Clamp

The first direct measurement of ACh current noise under voltage clamp was published in 1973 by Anderson and Stevens. The measurements were made on the sartorius nerve–muscle preparation of *Rana pipiens*. Excitation–contraction coupling was disrupted by hypertonic treatment with either glycerol or ethylene glycol. The latter was preferred because it gave larger resting potentials and input resistances. High Ca^{2+} and Mg^{2+} ringer and low pH (6.2–6.8) sometimes produced similar advantages.

Two glass microelectrodes filled with 2.5 M KCl were used to impale the muscle cell. Electrodes were kept 50 μm apart, one for passing current and the other for measuring potential. The electrodes were low impedance (1–5 MΩ) and must have been fairly large. A third microelectrode was placed nearby to iontophorese 2.5 M ACh extracellularly. A successful penetration gave spontaneous miniature end-plate currents (mepc) and evoked end-plate currents (epc) under voltage clamp. These were used as criteria for positioning the microelectrodes.

Before developing their voltage clamp experiments, Anderson and Stevens calculated the expected current noise spectrum from the measured voltage noise and impedance. The impedance was measured by injecting white current noise into the muscle fiber and measuring the resulting voltage noise spectrum, $\tilde{S}_V$. Under the assumption that the $\tilde{S}_V$ is sufficiently above the inherent voltage noise S_V, and still in the linear range of the membrane, the square modulus of the impedance of the muscle cell has the shape of $\tilde{S}_V$. Let the input current noise density be given by $\tilde{S}_I$; then (see Sections 35 and 38)

$$| Z |^2 = \tilde{S}_V / \tilde{S}_I = \tilde{S}_V \Delta f / I^2$$

where Δf is the frequency band over which $\tilde{S}_I$ is white and I^2 is the mean square value of the input current noise. The current noise spectrum is then given by

$$S_I = \frac{S_V}{| Z |^2} = \frac{S_V I^2}{\tilde{S}_V \Delta f}$$

where every quantity on the right-hand side is either measured or known. This method gave quantitative agreement with direct measurement of S_I under voltage clamping; however, its accuracy was limited and it was eventually dropped.

In the absence of ACh applied in the external bath, the background noise consisted of membrane $1/f$ noise below 200 Hz (Chapter 5, Section B). Above 200 Hz, most of the background noise was from the instruments. The membrane $1/f$ noise in the frog muscle had a reversal potential of about -10 mV. This suggested a contribution from a number of ionic components since none of the ionic equilibrium potentials are near this value. $1/f$ noise was not confined to the end-plate region, but occurred anywhere along the muscle fiber (see also DeFelice and Adair, 1973). Curious "humps" in the current spectra were reported at hyperpolarizing potentials up to -120 mV (rest $= -50$ mV). These spectral components, when separated from a $1/f$ component, had roughly the shape of a Lorentzian. It was suggested that these Lorentzian components might represent conductance fluctuations of the type predicted by Stevens (1972) for K channels in the excitable nerve, although the suggestion was not pursued. The voltage electrode noise and the amplifier noise are added to the membrane voltage and act as command voltages for the voltage clamp system.

PROBLEM.   Assume the muscle fiber is represented by a long RC cable with the end plate situated at the midpoint of the cable. Show that the input impedance is given by

$$Z = \tfrac{1}{2}[r''r_m/(1 + i\omega\tau)]^{1/2}$$

where r'' (in units of Ω/cm) is the resistance per unit length along the cable interior, r_m (in units of Ω cm) is the membrane resistance encountered per unit length along the cable, and $\tau = r_m c_m$, where c_m (in F/cm) is the membrane capacitance encountered per unit length along the cable (see Sections 28 and 29). Calculate $S_I = S_e/|Z|^2$ assuming reasonable values of r'', r_m, and c_m for the muscle fiber, and assuming $S_e = 10^{-13}$ V²/Hz. Plot S_I against f on a log–log scale.

Anderson and Stevens subtract background noise spectra from all the ACh-noise spectra; the basic assumption behind these different spectra is that the background noise and the ACh-induced noise are uncorrelated. For large ACh-induced currents, ACh noise is well above background. This is shown in Fig. 92.1. For a mean ACh current of about 100 nA, the ACh noise is about 1 nA peak-to-peak. This is also the size of a spontaneous mepc

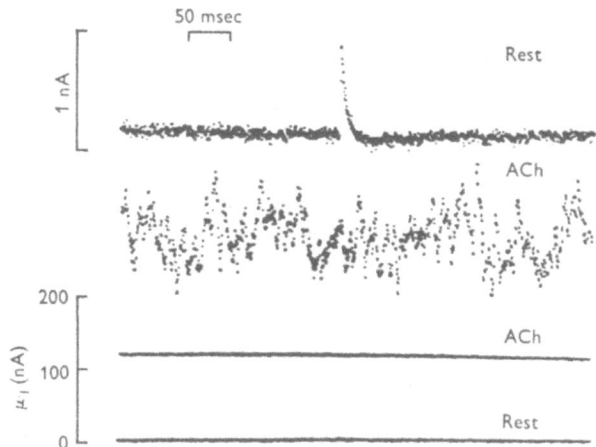

Figure 92.1. Current noise under voltage clamp from the postsynaptic membrane of the frog neuromuscular junction. Top trace: background noise (no ACh added) at rest. A spontaneous miniature end-plate current (mepc) is seen. Middle trace: ACh-induced noise. Bottom trace: the mean ACh-induced current corresponding to the ACh noise in the middle trace. Holding potential $V = -100$ mV. Temperature 8°C. In this experiment, rms noise increased from 0.07 nA at rest to 0.25 nA at -100 mV (1–500-Hz bandwidth). Note that the ACh noise is roughly the same amplitude as the mepc in this experiment. (The traces are spotty because the time sweep is digital.) Anderson and Stevens (1973).

(upper trace in Figure 92.1). The background noise is taken as the current at rest before the addition of ACh.

In Figure 92.1, $\mu_I = I$ is the mean current at a particular ACh concentration and holding potential. I is related to the mean membrane conductance by (Section 73)

$$I = G(V - E)$$

where

$$G = \gamma p N$$

and E is the reversal potential for the ACh current. Experimentally, $E \simeq 0$. Let σ_I^2 be the variance of the ACh noise at a particular I and σ_G^2 be the variance of the ACh-induced conductance fluctuations. From Section 73,

$$\sigma_G^2 = \frac{\sigma_I^2}{(V - E)^2}$$

For the two-state channel model,

$$\sigma_G^2 = \gamma^2 N p(1 - p) = \gamma G(1 - p)$$

Assume that the channels are nearly always closed, i.e.,

$$p \ll 1$$

In this case

$$\sigma_G^2 \simeq \gamma G$$

PROBLEM.   Anderson and Stevens argue that for appropriately small ACh concentrations, p is small compared to one (see Stevens, 1972, Equation 20). Compare this assumption with two models of ACh–receptor interaction. First assume that a receptor binds ACh and that p is the probability of the open configuration of this complex. Second, assume that p is the probability of an encounter between an ACh molecule and a receptor; when ACh jumps on the receptor the channel is open, when ACh jumps off the receptor the channel is closed. What does N represent in each of these models? To which does the assumption that small ACh concentrations imply $p \ll 1$ apply?

For G between 0 and 4 μS, and σ_G^2 between 0 and 10^{-16} S², σ_G^2 is proportional to G with a mean slope of 20.5 pS. Desensitization (the gradual repolarization during constant ACh application) did not affect the results.

Both G and $\sigma_G{}^2$ decrease proportionally. The spectral density of the ACh noise scaled down during desensitization, but was undistorted on the frequency axis. The spectral density of ACh noise is well fit by a single Lorentzian.

The spectral density of the noise depends on membrane voltage and temperature. This is shown in Figure 92.2: At constant temperature (a), depolarizing the membrane increases the cutoff frequency of the Lorentzian. At constant membrane potential (b), increasing the temperature increases the cutoff frequency of the Lorentzian. For both the voltage and the temperature effects shown in Figure 92.2, the change in the spectra is in the direction that tends to leave the variance (the area under the spectral density) constant. That is, when the cutoff frequency shifts to the right, the amplitude goes down. This is a property of the functional form

$$\frac{\theta}{1 + \omega^2\theta^2} = \frac{1/2\pi f_\theta}{1 + f/f_\theta}$$

where θ (or f_θ) depends on the parameter being changed (voltage or temperature.)

The end-plate current noise spectra were fit by the formula (double-sided spectral density; see Section 44)

$$SS(f) = \frac{2\gamma I(V - E)\theta}{1 + \omega^2\theta^2}$$

where $I \equiv \mu_I$ is the mean current induced by ACh, V is the holding potential, and E is the reversal potential for the current. The solid curves in Figure 92.2 are described by this equation. γ is interpreted as the conductance of a single open channel.

PROBLEM. Show that the above formula for $SS(f)$ follows from the two-state channel model developed in Sections 72 and 73 if $p \ll 1$. Note that Anderson and Stevens use α and β opposite to the way they are used in this book.

It was found that θ depends on voltage according to the relation

$$\theta = be^{-aV}$$

where b and a are positive constants. V is with respect to the outside bath. Depolarization leads to shorter θ. From Section 72,

$$\theta = \frac{1}{\alpha + \beta}$$

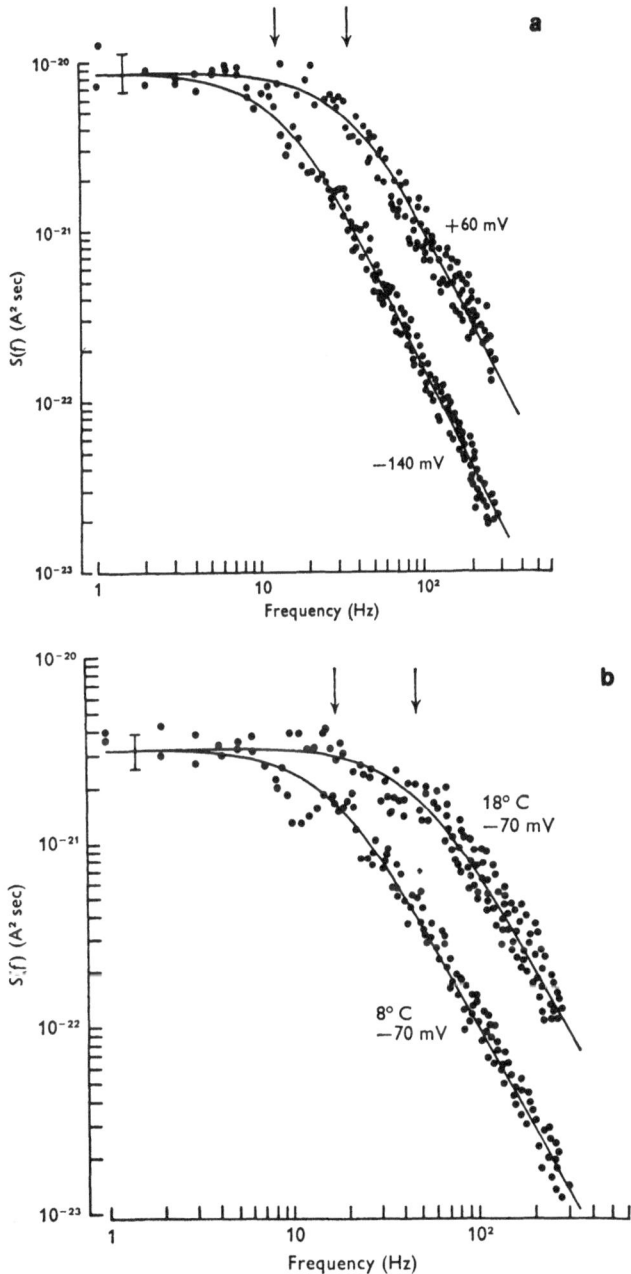

Figure 92.2. (a) The effect of membrane potential on ACh current noise spectra at 8°C. The 60-mV curve has been shifted up. Multiply by 0.008 to obtain the correct scale for the 60-mV spectrum. The cutoff frequencies that fit the two curves are $f_\theta = 34$ Hz for 60 mV and $f_\theta = 12$ Hz for -140 mV. (b) The effect of temperature on ACh current noise spectra at -70 mV. The 18°C curve has been shifted up. Multiply by 0.36 to obtain the correct scale for the 18°C spectrum. The cutoff frequencies are $f_c = 18$ Hz for the 8°C and $f_\theta = 49$ Hz for 18°C. Anderson and Stevens (1973).

Anderson and Stevens use α as the closing rate constant and use β as the opening rate constant—opposite to our notation. Their assumption that $p \ll 1$ implies a channel is open very little and therefore $\beta \ll \alpha$. In this case (their notation)

$$\frac{1}{\theta} \simeq \alpha \qquad (p \ll 1)$$

Therefore

$$\alpha = \frac{1}{b} e^{aV}$$

Depolarization leads to larger α, as shown in Figure 92.3. Figure 92.3 also

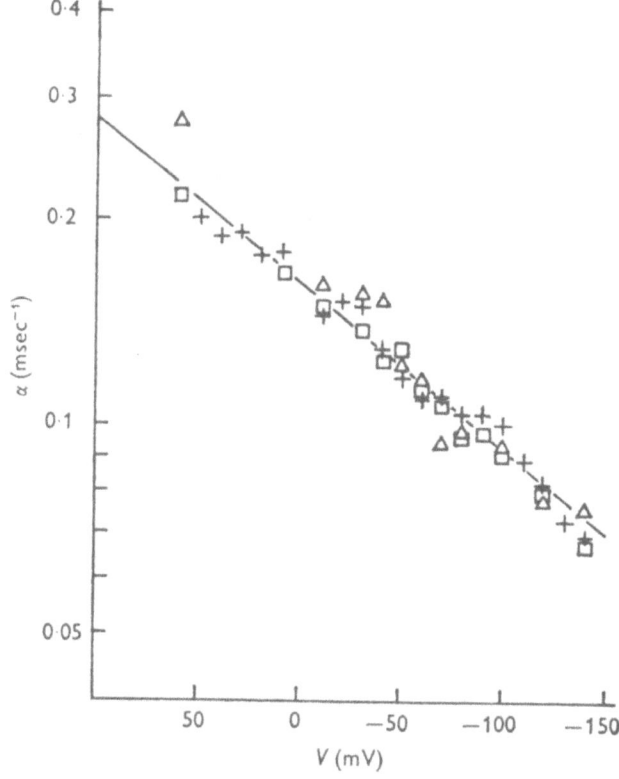

Figure 92.3. The dependence of the rate constant α on voltage at constant temperature of 8°C. ACh ($\square$) refers to estimates of α from noise measurements ($\alpha = 1/\theta$). E.p.c. ($+$) refers to nerve evoked end-plate currents, and M.e.p.c. ($\triangle$) to spontaneous miniatures. All obey the relationship (solid line) $\alpha = 0.17 \exp(0.0059V) \, \mathrm{msec}^{-1}$ (where V is in millivolts). Note that α is used oppositely to our notation in Section 72. Anderson and Stevens (1973).

shows that the time constant associated with the decay phase of spontaneous mepc's and epc's have the same voltage dependence as the noise time constant. The macroscopic and microscopic phenomena obey the same law.

The value of the open channel conductance γ is constant over the range of membrane potentials studied. Calculations from $SS(0)$ from the best experiments gave a mean value near 32 pS. The mean overall experiments was nearer 20 pS. The dependence of γ on temperature was not significant.

PROBLEM. Compare the mean open time of a channel at -140 mV to the mean open time at $+60$ mV at 8°C. ($p \ll 1$) Compare the mean open time of a channel at 8°C to 18°C at -70 mV. How many channels would have to be at the end plate to give a total possible conductance of 1% of the observed peak conductance of 4 μS?

93. Other Types of Chemically Induced Noise under Voltage Clamp

Colquhoun, Dionne, Steinbach, and Stevens (1975) have compared the effect of ACh and three other agonists on the ACh receptor under voltage clamp. The four drugs are shown in Figure 93.1. To analyze the noise, the same model described in Section 92 was used. Again, it was assumed that the probability of a channel being open is small, i.e., $p \ll 1$. It was found that all four of the drugs in Figure 93.1 induce different open-channel conductances that have different mean open-channel lifetimes under the same conditions of voltage and temperature. The mean single-channel conductance was estimated from the formula

$$\gamma = \frac{\sigma_I^2}{I(V - E)}$$

where σ_I^2 is the current noise variance between 1 and 500 Hz, I is the mean induced current, and E is the reversal potential for the current. Separate measurements showed that $E \simeq 0$ for all the drugs shown in Figure 93.1. The variance was measured between -60 and -80 mV and at 10 to 15°C. (Does this spread of values affect the comparison made below?) TTX was present to block membrane action potentials. The results are summarized in Table 93.1, reproduced from Colquhoun *et al.* (1975). The number in parentheses is the number of observations.

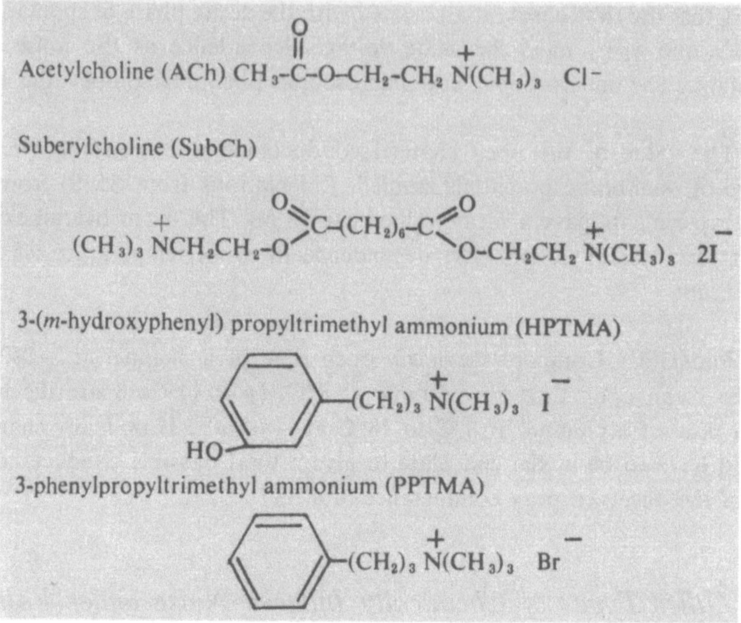

Acetylcholine (ACh) $CH_3-\overset{\overset{O}{\|}}{C}-O-CH_2-CH_2 \overset{+}{N}(CH_3)_3$ Cl^-

Suberylcholine (SubCh)

$(CH_3)_3 \overset{+}{N}CH_2CH_2-O-\overset{O}{\overset{\|}{C}}-(CH_2)_6-\overset{O}{\overset{\|}{C}}-O-CH_2CH_2 \overset{+}{N}(CH_3)_3$ $2I^-$

3-(*m*-hydroxyphenyl) propyltrimethyl ammonium (HPTMA)

$-(CH_2)_3 \overset{+}{N}(CH_3)_3$ I^-

HO–

3-phenylpropyltrimethyl ammonium (PPTMA)

$-(CH_2)_3 \overset{+}{N}(CH_3)_3$ Br^-

Figure 93.1. Drugs applied to *Rana* end-plate preparations to study induced current noise under voltage clamp. From Colquhoun *et al.* (1975).

The variance was taken as the integral of the current noise spectral density. ACh and SubCh spectra fit well to a single Lorentzian. The other two agonists, however, fit a two-Lorentzian model. The lifetimes given in Table 93.1 are from the higher frequency component. Compare the results for ACh and SubCh with those found by Katz and Miledi (1973), Section 91.

TABLE 93.1. *Single-Channel Properties of the Agonists Shown in Figure 93.1*

Drug	γ (pS)	θ (msec)
SubCh	$28.6 \pm 1.0(22)$	$5.6 \pm 0.3(8)$
ACh	$25.0 \pm 0.9(8)$	$3.2 \pm 0.3(6)$
HPTMA	$18.8 \pm 0.8(19)$	$1.0 \pm 0.06(13)$
PPTMA	$12.8 \pm 1.1(7)$	$0.83 \pm 0.02(3)$

PROBLEM. Assume there are two different types of channel present, each giving a Lorentzian noise spectrum. The two species have separate time constants τ_1 and τ_2 but the same $\gamma = \gamma_1 = \gamma_2$. Show that the noise variance of the total spectrum gives this value of γ.

PROBLEM. Calculate the average number of ions that flow through an open ACh channel at about -70 mV for the four agonists of Table 93.1. At -100 mV and -40 mV, do you expect these numbers to increase or decrease?

PROBLEM. Show that the formula $\gamma = \sigma_I^2/I(V - E)$ follows from the two-state channel model (Sections 72 and 73) if $p \ll 1$.

94. Effect of Procaine on ACh Noise

Katz and Miledi (1975) studied the effect of procaine on the action of ACh at the frog end plate. Of the two extreme forms of antagonist, those that block the access of ACh to the receptor and those that change the kinetics of the ACh–receptor interaction, procaine is an example of the latter.

Procaine prolongs the end-plate response (after an initial fast repolarization phase) much beyond the normal duration of the end-plate potential. The end-plate response is reduced to a few millivolts in the presence of procaine and the spontaneous mepp's are hardly discernible. Not surprisingly, the ACh-induced noise is also greatly diminished. The relative size of the elementary event is reduced by about 20% in procaine. (Katz and Miledi now use $a = \sigma^2/\Delta V$ rather than $2\sigma^2/\Delta V$ to estimate the size of the elementary voltage event, depreciating the shape of the event because of lack of specific information about it, see Section 90.) More importantly, the shape of the spectrum of the fluctuations no longer corresponds to that expected for simple exponential events, or on–off events whose intervals are exponentially distributed. Rather, the spectra are complex and may contain several components. The intracellular noise spectrum is shown in Figure 94.1. Notice that the frequency scale extends below that in previous studies (Sections 90 and 91).

The extracellular focal recordings of both the noise and the end-plate potentials are more revealing. The end-plate current after exposure to procaine can show an initial fast phase followed by a slow repolarization extending out to 100 msec. The exact shape of the fast phase depends on the

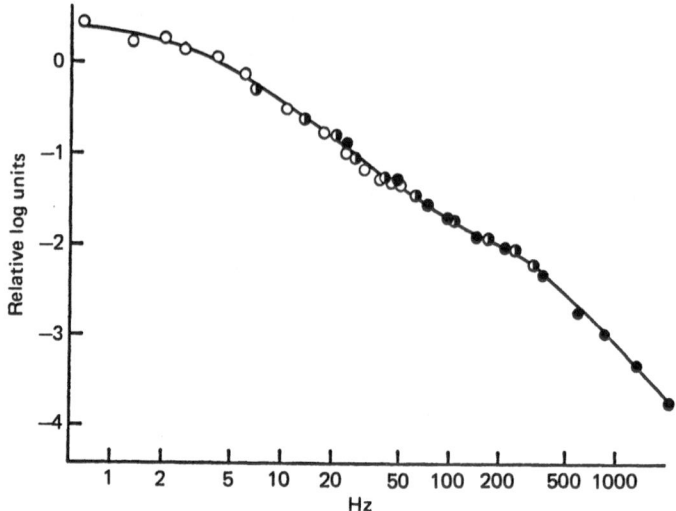

Figure 94.1. Intracellular ACh-noise spectrum from frog end plate in 0.37 mmol procaine. Temperature = 5°C. Symbols refer to separate segments of computation of the spectrum. From Katz and Miledi (1975).

effective summation of the quantal extracellular mepp's. Those mepp's occurring spontaneously also have a fast initial phase and a slow tail, although individual events are barely visible. In spite of the low noise intensity, noise spectra could be taken. Figure 94.2 contrasts control ACh noise and ACh noise in the presence of procaine.

It is evident that the effect of procaine is to introduce higher frequency components in the spectrum. These correspond to the shorter relaxation times of 0.3-0.5 msec. At this temperature, the ACh noise has a relaxation 10 to 20 times longer in the absence of procaine. Values of $\sigma_V{}^2/\Delta V$ up to 0.45 μV were obtained. These noise data are expected from the fast initial phase in the mepp's in the procaine-treated end plate. The conclusion is that the effect of procaine is to shorten the action of the transmitter, and not to reduce its peak activity. Although there was evidence for a dual time course for the macroscopic events, and for the noise, several questions remained unanswerable. For example, noise analysis will not allow discrimination between dual kinetics for each channel or two populations of channels with different kinetics. The former view would imply "that the closure of channels, at least after procaine, is not a simple 'switching' process, with the conductance moving sharply between two fixed levels." (Katz and Miledi, 1975). Such a statement is probably impossible to make from noise analysis and requires observation of discrete single events (Section 97)

PROBLEM.   Assume that each molecular event is given by the relation

$$f(t) = a_1 e^{-t/\theta_1} + a_2 e^{-t/\theta_2}$$

Show from Campbell's theorem (Chapter 5, Section D) that the expected mean voltage from the random sum of such events is given by

$$V = \nu(a_1\theta_1 + a_2\theta_2) = V_1 + V_2$$

and

$$\sigma^2 = \tfrac{1}{2}\nu(a_1{}^2\theta_1 + a_2{}^2\theta_2 + 4a_1a_2\theta')$$

where $\theta' = \theta_1\theta_2/(\theta_1 + \theta_2)$ is the parallel combination of the time constants. Show also that

$$S(0) = 2\nu\,(a_1\theta_1 + a_2\theta_2)^2 = \frac{2}{\nu}\,(V_1 + V_1)^2$$

where $S(0)$ is the low-frequency asymptote of the spectral density of the random sum. What is $S(f)$ for this case?

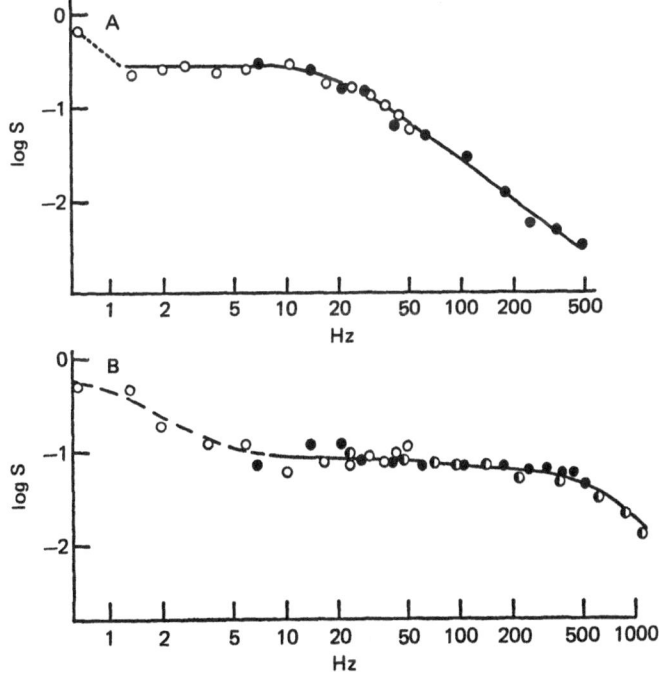

Figure 94.2. Frog end-plate extracellularly recorded ACh-induced noise (top) and ACh-induced noise in the presence of 0.37 mmol procaine (bottom). Temperature: 9°C. From Katz and Miledi (1975).

PROBLEM.   Assume that each molecular event is given by a random switch, but that there are two populations of switches with average time constants θ_1 and θ_2 and amplitudes a_1 and a_2. Calculate V, σ^2, and $S(f)$. Compare these with the same quantities in the problem above.

95. Effect of Dithiothreitol on ACh Noise

Ben-Haim, Dreyer, and Peper (1975) studied the effects of modifying the ACh receptor. The experiments were done on normal *cutaneus pectoris* muscle of *Rana esculenta* under voltage clamp. ACh was released from an electrode 50–100 μm above the end plate and current noise from the postsynaptic junction was measured before and after the modifying agent was introduced.

The modifying drug was dithiothreitol (DTT) which is a specific reducing agent. DTT breaks the disulfide bond located several angstroms from the anionic site on the ACh receptor. The reduction is reversible.

The effect of DTT is to decrease γ. The normal value of γ in these preparations was 18.5 pS. This was reduced by more than half for the reduced ACh receptor. The mean channel time constant τ was also reduced by DTT. However, the normal value obtained for τ was 3–4 times smaller than reported by Anderson and Stevens (1973).

Ben-Haim *et al.* (1975) confirmed the voltage noise studies of Landau and Ben-Haim (1974). They also showed from dose–response curves that one effect of DTT is to reduce the affinity of ACh to the receptors. The peak end-plate conductance was measured against the iontophoretic charge released from the ACh pipette. One millimole of DTT decreased the conductance by about 3 for the same amount of released ACh. This was interpreted as a decrease in γ with DTT with no change in the number of ionic channels activated by ACh.

96. Current Noise from Denervated Skeletal Muscle

Neher and Sakmann (1976b) voltage clamped denervated frog muscle fibers and used the autocorrelation function to analyze ACh-, Carb-, and SubCh-induced noise. Denervated and normal muscle from *Rana esculenta* and *temporaria* was used. Frogs were kept 40–70 days after denervation of the *cutaneus pectoris* before an experiment. TTX was added to a normal Ringer's solution whose temperature was kept at 8 or 18°C. Former end-

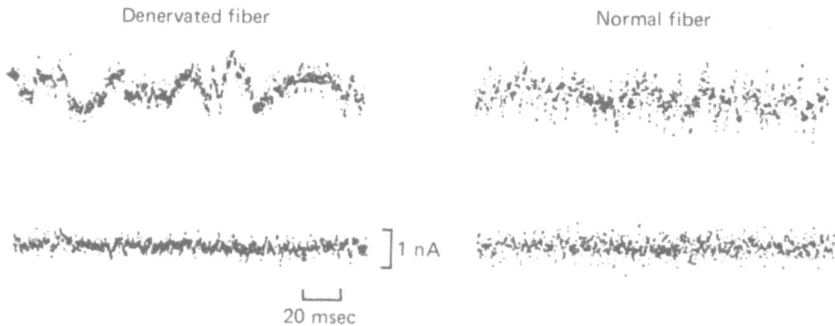

Figure 96.1. Left side: ACh-induced noise from an extrajunctional region (upper trace) compared with the background noise (lower trace) before the drug was applied. Right side: ACh-induced noise from the end-plate region of a normal fiber. In both experiments, the mean ACh-induced plateau current was $I = 40$ nA. Temperature: 18°C. Holding potential: -80 mV. The traces are digitized records of filtered data (0.5–2500 Hz). Neher and Sakmann (1976b).

plate regions were identified before an experiment by ACh sensitivity and after an experiment by AChE staining. Extrajunctional recordings were defined as recordings at least 0.3 mm from the former end-plate regions. Intact end-plates were identified by the rise time of spontaneous miniature end-plate currents, ACh sensitivity, and anatomical features.

A standard two-microelectrode clamp was used. The electrodes were kept 40–60 μm apart. K_2SO_4-filled microelectrodes were preferred for current injection; the holding currents before and after drug application were more constant than for KCl-filled microelectrodes. ACh, Carb, and SubCh released iontophoretically 20–30 μm from the muscle fibers. The drug-induced plateau currents were kept to 30–70 nA. Plateau currents lasted only 10–25 sec owing to desensitization, and sometimes repeated applications of the drugs were necessary. The autocorrelation of the drug-induced noise was taken as the difference between the correlation functions before and after the application of the drug.

Figure 96.1 shows the effect of ACh on the current noise measured from the normal end-plate region in an intact fiber, and from an extrajunctional region of a denervated fiber.

The ACh-induced noise from extrajunctional regions of denervated fibers is similar in magnitude to normal end-plate noise but it is slower. The autocorrelation functions of both kinds of noise are well fitted by the expression

$$C_I(\tau) = \sigma_I{}^2 e^{-\tau/\theta}$$

This correlation function is expected from the two-state channel model of Section 72. $\sigma_I^2 = C_I(0)$ is the variance of the noise. The value of θ from the denervated fiber in Figure 2 is $\theta = 5.6$ msec; $\theta = 1.3$ msec for the normal fiber. θ increases at lower temperatures.

The value of θ from extrajunctional regions of denervated fibers depends on the applied drug. At 8°C and -80 mV holding potential:

$$\theta = 3.9 \pm 0.4 \text{ msec}, \quad \text{Carb}$$
$$\theta = 11 \pm 1.6 \text{ msec}, \quad \text{ACh}$$
$$\theta = 19 \pm 2.5 \text{ msec}, \quad \text{SubCh}$$

θ increases at hyperpolarizations.

The elementary current pulse for each drug–receptor interaction is calculated from the formula

$$i = \frac{\sigma_I^2}{I}$$

Single-channel conductance is calculated from (Section 73)

$$\gamma = \frac{i}{V - E}$$

where E is the reversal potential for the induced current I. E was measured from the instantaneous current–voltage relationship during a drug-induced plateau current of 25–45 nA. $E = 2.6 \pm 2.7$ mV was taken as zero in the calculations. The value of γ measured from extrajunctional regions was about the same for all three drugs. However, γ was smaller for extracellular regions of denervated fibers compared with end-plate regions of normal fibers:

$$\gamma = 15 \pm 1.8 \text{ pS}, \quad \text{extrajunctional, denervated}$$
$$\gamma = 23 \pm 2 \text{ pS}, \quad \text{end plate, normal}$$

PROBLEM.   Show that the formula $\sigma_I^2 = iI$ follows from the two-state channel model developed in Sections 72 and 73 for the case

$$p \ll 1$$

where p is the open-channel probability. How does this approximation affect the estimation of γ?

Neher and Sakmann also studied the properties of the former end-plate regions in denervated muscle fibers. In 11 out of 15 experiments, two time constants were needed to fit the autocorrelation functions of the drug-induced noise:

$$C_I(\tau) = \sigma_1^2 e^{-\tau/\theta_1} + \sigma_2^2 e^{-\tau/\theta_2}$$

Figure 96.2 shows the autocorrelation function of ACh-induced noise from a former end-plate region. $\phi(T)$ is identically $C_I(\tau)$.

PROBLEM. Assume 1.0 sec of noise data was used to calculate the correlation function shown in Figure 96.2. Estimate the error in the correlation function at $\tau = 0$ and large τ (see Section 55).

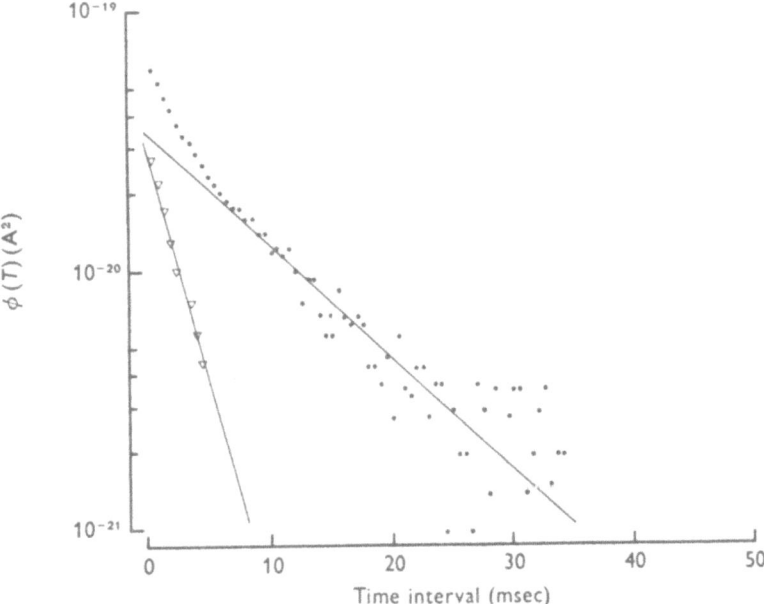

Figure 96.2. Autocorrelation function of ACh-induced current noise. 62-day denervated fiber. -80 mV, $8°C$, $I = 60$ nA. Triangles represent the difference between a straight line fitted to the slow component and the first points of the correlation function. The two ACh time constants extracted from Figure 96.2 are $\theta_1 = 2.6$ msec, $\theta_2 = 11.0$ msec. The variances for the slow and fast component were about equal. The fast time constant is about equal to that for end-plate regions of normal fibers. The slow time constant corresponds to the extrajunctional regions of denervated fibers. In the former end-plate regions, both θ_1 and θ_2 are smaller for Carb and larger for SubCh compared to ACh. When only a single time constant was found in a former end-plate region, it corresponded to either the normal value or to the extrajunctional denervated value. From Neher and Sakmann (1976b).

PROBLEM.   In Section 52 we showed that the correlation function of $1/f$ noise is

$$C(\tau) = \frac{Bf_2^2}{2(f_2^2 - f_1^2)} [E(2\pi\tau f_2) - E(2\pi\tau f_1)]$$

Use $f_1 = 0.5$ Hz and $f_2 = 2500$ Hz. Plot $C(\tau)$ on semilog paper using the values of $E(x)$ given in Table 53.1. Compare this plot with the data shown in Figure 96.2.

The assumptions used to derive the elementary conductance data are the same as those used by Anderson and Stevens (1973) (Section 92). The drug-induced membrane noise is due to the fluctuations of the drug–receptor complex between an open and a closed conductance state. The formation of the complex is assumed to occur much faster. The relations used require that only a small percentage of all channels are in the open state. The highest plateau currents used were about $I = 70$ nA at -80 mV. If $i = 10^{-19}/I \simeq 1.4$ pA is the ACh-induced current, then 70 nA/1.4 pA $\simeq 5 \times 10^4$ open channels are implied.

PROBLEM.   Show that the variance in the current noise is related to the mean current by (Section 73)

$$\sigma_I^2 = \frac{1}{N_0} I^2 \qquad (p \ll 1)$$

where N_0 is the number of parallel current gates through which the current passes. Compare this formula with the calculation.

PROBLEM.   Assume a muscle fiber is 30 µm in diameter and assume the upper half of a 200-µm length of the fiber contributes to I. Calculate the receptor density necessary if only 10% of the receptors are involved in a typical drug-induced current. Is this density reasonable?

97. Single-Channel Currents

Neher and Sakmann (1976a) used their denervated preparations to study single-channel currents; theirs was the first report of direct channel measurements from a biological membrane. The events were recorded from a micropipette pressed against a muscle fiber treated with enzymes to digest the connective tissue and basement membrane. The micropipette is a current electrode. The recording pipette contained the agonist for the receptor; the agonist is effectively isolated on the fiber membrane by the 3–5-μm pipette tip diameter. The membrane near the recording pipette was kept under voltage control by a conventional two-microelectrode voltage clamp circuit. The small area of membrane isolated by the tip increased the signal-to-noise ratio, and the long duration of the extrajunctional channels enhanced the chances of their detection. To increase open channel duration further, the membrane was hyperpolarized, low temperatures were used, and in most experiments SubCh was the agonist (Section 93).

PROBLEM.   Draw an equivalent circuit of the current electrode pressed against the muscle cell membrane. Include the shunt between the electrode and the membrane and estimate the resistance of the membrane area under the tip relative to the rest of the cell. What is meant by the claim that a small area of isolated membrane increased the signal-to-noise ratio (S/N)?

Figure 97.1 shows the results for the three drugs SubCh, ACh, and Carb. Current noise is shown under the voltage clamp for typical experiments. To demonstrate that the small rectangular events are related to the drug–receptor interaction, Neher and Sakmann show that the frequency of SubCh events is proportional to the membrane's sensitivity to the agonist (Figure 97.1, bottom left). The sensitivity was determined by mean change in membrane potential achieved for a given d of SubCh. All of the experiments were done on hypersensitive fibers. The first peak of the histogram (Figure 97.1, bottom right) is background noise with a standard deviation of less than 1 pA. The next three peaks are separated by 3.4 pA; the small peak near 10 pA is interpreted as the sum of three simultaneous elementary events.

One would expect the size of the elementary current pulse to be dependent on the membrane potential. The two values reported for SubCh channels are

$$i = 3.4 \text{ pA}, \quad -120 \text{ mV}$$
$$i = 2.2 \text{ pA}, \quad -80 \text{ mV}$$

These are consistent with the relation (Section 73)

$$i = \gamma(V - E)$$

when $\gamma = 28$ pS and $E = -7$ mV. This is in close agreement with the autocorrelation analysis of current noise measured by Neher and Sakmann (1976b), Section 96.

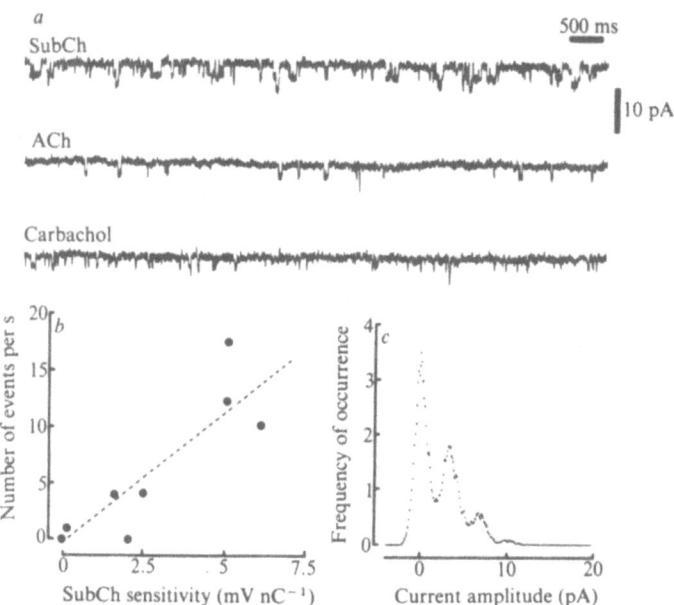

Figure 97.1. The top three traces show records of current noise from 10 μm² of extrajunctional denervated frog muscle fiber. The concentration of the agonist in the recording pipette is 2×10^{-7} M (SubCh), 2×10^{-6} M (ACh), and 6×10^{-5} M (Carb). The traces are pen recordings with an upper cutoff of 100 Hz. Holding potential: -120 mV. Temperature: 8°C. The bottom left graph plots the number of SubCh events versus the SubCh sensitivity in mV/nC. The bottom right graph is a histogram of SubCh events. Neher and Sakmann (1976a).

The mean open time for SubCh channels was measured directly from 40–50 individual events, and from power spectra of records like that shown in Figure 97.1. The two methods agree to within $\pm 30\%$; this may be the expected statistical error although the authors do not discuss the discrepancy. For SubCh events at 8°C,

$$\theta = 45 \pm 3 \text{ msec,} \qquad -120 \text{ mV}$$
$$\theta = 28 \pm 3 \text{ msec,} \qquad -80 \text{ mV}$$

These values are also in close agreement with those obtained from the autocorrelation analysis of millions of channels (Section 96). The expected change in θ due to different agonists was also found. This is seen qualitatively in Figure 97.1.

The records of individual events had a profound impact on membrane noise analysis. They supported the statistical approach that had been applied to membranes during the previous decade, and the hypothesis that channels have two conductance states.

A detailed technical description of the methods used to record single-channel currents is given in Neher, Sakmann and Steinbach (1978).

98. Ion Flow through the ACh Channel. A Noise Analysis

Dionne and Ruff (1977) used noise analysis to show that the current induced by ACh at the frog end plate flows through one type of channel even though the current is carried by more than one type of ion.

The current through an open channel is given by (Section 73)

$$i = \gamma(V - E)$$

where γ is the open-channel conductance and E is the reversal potential for that channel. The mean current and the variance in the current from many channels in parallel are given by the equations

$$I = pNi$$

$$\sigma_I^2 = p(p - 1)Ni^2$$

If one type of channel is involved, the equations are

$$I = pN\gamma(V - E)$$

$$\sigma_I^2 \simeq I\gamma(V - E) \qquad (p \ll 1)$$

If two types of channel are involved (with appropriately signed equilibrium potentials, e.g., $E_1 < 0$ and $E_2 > 0$) the equations are

$$I = p_1 N_1 \gamma_1 (V - E_1) + p_2 N_2 \gamma_2 (V - E_2)$$

$$\sigma_I^2 \simeq I_1 \gamma_1 (V - E_1) + I_2 \gamma_2 (V - E_2) = p_1 N_1 \gamma_1^2 (V - E_1)^2 + p_2 N_2 \gamma_2^2 (V - E_2)^2$$

σ_I^2 vs. V is approximately the sum of two parabolas with $(V - E)^2$ dependence. (This is only approximate because p is also a function of voltage.)

Consider the voltage V for which $I = 0$. This may occur for two cases: (1) Channels all of one type, $V = E_1 = E_2$ and I is zero in each channel. In this case, σ_I^2 is also zero. (2) Channels of two types, V such that $I_1 = -I_2$. In this case, current is flowing in the channels and σ_I^2 is finite.

PROBLEM.   Make a schematic drawing of I vs. V and σ_I^2 vs. V for the two cases discussed above. Although $I_1 + I_2 = 0$ may equal zero on average, a fluctuation in I exists owing to the random opening and closing of channels.

PROBLEM.   Draw a circuit equivalent with random switches to illustrate the two cases discussed above. Include a parallel pathway with a resistor and battery in series to represent the input resistance and resting potential in the absence of ACh. Give an expression for the closed-circuit current in each case.

Dionne and Ruff (1977) showed that when $I = 0$, $\sigma_I^2 = 0$. This implied that the channels in the endplate are all of one type. Specifically the endplate current is due to K and Na ions that move through the same pathway. At $I = 0$, inward Na balances outward K on average. This occurs at $V = -2$ mV, far from the equilibrium potential of either ion.

PROBLEM.   Does $I = 0$ at $V = -2$ mV imply that $\gamma_1 = \gamma_2$? Can σ_I^2 really be zero for any case?

The generation of noise from the movement of ions of different species has been considered previously in Section 70. Voltage noise from a KCl-filled glass microelectrode tip is approximately described by the sum of two parabolas centered on the equilibrium potentials for K and Cl. The two parabolas add to form a new parabola finite at every voltage. In this case, the noise-generating mechanism is the movement of ions through the aqueous environment in the microtip. K and Cl generate noise independently even though they move together on average. This is different than the case considered above, where moving together through the same channel eliminated the source of noise. Noise from the squid giant axon fits a model in which the noise is generated by K and Na moving through separate channels (Section 106). In this case, the variance would be approximately the sum of two parabolas centered on E_K and E_{Na}, for the very reason eliminated by the experiments of Dionne and Ruff for the ACh channel.

PROBLEM.   At the frog end plate, the actual relationships between the mean, the variance, and the voltage are

$$I = B(V - E)e^{AV}$$
$$\sigma_I^2 = \gamma B(V - E)^2 e^{AV}$$

Plot I and σ_I^2 vs. V for $B = 144$ nS, $A = -15.36$ V^{-1}, and $E = -2$ mV. These are the values obtained by Dionne and Ruff. (Note: Dionne and Ruff have nA units for B in the legend of their Figure 1.) From $I = pN\gamma(V - E)$, $B = pN\gamma \exp(AV)$, calculate γ for $\sigma_I^2 = 10^{-20}$ A^2 at $V = -40$ mV. Answer: $\gamma = 26$ pS. Calculate the average number of open channels at this voltage. Answer: 2996.

Assume N ACh/ACh-receptor complexes flip between an open and a closed state (Section 72); in our notation

$$N_c \underset{\beta}{\overset{\alpha}{\rightleftharpoons}} N_o$$

The probability of a channel being open is

$$p = \frac{\alpha}{\alpha + \beta} \simeq \frac{\alpha}{\beta} \quad (p \ll 1)$$

and the channel time constant is

$$\theta = \frac{1}{\alpha + \beta} \simeq \frac{1}{\beta} \quad (p \ll 1)$$

Dionne and Ruff, and Anderson and Stevens (Section 92), use α and β exactly opposite to our notation. β (Anderson and Stevens's α) depends on voltage according to (Figure 92.3)

$$\frac{1}{\beta} = be^{-aV}$$

As the membrane depolarizes, θ becomes shorter (V with respect to the external bath and a is positive).

PROBLEM. Assume α has no voltage dependence. Find α from the values given in the above problem and b (given in Figure 92.3). Dionne and Ruff actually use a more complex formulation; both α and β are voltage dependent and p depends on ACh concentration.

PROBLEM. Compare σ_I^2 vs. V and I vs. V for two cases: (1) $A = 0$ and (2) $A = 15.36$ V^{-1}, for V between -60 and $+120$ mV. In case (2), both σ_I^2 and I drop off at high depolarization even though the driving force ($V - E$) increases.

99. *Glutamate Noise I*

Voltage noise from membranes with glutamate receptors was studied by Crawford and McBurney (1976). The giant muscle fiber of the spider crab *Mais squinado* was used. The fibers, usually 1–2 mm in diameter, are large enough to insert a 250-μm AgAgCl electrode along their length. The muscle fibers are about 2 cm long. Typical areas were 0.6 cm². The value of 360 Ω cm² is given for the specific membrane resistance for these fibers, implying an input resistance of 600 Ω. This is a very low value compared to other preparations used to study noise. Usually it is desirable to increase the input resistance of the noise source to improve the signal-to-noise ratio. A bipolar transistor input stage amplifier was used. These devices may have low voltage noise but large current noise compared to FET input devices. This may be an advantage when measuring voltage noise from low input resistances.

Extracellular noise was recorded focally with a voltage electrode. The external electrode was a 10–50-μm diameter glass pipette filled with crab saline. The sealing resistances obtained by pressing the pipette against the fiber were about 50 kΩ.

A monosodium salt of L-glutamate was added to the entire external solution for the intracellular recordings, and added locally in the extracellular recordings. Intracellular voltage noise, and extracellular voltage fluctuations proportional to current noise, were measured at room temperature before and after glutamate application.

The low input resistance of the giant muscle fibers makes the spontaneous miniatures too small to observe with relatively noisy intracellular glass microelectrodes. The intracellular wire reduces background noise to 0.5 μV rms from 1–1000 Hz. The miniatures are about 5 μV in amplitude, about one-hundredth the size of the mepp's at the frog neuromuscular junction. When glutamate is applied the fiber depolarizes, the amplitude of the miniatures decreases, and the voltage noise increases. Similar effects are absent with γ-aminobutyric acid (GABA) at concentrations up to 1 mmol. The glutamate effects occur at 5×10^{-4} M. The variance of the glutamate-induced noise was proportional to the mean depolarization. The ratio of variance to mean was 218 pV; in the range of 0–7-mV depolarization, from a resting level of -50 mV, the ratio was constant. If a random sum of exponentially decaying pulses is assumed, as in Katz and Miledi, Section 90, each elementary pulse has an amplitude of about

$$a = 0.44 \text{ nV}$$

PROBLEM. Estimate the peak-to-peak noise observed near -50 mV in the presence of 5×10^{-4} M glutamate. Compare this with a background noise of 0.5 μV rms. If the power spectral density of the voltage noise has a Lorentzian shape with a cutoff frequency of 20 Hz, calculate the mean frequency of arrival of the elementary events responsible for the induced glutamate noise. Assume that the Lorentzian shape of the voltage noise spectrum is due to the RC properties of the membrane. Calculate the capacitance for 1 cm² of membrane; is this a reasonable value?

Voltage noise from the crab muscle fiber in the presence of glutamate presumably arises from discrete points on the cell surface. Extracellular focal recordings of glutamate-induced noise showed quiet regions of the membrane. Only when the focal electrode picked up extracellular miniatures, near nerve twigs on the muscle surface, could glutamate-induced noise be observed.

Figure 99.1 shows a sample of extracellular noise measured with and without glutamate in the external bath. The extracellular dc response is also shown.

PROBLEM. Assume a two-conductance state model of the glutamate–receptor interaction. Calculate the mean open time of a single channel. Estimate the variance of the extracellular noise from Figure 99.1 (V²). From the value of the sealing resistance for the focal electrode, sketch the current spectral density (A²/Hz) on an absolute scale. Estimate the elementary current associated with each open channel.

An enzymatic system that will degrade glutamate at the crab neuro-muscular junction is unknown. The intracellular noise experiments gave a value of 0.44 nV for an elementary voltage event; since a spontaneous miniature is about 5 μV, 10,000 events are implied. This is larger than the number of elementary events estimated for a miniature at the frog neuro-muscular junction. The synaptic vesicles in crab muscle may contain more transmitter molecules, or the release of a glutamate molecule may have a greater probability than an ACh molecule of reaching its receptor.

PROBLEM. In the absence of an enzymatic mechanism to remove the glutamate from the synaptic cleft, estimate the transport flux necessary to clear glutamate from the cleft. Compare this figure to 0.01 pmol/cm² sec, the value estimate from the uptake of radioactive glutamate by the nerves of the crab muscle.

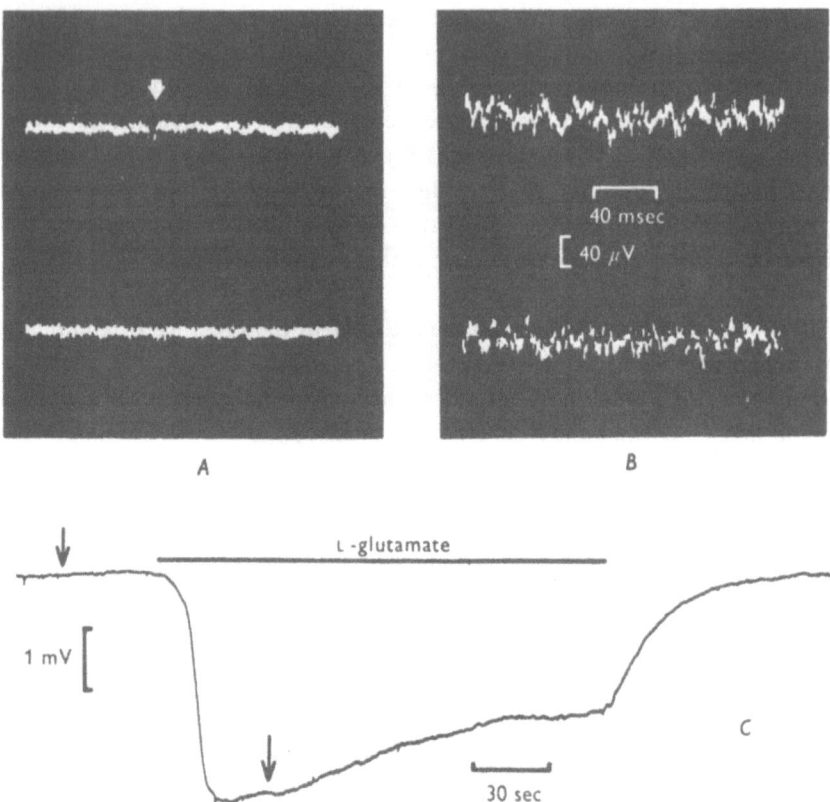

Figure 99.1. Focal records of the extracellular potential from the surface of crab muscle. A: Without L-glutamate. The arrow points to a spontaneous extracellular miniature potential. B: With L-glutamate in the external bath near the focal electrode. All traces were ac coupled with a time constant of 1 sec. The extracellular dc response is shown in C. L-glutamate was applied at the start of the bar. The focal electrode was removed at the end of the bar. The arrows in C show where A and B were recorded. The ratio of the variance to the mean extracellular glutamate response was 2.6×10^{-8} V. This result was independent of desensitization. The spectral density of the glutamate-induced noise was fitted to a Lorentzian curve with a mean cutoff frequency of 111 Hz. From Crawford and McBurney (1976).

Crawford and McBurney conclude the following:

... analysis of transmitter noise yields yet another criterion for transmitter identification; for it must surely be a requirement of any putative transmitter substance that it produces the same "elementary event" in the postsynaptic membrane as does the natural transmitter itself.

They also show that the time constants of conductance changes induced by the natural transmitter and released quantally agree with those estimated from L-glutamate noise. Other substances that produce conductance changes in these preparations, such as aspartate, have different kinetics and are not regarded as the natural transmitter.

100. *Glutamate Noise II*

Glutamate noise was studied by Anderson, Cull-Candy, and Miledi (1976, 1978) under voltage clamp. The extensor tibiae of adult locusts, *Schistocerca gregaria*, were used. The muscle fibers are about 1.5 mm long and 100 μm in diameter. In Cl-free medium the input resistance is 5–8 MΩ and the space constant is about six times the length of the fiber. The hyperpolarizing glutamate induced channels are not effective in Cl-free solutions. L-glutamate was applied iontophoretically from a micropipette located between the current and the voltage intracellular microelectrodes. Only the depolarizing response was studied. The intracellular microelectrodes, separated by 100 μm, were rotated near junctional regions. An analog of L-glutamate, quisqualic acid, was also tested.

The variance of the glutamate-induced current noise is proportional to the mean induced current. The spectrum of the noise was well described by a Lorentzian (Section 92). The double-sided spectral density was used in this paper (Section 44):

$$SS(f) = \frac{2\gamma I(V-E)\theta}{1 + \omega^2\theta^2}$$

The reversal potential (E) for the mean glutamate-induced current (I) is approximately zero. Their results are summarized in Figure 100.1. Control spectra before application of glutamate have been subtracted.

PROBLEM. Calculate the mean open time of the channels, and the single-channel conductance for all four curves shown in Figure 100.1. What is the conversion factor that will change the current spectral densities of the top graph of Figure 100.1 into conductance spectral densities?

Glutamate-induced channels are open longer, on the average, at lower temperatures. The Q_{10} is about 1.5, or 8.4 kcal/M. This is lower than ACh-induced channels but much higher than the similar quantity observed in voltage controlled K channels in squid giant axon (Section 106). The average single conductance was originally reported by Anderson *et al.* (1976) as

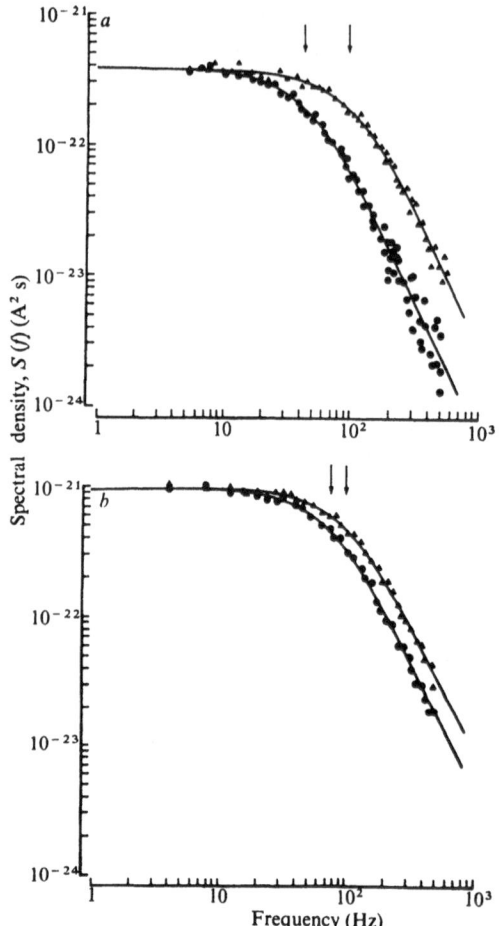

Figure 100.1. Current noise of glutamate-induced noise from locust muscle. The top graph compares two temperatures at −104-mV holding potential. The bottom graph compares two holding potentials at 23°C. The solid lines represent Lorentzian-shaped spectra. Top: The two curves have cutoff frequencies of 45 Hz for 8.3°C and 98 Hz for 22.5 °C. The mean currents were 9 and 11 nA. The curve with the 45-Hz cutoff has been shifted up; it should be divided by 1.5 to obtain its absolute scale. Bottom: The two curves have cutoff frequencies of 76 Hz at −60 mV and 100 Hz at −100 mV. The mean currents are 16 and 12 nA. The curve with the 100-Hz cutoff has been shifted down slightly; it should be divided by 0.97 to obtain its absolute scale. Anderson, *et al.* (1976).

231 pS at 23°C; the latest estimate of the average value is*

$$\gamma = 122 \text{ pS at } 23°C$$

γ decreases at lower temperature, for example, $\gamma = 39$ pS at 8.5°C. Recall that ACh-induced channel conductance has no significant temperature dependence (Section 92). γ_{glu}, as γ_{ACh}, is practically voltage independent.

Glutamate induced-channels are open longer, on the average, at depolarized potentials. For example, at −50 mV, $\theta = 2.12$ msec, and at −110 mV, $\theta = 1.56$ msec (23°C). This effect is opposite to ACh-induced

---

* Cull-Candy, personal communication (April 1977).

channels at the frog neuromuscular junction, where channels are open more briefly as the membrane is depolarized. (θ is approximately the mean open time only if $p \ll 1$; Section 98.)

Quisqualic acid is a potent agonist in other preparations where glutamate is the presumptive natural transmitter. Quisqualic-induced current noise has the same qualitative features as glutamate noise. Generally, quisqualic acid produces channels that stay open longer, but have about the same single-channel conductance.

101. *Electrical Noise from Motoneurons*

In 1952, Brock, Coombs, and Eccles successfully introduced glass microelectrodes into motoneurons in the spinal cords of anesthetized cats. Graham and Gerard (1946) and Ling and Gerard (1949) developed the new technique and it was soon applied to a variety of preparations: Ling and Gerard (1949), Nastuk and Hodgkin (1950), Nastuk (1950), and Fatt and Katz (1950), the resting and action potentials of frog muscle fibers; Woodbury and Crill (1951), frog heart muscle; Draper and Weidmann (1950), mammalian heart muscle; Weidmann (1951) studied the giant axons of *Sepia*, and showed that the new technique yielded the same results as the earlier method of inserting a large internal electrode from a cut end.

Great care was given in these earlier papers to noting every detail and variation that was observed in an experiment. An example of this style of reporting is provided by the two kinds of electrical noise noted by Brock, Coombs, and Eccles in their 1952 paper. The first was called "surface membrane noise"; it was observed whenever a microelectrode came into close contact with a neuron and was just on the verge of penetrating the cell body. The second was called "synaptic noise" and it was observed across the membrane of the cell body after penetration. Figure 101.1 shows the two types of noise (B and C) compared to the background noise from the microelectrode placed in the bath far from the cell. [Relevant here are two articles that treat synaptic noise as a source of variability in the interval between motoneuron action potentials (Calvin and Stevens, 1967, 1968). See also Holden (1976b).]

The surface membrane noise shown in Figure 101.1B has peak-to-peak voltages as large as 1 mV. One possible cause of the noise increase is that the tip of the microelectrode is sealed off as it approaches the membrane. To test this possibility one could approach an inert membrane with a microelectrode. Instead, Brock, Coombs, and Eccles compared the noise from a

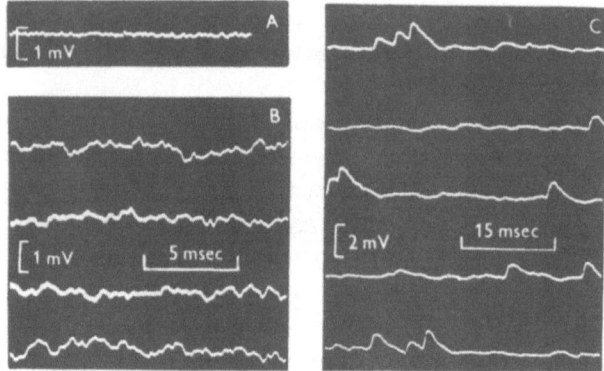

Figure 101.1. Electrical noise from motoneurons. From Brock, Coombs, and Eccles (1952). A: Extracellular noise far from a microelectrode placed far from the motoneuron. B: A series of records from the same microelectrode placed next to the neuronal surface membrane. C: Noise from the same electrode after a successful penetration of the nerve membrane.

10 MΩ resistor (approximately the value of microelectrode) with the noise from a 10^5-MΩ resistance. It is doubtful that this is an adequate test; the noise from a microelectrode is much larger than resistor noise (DeFelice and Firth, 1971). Also, a 10^5-MΩ input resistance limits the bandwidth of the input circuit due to stray capacitance and this would limit the change in noise at the higher resistance values.

Although the control was probably inadequate, it is likely that much of the surface membrane noise was actually from the motoneurons. They attributed the noise to changes in ionic flux through the membrane area defined by the tip of the microelectrode (approximately 10^{-9} cm²). It was known that a flux of 20 pmol/cm² sec is expected from nerve membranes. This gives about 10^4 ions/sec across the membrane area seen by the microtip. For a capacitance of 1 μF/cm², one ion would produce a voltage change of 0.16 mV. These calculations are given below:

$$20 \times 10^{-12} \frac{\text{mol}}{\text{cm}^2 \text{ sec}} \times 10^{-9} \text{ cm}^2 \times 6.03 \times 10^{23} \frac{\text{ions}}{\text{mol}} = 1.2 \times 10^4 \text{ ions/sec}$$

$$V = Q/C = \frac{1.6 \times 10^{-19} \text{ C}}{(10^{-6} \text{ F/cm}^2)(10^{-9} \text{ cm}^2)} = 1.6 \times 10^{-4} \text{ V}$$

This looks reasonable, since the observed fluctuations were about 0.5 mV. However, Brock, Coombs, and Eccles point out that the input capacitance of the measuring circuit was about 4 pF, which would require 12,000 ionic charges to produce 0.5 mV. They postulated that ionic flux would have to occur in bursts across the membrane to account for the observed noise shown in Figure 101.1B. This could occur normally, but at least it must occur under the conditions of measurement with the microtip next to the surface membrane.

The synaptic noise shown in Figure 101.1C is composed of irregular potential waves between 0.5 and 1.5 mV high. Exceptionally small waves of 0.2 mV were also observed. This noise is attributed to the random synaptic bombardment of motoneurons by a single afferent. Approximately ten convergent presynaptic impulses, occurring at the same time but at different (nearby) endings, are required to evoke a reflex discharge in the motoneuron.

Brock, Coombs, and Eccles compare their synaptic noise width with that previously reported by Fatt and Katz from the frog neuromuscular junction (see Section 89). At that time, the two phenomena were regarded as different because the noise at the neuromuscular junction was localized to the motor nerve terminals: Analogous noise (to the mepp's) also exists from synaptic knobs (in motoneurons) but it would be of too low a voltage for identification by the present techniques. In retrospect, it is rather difficult to understand this position.

PROBLEM. Argue that it is improbable that the resting level of inward flux of Na could be so high that its suppression would cause a voltage change of 100 V/sec under inhibitory synaptic knobs.

Brock *et al.* speculate that synaptic excitatory and inhibitory action require two different transmitters, one to stimulate the inward Na current (excitatory) and another to stimulate the Na pump (inhibitory). Fatt and Katz (1951) had shown a nonspecific, ACh-induced increase in permeability at the neuromuscular junction.

102. Excitability Noise in Neurons

Noise from neural elements was recognized long before the work of Brock *et al.* (1952); however, the fluctuations studied were of a higher order and dealt primarily with the suprathreshold response of the nerve rather than subthreshold transmembrane events. The link between membrane noise and macroscopic fluctuations was made much later. Although the connection between these two levels of biological noise is not the main subject of this book, it is of interest because many of the early workers on membrane noise began their studies with a view to understanding macroscopic randomness such as the apparent fluctuation in excitability in nerve (see also del Castillo and Suckling, 1957). Some of these early studies will be discussed below. Recent papers relevant to noise and excitability include Clay (1976, 1977) and Sigworth (1979a).

A. A. Verveen submitted his thesis on "Fluctuation in Excitability" in 1961. The work was influenced by earlier studies of Pecher (1936, 1937, 1939). The basic phenomenon studied by Verveen is as follows. If an electrical stimulus of a certain strength is applied again and again to a nerve fiber, although the stimulus strength remains constant, the fiber may respond differently each time the stimulus is presented. At high stimulus intensity a response occurs every time. At low intensity no response occurs. These extremes, however, are not separated by a sharp threshold for there is an intermediate condition for which a response may or may not occur. This effect is shown in Figure 102.1. Within the range of threshold, the probability of firing increases from zero to one. For low frequencies of stimulation, the probability is approximately Gaussian. Long- and short-duration stimuli give the same probability distribution when normalized about threshold (defined in Figure 102.1). The relative spread (*RS*) is approximately constant for varying conditions of stimulation. *RS* determined for 80 frog nerve fibers was

$$RS = 0.011 \pm 0.005$$

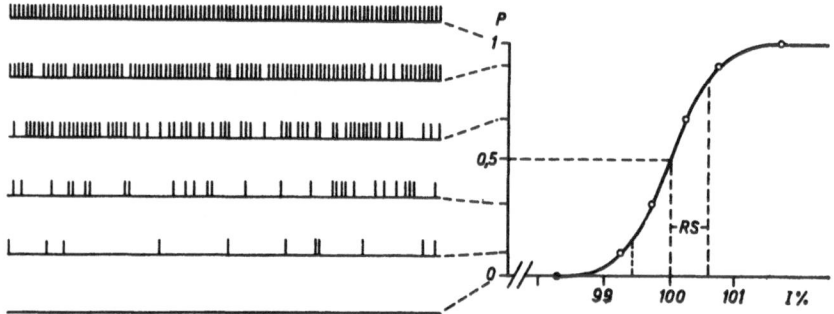

Figure 102.1. Action potentials (left) from a frog nerve in response to a constant stimulus given every 2 sec. The intensity of the stimulus (*I*) is lowered for each trace from top to bottom. The probability of a response (right) for a given stimulus strength normalized about the treshold stimulus. Threshold is defined as the stimulus at which, on average, the nerve fires once for two trials. Verveen (1961).

Unmyelinated crayfish fibers had a much smaller variability. From 15 fibers,

$$RS = 0.0012 \pm 0.0003$$

In both cases, the histograms of RS are skewed to the right.

Fluctuation in excitability was studied earlier by Blair and Erlanger (1932, 1933), who were able to show that the fluctuation was inherent to the neurons, and by Monnier and Jasper (1932). Pecher (1936, 1937, 1939) proved that external causes played a minor role in the fluctuation of excitability in nerve fibers by showing independence of the variance of firing in two fibers from the same preparation. His studies supported the concept that the fluctuation of excitability in neural elements is an inherent property of the neurons themselves.

Verveen's thesis studies involved extracellular recordings from myelinated and unmyelinated nerve fibers. Single axons from frog and crayfish preparations were distinguished electrically by their large size. Responses from signal nodes of Ranvier could be isolated; this was demonstrated earlier by del Castillo and Stark (1952), who state the following:

> ... local response becomes particularly clear at threshold strength when successive stimuli produce a random display of potential changes, the hump varying continually in amplitude and duration, and occasionally giving rise to a more or less delayed spike.

Since fluctuation seemed intrinsic to neurons, the next logical step was to search for them at the level of the nerve membrane. Inherent membrane noise had already been reported by Brock *et al.* (1952), Section 101, but its origin was uncertain; the noise they observed might have been due to local activation of the neuronal soma by transmitters. The only proof of an intrinsic membrane noise is the direct measurement of isolated transmembrane events themselves. This work was started by Verveen and Derksen in the early 1960s.

103. Nerve Membrane Noise

In 1964–1965 Verveen and Derksen began to measure membrane noise directly. The node of Ranvier was chosen because its electrical properties were well known and because "the fluctuations in excitability and in latency, which are directly associated with fluctuations in the resting potential membrane" (Derksen, 1965, p. 390) were known from the earlier work of Pecher and Verveen.

Verveen and Derksen presented their first results at the June 1964 meeting of the International Union of Pure and Applied Biophysics in Paris, published in *Kybernetik* in 1965. In these experiments, the voltage was measured between two nodes isolated by air gaps or ion-free sucrose solution. The voltage noise was amplified and passed through narrowband ($Q = 10$, see Section 38) filters and integrated 5–10 min to obtain an estimate of the power spectral density.

Voltage noise spectra were proportional to $1/f$ between 10 and 1000 Hz although some spectra became flat above and below this range. The autocorrelation function was calculated from the voltage noise spectral densities (Section 44) and compared with Pecher's observation that at least 2 sec between stimuli are necessary for independence. The similarity of noise spectra from the node and noise spectra from carbon resistors and semiconductors was noted (see Section 64). Two theoretical approaches to explain the $1/f$ spectrum are mentioned in these early works. One suggested the possibility "that ion conduction takes place through channels in the membrane and that a channel could be blocked by absorption of an ion ... waiting for passage..." (Verveen and Derksen, 1965, p. 155). A distribution of absorption times in the channel could account for a $1/f$ spectrum. The second explanation was that the $1/f$ noise came from the internodal segments used as access pathways to the node. In order to test these hypotheses, a symmetric three-node cross-correlation method was developed. Voltage noise from the internodal pathways was assumed to be uncorrelated and therefore the cross-correlation function is approximately the autocorrelation function of the membrane noise from a single node. Membrane noise and its possible relationship to coding information in nerve action potentials are discussed in detail. Verveen and Derksen emphasized in their early work that the inherent noise in nerves is much larger than that predicted by Nyquist's theorem (Section 60) used by others in most earlier calculations.

The three-node, cross-correlation method was used in all later experiments. These experiments were done jointly by Verveen and Derksen and were published in detail by Derksen (1965) as his thesis, and summarized by Derksen and Verveen (1966) in *Science* (see Section 103). Figure 103.1 is taken from Derksen's thesis. Abolishing Na in the external medium did not change the spectrum. Replacing NaCl by KCl immediately abolished the $1/f$ spectrum and the effect was reversible.

PROBLEM. Assuming equilibrium potentials of K and Cl of -90 and -55 mV, argue from the data of Figure 103.1 that K flux, and not Cl, is probably responsible for the $1/f$ noise (see Sections 98–100).

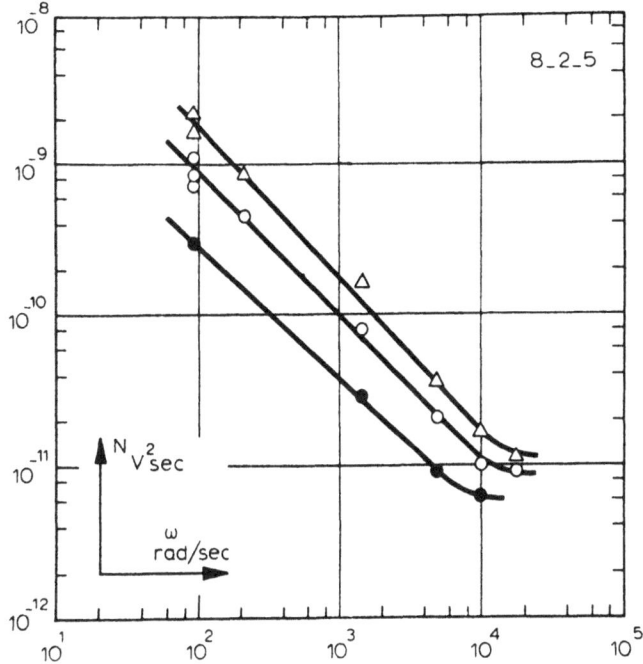

Figure 103.1. The voltage noise spectra densities at rest (open circles), 10-mV depolarization (open triangles) and 10-mV hyperpolarization (filled circles), from frog node at room temperature in normal Ringer. Derksen (1965).

Another type of noise was described at low frequencies. This is shown in Figure 103.2. The bursts occur in both output voltages of the symmetric three-node preparation, indicating that they probably originate in the middle node membrane. Bursts were separated by 1–10-min quiet intervals. They occur in groups of 10–100 separated by 5–100 msec. Bursts are less frequently observed in fibers stored at 0°C in Ringers for 10–12 hr, so they do not signify deterioration. The amplitude distribution of the burst noise was studied by Verveen, Derksen, and Schick (1967). The third moment was calculated, indicating the degree of skew in the amplitude distribution. Positively skewed distributions were found in hyperpolarized states. Interfering with active transport did not influence the burst noise. Partial replacement of external Na shifted the range in which bursts are common to more hyperpolarized levels. The effect is reversible. At 25% normal Na, skewness begins near −100 mV instead of −70 mV in normal Na. Verveen and Derksen conclude that the burst noise is related to the flux of Na ions through passive sites in the membrane.

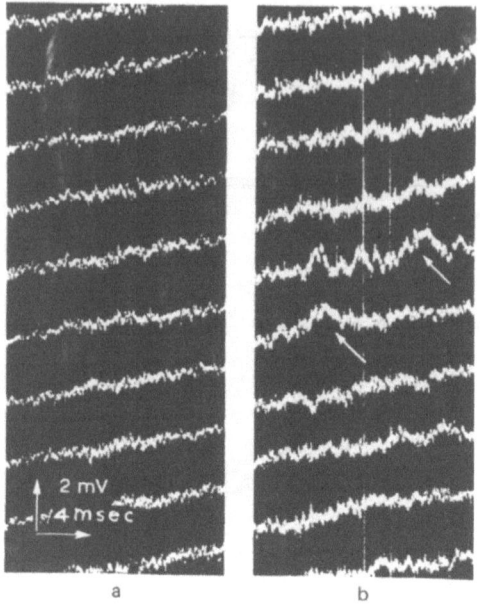

Figure 103.2. Unfiltered amplifier output voltage monitored over a long duration (30–60 min) (a) Node voltage noise from a relatively quiet period $(1/f)$ spectrum. (b) Node voltage noise with burst activity (arrows). Derksen (1965).

Verveen (1962) also studied the influence of axon diameter of fluctuation in excitability. It was found that the coefficient of variation (RS) varied with the axon diameter according to the relation

$$\log RS = -1.5 - 0.8 \log d$$

where d is in μm.

PROBLEM. Assume 20 mV between rest and the critical depolarizing level for firing. Show that the rms noise (σ) is given by $\log \sigma = -3.2 -0.8 \log d$. Compare this with the expression given earlier (Section 89) for Johnson noise from a nerve axon in a 10-kHz bandwidth [$\log \sigma = -4.2 - (3/2) \log d$]. A good review of this early work is given in Verveen and Derksen (1968).

104. Voltage Noise from the Node of Ranvier

This early work by the Leyden noise group was interrupted by the death of Hans Derksen. It was resumed under Verveen with his student, Elias Siebenga, and is summarized in Siebenga's thesis (1974) entitled "Some Investigations of the Voltage Fluctuations in the Node of Ranvier of *Rana temporaria*."

The main result of the new work on voltage noise from the node was the appearance of a second component in the spectra of the Gaussian noise at large depolarizations. This was published in 1970 and reported to the First European Biophysics Congress in September 1971 in Baden, Austria (Siebenga and Verveen, 1970, 1971). Subsequent papers are Siebenga and Verveen (1972) and Siebenga and Meyer (1973). $1/f$ noise was present in all spectral densities, but at membrane potentials held by current clamp above -30 mV an additional noise source was found with an approximate spectrum proportional to

$$\frac{1}{1 + \omega^2\theta^2}$$

This Lorentzian noise component was temperature sensitive, whereas $1/f$ noise was temperature insensitive.

The I–V relationship for the node was measured. Figure 104.1 shows the intensity of the $1/f$ noise component (left-hand side, filled circles) and the K current through the node (left-hand side, open circles) as a function of membrane voltage. I_K was estimated by subtracting the steady-state leakage from the total current. The $1/f$ noise component obeys the relationship (Figure 104.1, right-hand side)

$$\frac{S_V}{R} = A + BI_K(V - E_K)$$

where $1/R$ is the steady-state K slope conductance.

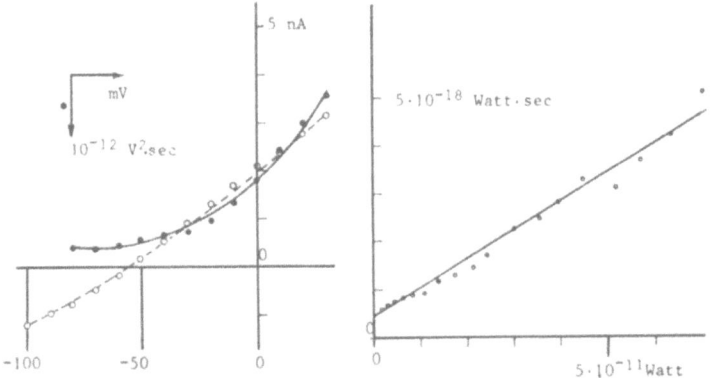

Figure 104.1. Left: $1/f$ voltage noise intensity at 100 Hz, 7°C against membrane potential (filled circles). Steady-state I–V relationship for the same node (open circles). Right: Power fluctuations (S_V/R) against $I_K(V - E_K)$ calculated from the date on the left. R is the slope resistance. Siebenga (1974).

The Lorentzian component was studied in more detail by Siebenga, Meyer, and Verveen (1973) and by Siebenga, de Goede, and Verveen (1974). The Lorentzian component increases with depolarization, as shown in Figure 104.2. The top three curves in the figure are theoretical $1/f$ and Lorentzian lines.

A random switch model of conductance fluctuations of K channels was assumed to underlie the Lorentzian component. Conductance fluctuations were estimated from the measured voltage fluctuations under the assumption that the impedance of the nodal membrane is independent of frequency. An order of magnitude calculation, based on the ratio of the variance of the Lorentzian voltage noise component to the mean membrane voltage (see Section 84), led to an estimate of the number of K channels per node:

$$N \simeq V^2/\sigma_{V'}{}^2 \simeq 10^1$$

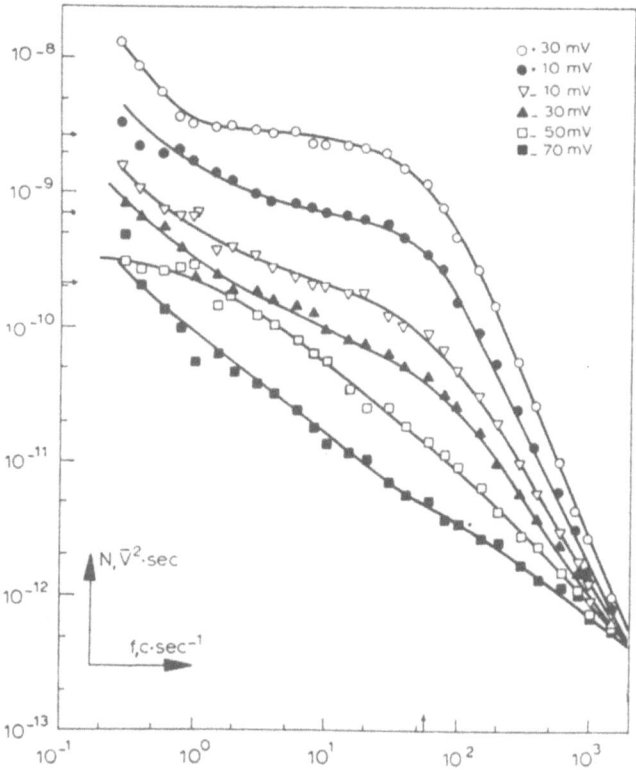

Figure 104.2. Voltage noise spectra of a node of Ranvier in normal Ringer's at room temperature. Siebenga (1974).

No relationship between N and membrane potential was found. Because TEA reduced the Lorentzian component, whereas Na replacement, TTX, and metabolic inhibitors had little effect, this noise component was assigned to the K system. The time constant of the Lorentzian component had no voltage dependence. From the macroscopic K conductance, γ_K was estimated between 10–40 pS. This estimate was lowered to 2–3 pS in later experiments by van den Berg, Siebenga, and de Bruin (1977).

PROBLEM. The expression given in Lee (1960, p. 224) for the spectral density of a random telegraph signal is

$$\Phi(\omega) = \frac{E_m{}^2}{\pi} \frac{2k}{(2k)^2 + \omega^2}$$

This is the double-sided power spectrum [$SS(f)$ in our notation]. In this expression k is the average number of zero crossings per second and the random wave flips between the values E_m and $-E_m$. The average time between transitions is therefore

$$\theta = \frac{1}{2k}$$

The definition Lee uses for Φ (Lee, 1960, p. 57) is

$$\Phi(\omega) = \frac{1}{2\pi} \int_{-\infty}^{\infty} \phi(\tau)e^{-i\omega\tau}\, d\tau$$

(The symbol ϕ stands for the correlation function, C in our notation). The definition for the measured, one-sided spectral density is (Section 44)

$$S(f) = 4 \int_0^{\infty} C(\tau)\cos(\omega\tau)\, d\tau$$

Show that Lee's expression for the $\Phi(\omega)$ of the random wave reduces to

$$S(f) = \frac{4E_m{}^2\theta}{1 + (\omega\theta)^2}$$

R. van den Berg of the Leyden noise group measured K and Na channel noise and $1/f$ noise under voltage clamp. Burst noise, originally observed by Verveen and Derksen in their early experiments, was also studied. Of particular interest is the group's concern with noise analysis under long and short voltage clamp pulse regimes. Interested readers are referred to van den Berg *et al.* (1975, 1977), van den Berg and Beekman (1977), and van den Berg's thesis (1978) "Electrical Fluctuations in Myelinated Nerve Membrane."

105. The Squid Giant Axon

In 1973 Fishman, Moore, and Poussart initiated the study of electrical noise from the squid giant axon. The advantage of the giant axon is its size, which allows relatively large electrodes to be inserted into the cytoplasm. Good electrical access to the membrane means that the membrane voltage can be well controlled and that the access resistance is not an important source of spurious noise. (Fishman, 1973a,b, 1975a,b,c; Fishman, Poussart, and Moore, 1975, 1977; Fishman, Moore, and Poussart, 1975, 1976, 1977; Poussart, Moore, and Fishman, 1976, 1977)

PROBLEM. Assume that a current amplifier has a voltage noise source e_a and a separate current noise source i_a at its input (Section 31). Show that the measured current noise spectral density from a membrane with impedance Z is

$$S_I(f) = S_I{}^m(f) + S_I{}^a(f) + \frac{S_V^a(f)}{|Z|^2} \qquad (105.1)$$

where $S_I{}^m$ is the spectral density of the current noise from the membrane, $S_I{}^a$ is the spectral density of $i_a(t)$, and $S_V{}^a$ is the spectral density of $e_a(t)$. The ratio of $S_I{}^m$ to the second two terms is called the signal/noise ratio. (See Haus *et al.*, 1959; Poussart, 1969, 1970, 1971, 1973; Motchenbacher and Fitchen, 1973.)

PROBLEM. Show that the signal-to-noise ratio (S/N) is related to membrane area A by

$$\frac{S}{N} = \frac{A(S_I{}^m/A_0)}{S_I{}^a + A^2(S_V{}^a/|Z_m|^2)}$$

where Z_m is the specific membrane impedance (Ω cm²; see Section 27). The quantities in parentheses are constants for a particular membrane at constant voltage. Show that S/N has a maximum value when

$$A = \frac{S_I{}^a(f)}{S_V{}^a(f)} |Z_m|$$

Show that this same formula holds if one assumes $i_a(t)$ and $e_a(t)$ at the input of a voltage amplifier, and measures voltage noise instead of current noise.

The criteria for a best (maximum) signal-to-noise ratio are frequency and voltage dependent, since in general $S_I{}^a$ and $S_V{}^a$ are not white noise sources and Z_m depends on voltage.

Fishman's group obtained their early measurements from small areas of the squid giant axon surface isolated by pressing an electrode on the membrane and causing sucrose to flow around the patch. Membrane areas so isolated were between 10^{-4} and 10^{-5} cm². The shunt resistance through the sucrose and the interstitial space of the Schwann cell layer of the axon is about 1 MΩ.

PROBLEM.   Calculate the dc resistance of a patch of membrane 10^{-5} cm² assuming a specific membrane resistance of 2000 Ω cm² (a typical value near rest; see Section 27).

PROBLEM.   Assume the circuit of Figure 81.1 (with C in parallel) for the axon membrane. Use the values given by Conti in Section 87 and plot $|Z|^2$ for 10^{-5} cm². Repeat the calculation with a 1-MΩ shunt resistor in parallel with Z.

Voltage noise, current noise, and impedance were studied using the patch method. Because the shunt resistance dominates the membrane impedance for a wide range of frequencies and voltages, current noise spectral densities and voltage noise spectral densities tend to have the same shape when measured from the patch. This is shown in Figure 105.1b.

PROBLEM.   Assume a shunt impedance Z_s is in parallel with the membrane impedance Z. Derive the relationship between S_I, S_V, Z, and Z_s. Compare this relationship with Figure 105.1b (in the figure Z refers to the total measured impedance, membrane and shunt in parallel).

Their early experiments were instrumental in stimulating further research in nerve membrane noise. They suggested, as did Siebenga's work (Section 104) that there is a measurable noise component in nerve other than $1/f$ noise and that this component reflects channel kinetics.

In 1973 Wanke, DeFelice, and Conti (Section 106) showed that to a good approximation the relationship

$$S_V = S_I \, |Z|^2 \qquad\qquad (105.2)$$

holds in the squid giant axon where Z is the small-signal impedance (Section 87). Fishman compared his patch method (Figure 105.1b) with an axial wire method similar to Wanke's (Figure 105.1a). Noise and impedance measured by the axial wire method show the expected difference in the shapes of S_I and S_V and the resonance in Z.

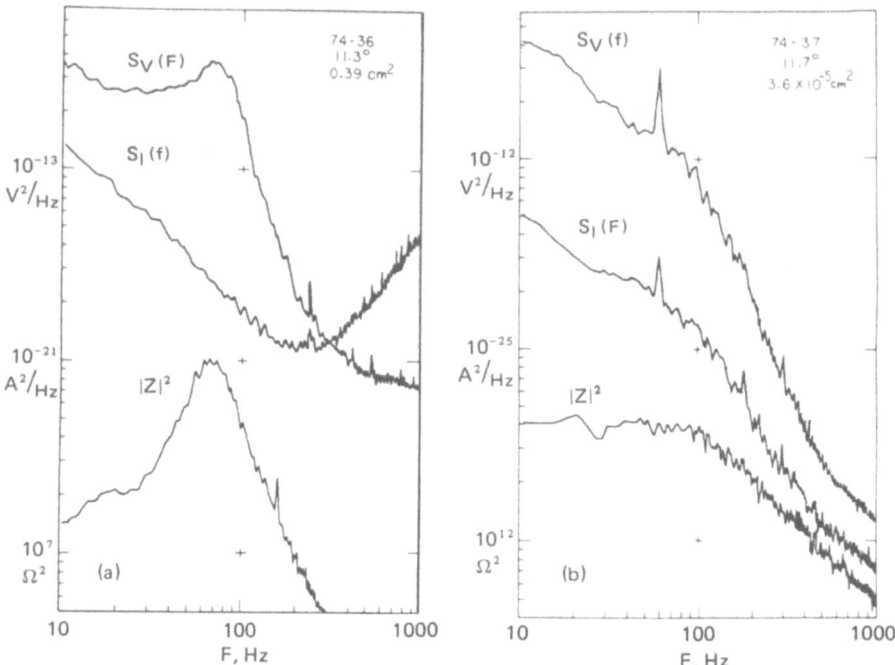

Figure 105.1. Current and voltage noise spectral densities and impedance (modulus square) from squid giant axon at rest. (a) Large area axial wire technique. (b) Small area sucrose patch technique. *Loligo paelei.* Fishman *et al.* (1975).

PROBLEM. Compare Equation (105.2) with Figure 105.1a. Explain the tailing up in $S_I(f)$ in terms of the amplifier noise model used to derive Equation (105.1). Estimate $S_V{}^a(f)$ at 1000 Hz, assuming $S_I{}^a$ is zero. Compare this with Wanke *et al.* (1974, Figure 3).

PROBLEM. Assume Z_s in Figure 105.1a is infinite. Sketch Z_s by comparing $|Z|^2$ in (b) with $|Z|^2$ in (a).

106. Current Noise from the Squid Giant Axon

In 1973 Conti, DeFelice, and Wanke (Wanke *et al.*, 1974; DeFelice *et al.*, 1975; Conti *et al.*, 1975) began a series of experiments designed to test the relationship between membrane current noise, voltage noise, and impedance and to measure current noise under voltage clamp for a variety of conditions. Large areas were isolated by air gaps and an axial wire was used to space clamp the membrane. Membrane area could be varied between 0.03 and 0.3 cm². The preparation was similar to that used by Hodgkin and Huxley in 1952.

PROBLEM. Assume the peak-to-peak current noise amplitude from 0.03 cm² of axon membrane is 10 nA. What is the amplitude from 0.3 cm² of membrane assuming all other conditions are the same?

The basic observations are summarized in Figure 106.1. The curve labeled ASW (artificial sea water) is the normal spectrum. The curve labeled TTX is from the same axon exposed to tetrodotoxin (TTX), a se-

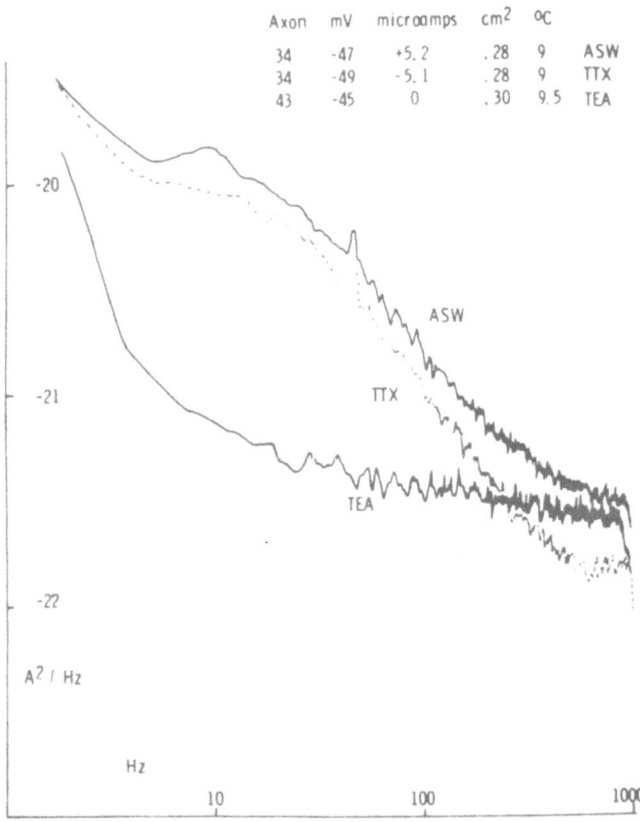

Figure 106.1. Current noise spectral density from squid giant axon. Large area axial wire technique. The effect of TTX and TEA on the normal (ASW) spectrum is shown for comparable situations. *Loligo vulgarus.* From DeFelice *et al.* (1975).

lective blocker of Na channels. Thus the TTX curve represents K channel noise plus background. The curve labeled TEA is from a different axon under similar conditions; tetraethylammonium (TEA) has been added inside the axon where it is an effective blocker of K channels. Thus the TEA curve represents Na channel noise plus background.

An immediate conclusion from Figure 106.1 is that current noise near rest is due primarily to the K system. Note that the K noise (TTX curve) is also given by

$$S_{TTX} \simeq S_{ASW} - S_{TEA}$$

PROBLEM. Estimate the peak-to-peak noise in the bandwidth 1–1000 Hz expected from 1 cm² of squid giant axon near rest. Do the same calculations for the case that TEA is present.

PROBLEM. Calculate the Johnson noise expected from the axon membrane using the impedance model given in Section 82. Compare this noise to the curves in Figure 106.1. (This comparison is only approximate since the noise in Figure 106.1 is from a depolarized membrane and the impedance data in Section 87 is from a membrane at rest near −60 mV.)

To a good approximation, K noise can be modeled as the sum of Lorentzian and $1/f$ noise spectral densities. This is shown in Figure 106.2. The solid line through the experimental curve is

$$S_I(f) = \frac{b}{f} + \frac{c}{1 + (f/f_\theta)^2}$$

where b, c, and f_θ are given in the figure. The theoretical Lorentzian and $1/f$ curves are also shown separately. There is reasonable agreement between the theoretical curve and the data below about 300 Hz. The tailing up of the measured spectral density above 300 Hz is due to membrane impedance and amplifier voltage noise [see Equation (105.1)]. The four open data points are measured values of this spurious noise component; the two solid points are the difference between total noise and spurious noise.

PROBLEM. Compare the magnitude of $1/f$ noise from nerve membrane with $1/f$ noise from carbon composition resistors (Section 64) and doped black lipid membranes (Section 69).

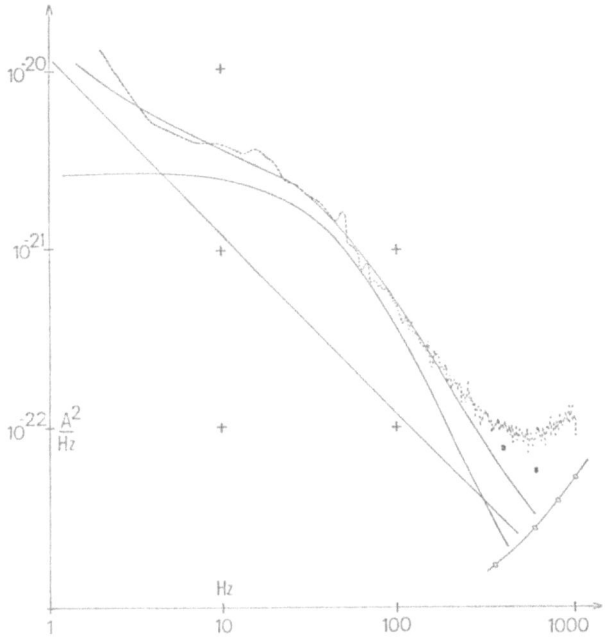

Figure 106.2. Current noise spectral density from a depolarized axon in TTX. The solid line through the data is the sum of b/f and Lorentzian noise components shown separately beneath. $T = 8.6°C$; $A = 0.28$ cm²; $b = 1.3 \times 10^{-20}$ (at 1 Hz); $c = 2.7 \times 10^{-21}$; $f_c = 38$ Hz; $V = -53$ mV; $I = 1.9$ μA; $V_{rest} = -61$ mV. DeFelice *et al.* (1975).

PROBLEM. Assume the two-state channel model developed in Sections 72 and 73. Let $E_K = -80$ mV and $p = 1/2$. Estimate $N_K \gamma_K$ from Figure 106.2. Assume the current given in the figure is K current; estimate N_K and γ_K independently.

Conti *et al.* (1975) fit their data to the Hodgkin and Huxley n^4 kinetics described in Section 76. Their estimations for K channel density and single-channel conductance are

$$N_K = 60/\mu m^2$$

$$\gamma_K = 12 \text{ pS}$$

PROBLEM. Estimate the minimum resistance for 1 cm² of giant axon membrane assuming all channels are open. Compare this value to $\bar{g}_K$ for the giant axon used by Hodgkin and Huxley (1952).

Na noise was obtained directly from TEA spectra (Figure 106.1) or from difference spectra

$$S_{Na} = S_{ASW} - S_{TTX}$$

Assuming m^3h kinetics, it may be shown that

$$S_{Na}(0) \simeq \frac{4I_{Na}^2}{N_{Na}} \frac{\theta_m}{h} \sum_{i=1}^{3} \frac{1}{i} \binom{3}{i} \left(\frac{1-m}{m}\right)^i$$

(see Section 77). Figure 106.3 shows a comparison between this theory and the measured $S_{Na}(0)$. The TEA experiments are shown as filled triangles and the difference experiments (ASW−TTX) are shown as open triangles. Using these data, Conti *et al.* calculated that

$$N_{Na} = 330/\mu m^2$$

$$\gamma_{Na} = 4 \text{ pS}$$

PROBLEM.   Compare N_{Na} and γ_{Na} with $\bar{g}_{Na}$ for the *HH* axon.

The cutoff frequency for the Lorentzian curves used to fit the K noise spectra (Figure 106.2) are temperature dependent. Figure 106.4 shows this effect. The membrane time constant

$$\theta = \frac{1}{2\pi f_\theta}$$

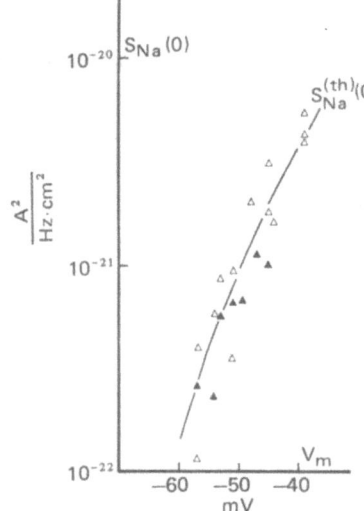

Figure 106.3. The low-frequency limit $S_{Na}(0)$ of the Na Lorentzian noise component from three TTX experiments (filled triangles) and three TEA experiments (open triangles) as a function of membrane potential. Data scaled to 6°C using a Q_{10} of 3, and to 1 cm² of membrane. The solid line is a theoretical fit using $330/\mu m^2$ as the density of Na channels. Conti *et al.* (1975).

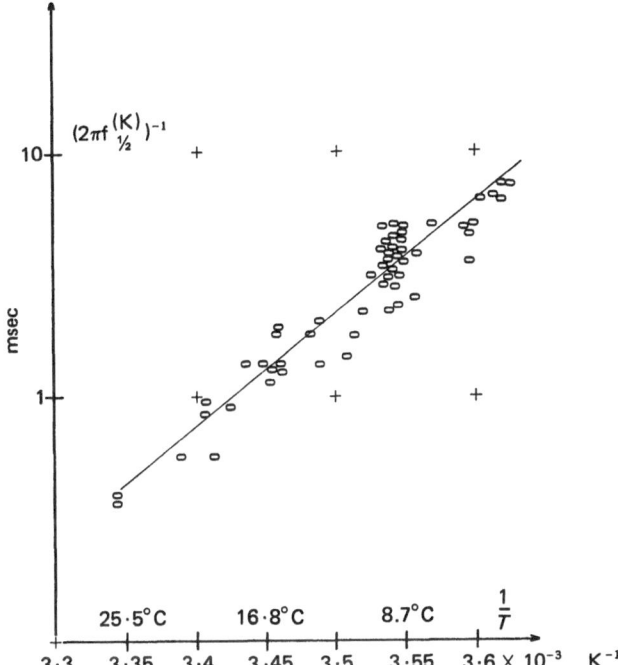

Figure 106.4. The time constant of the K Lorentzian noise component against reciprocal temperature (the actual temperature in °C is shown above the axis). Data from seven axons at membrane potentials between −45 and −55 mV. The straight line indicates an actuation energy of 0.9 eV. Conti *et al.* (1975).

is plotted against $1/T$, where T is the absolute temperature (the temperature in °C is shown above the horizontal axis).

PROBLEM. Show that the straight line fit to the data points in Figure 106.4 corresponds to the line

$$\ln \theta = \frac{\varepsilon}{kT} + \text{const}$$

where $\varepsilon \simeq 0.9$ eV. This implies a Q_{10} of about 3.3.

PROBLEM. Show that the variance of K noise is approximately independent of the temperature dependence of θ.

107. Current Noise from the Node of Ranvier I

In 1976 Conti, Hille, Neumcke, Nonner, and Stämpfli (Nonner et al., 1976; Conti et al., 1976a,b) measured current noise under voltage clamp from the node of single myelinated fibers of *Rana esculanta*. There are several important differences between this work and that described in Section 106. First, the node is a small, naturally isolated patch of membrane. The radius of the myelinated fiber they used had an average value of 7.7 µm. At the node, where the myelination is interrupted, the average radius was 5.3 µm.

PROBLEM. Although the nodal area is difficult to know exactly, Conti et al. (1976a,b) used 53 µm². This gave the same absolute Na permeability as previously published work. Calculate the nodal width (Section 20). The average dc resistance of the node is 77.5 MΩ. Calculate the specific membrane resistance in Ω cm² (Section 27). Compare this with squid giant axon.

Second, the voltage clamp protocol is entirely different. Conti et al. (1975) used 30-sec to 1-min voltage clamp steps to obtain steady-state spectra. Conti et al. (1976a,b) used repetitive 152-msec voltage clamp pulses. The spectra were obtained from the Fourier transform of the current noise during each pulse after a settling time of 40–80 msec. The interval between each pulse was 1–3.5 sec; 75–100 pulses were used to obtain the average spectral density at a particular voltage. The advantage is that the membrane is not kept at the depolarized voltages for long times so the experiments can be extended to larger depolarizations than were possible in Section 106. This method of noise analysis was introduced by Begenisich and Stevens (1975).

Currents were separated by the method already described in Section 106. TTX (150 nmol) and TEA (10 mmol) were used separately or together to isolate Na or K currents. This is summarized in Figure 107.1. The four cases are (1) normal Ringer's solution, (2) TEA–Ringer's, (3) TTX–Ringer's, and (4) TEA–TTX–Ringer's. The spectra are taken at −38 mV (32 mV depolarized from a nominal −70 mV resting potential). The expected result for each case is (1) Na and K channel noise plus background noise, (2) Na channel noise plus background, (3) K channel noise plus background, and (4) background noise.

PROBLEM. In Figure 107.1, TEA lowers the normal spectrum only slightly; in Figure 106.1, TEA changes the normal spectrum drastically. Give a qualitative explanation for this difference.

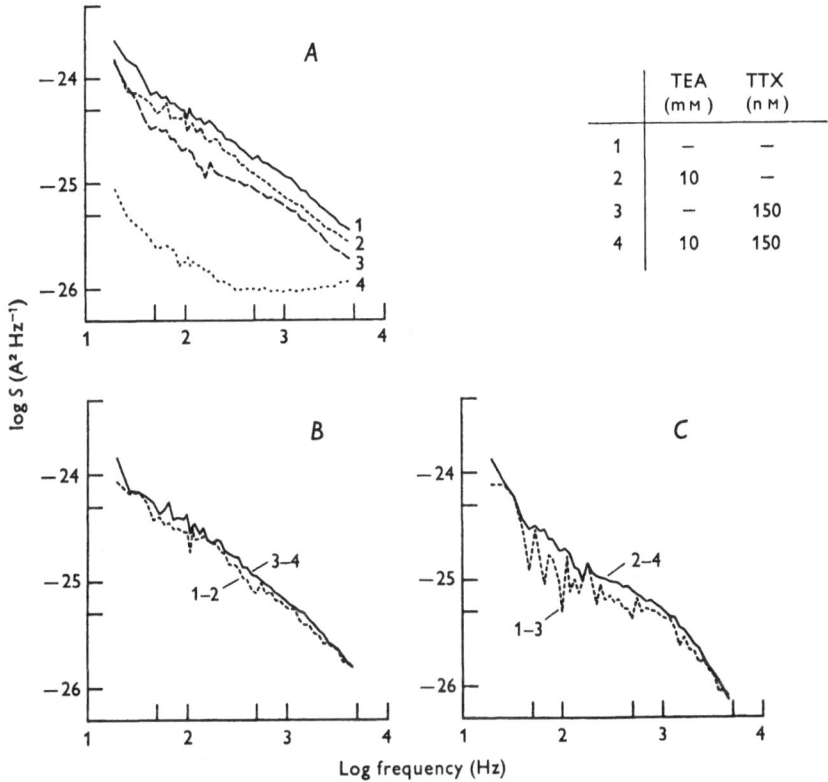

Figure 107.1. Current noise spectra from the node of Ranvier depolarized 32 mV from a nominal resting potential of -70 mV. A: Spectra obtained in the presence of solutions containing concentrations of TEA and TTX listed on the right. B: Difference spectra designed to emphasize K channel noise. C: Difference spectra designed to emphasize Na channel noise. From Conti *et al.* (1976a).

Conti *et al.* (1976a) model the difference curve 2–4 in Figure 107.1 with the equation

$$S_{2-4}(f) = \frac{B}{f} + S_{Na}(f)$$

where B/f is due to current through open Na channels and S_{Na} is due to the steady-state current chopped by the random opening and closing of Na channels.

PROBLEM.   It was found that $B = (1.1 \times 10^{-4})I_{Na}$. At rest ($-70$ mV) the steady-state Na current was between -17 and -31 pA. Compare the intensity of $1/f$ noise from a node of Ranvier at rest to a carbon composition resistor (Section 64) and to a BLM (Section 69).

PROBLEM. Assume an ohmic Na channel (invalid for the node; see below). Let $E_{Na} = 50$ mV. Estimate the average number of open Na channels in a node at rest.

PROBLEM. Ideally, the subtraction procedure that results in S_{2-4} in Figure 107.1 leaves only Na channel noise. However, if amplifier noise is considered, the measured current noise will depend on membrane impedance Z. Since Z in Solution 2 differs from Z in Solution 4, the subtraction is invalid. Assume amplifier noise as described by Equation (105.1). Show that the error introduced by the subtraction procedure is

$$S_V{}^a \frac{|Z_4|^2 - |Z_2|^2}{|Z_4 Z_2|^2}$$

where Z_4 and Z_2 are the membrane's impedance in solutions 4 and 2. A more detailed model, including access and sealing resistances of the node, is given in Conti *et al.* (1976a). In their model, $S_I{}^a$ does not subtract as it does in the above case (see their Equation 16).

The Na current noise was compared with the m^3h kinetic model described in Section 77. If $\theta_h \gg \theta_m$ (see Section 77), then

$$S_m(f) = \frac{4I^2}{N} \frac{\theta_m}{h} \sum_{i=1}^{3} \binom{3}{i} \left(\frac{1-m}{m}\right)^i \left(\frac{1}{i}\right) \frac{1}{1 + \omega^2(\theta_m/i)^2}$$

and

$$S_h(f) = \frac{4I^2}{N} \left(\frac{1-h}{h}\right) \frac{\theta_h}{1 + \omega^2\theta^2}$$

I and N are the Na current and total number of Na channels (open plus closed). In this approximation, $S_{Na} = S_m + S_h$.

PROBLEM. Show that the current through a single Na channel is given in this theory by

$$i = \frac{S_m(0)}{4m^3 I\theta_m \sum_{i=1}^{3} \binom{3}{i} \left(\frac{1-m}{m}\right)^i \left(\frac{1}{i}\right)} \qquad (107.1)$$

PROBLEM. Show by direct integration that

$$\int_0^\infty S_m(f)\, df = (1 - m^3)iI$$

and

$$\int_0^\infty S_h(f)\, df = (1 - h)m^3 iI$$

These are the variances of the m and h processes. The variance of the total Na current noise is the sum of the individual variances, or

$$\sigma_I^2 = (1 - m^3 h)iI \qquad (107.2)$$

where

$$p = m^3 h$$

is the probability that a Na channel is open.

The instantaneous Na current used in this work is given by the constant field equation (Goldman, 1943):

$$I(V) = P[\text{Na}]_o \frac{F^2}{RT} V\left(\frac{1 - \exp[(V - E)F/RT]}{1 - \exp(VF/RT)}\right)$$

where P is the Na permeability and $[\text{Na}]_0$ is the external concentration of Na (moles/liter). E is the Na reversal potential.

PROBLEM. Show that P has the units of liter/sec. Let c be the actual number of Na ions per liter in the external solution. Show that an equivalent expression for $I(V)$ is

$$I(V) = \frac{e}{kT} (cPeV)\left(\frac{1 - \exp[e(V - E)/kT]}{1 - \exp(eV/kT)}\right)$$

Show that

$$I(0) = -cPe(1 - e^{-eE/kT})$$

The current through a single Na channel is

$$i(V) = \frac{e}{kT} (c\pi eV)\left(\frac{1 - \exp[e(V - E)/kT]}{1 - \exp(eV/kT)}\right) \qquad \text{(node)}$$

where π is the single-channel permeability. Recall that in the squid giant axon

$$i(V) = \gamma(V - E) \qquad \text{(giant axon)}$$

where γ is the single-channel conductance.

PROBLEM. Let $c\pi$ and γ be arbitrary constants. Plot $i(V)$ for the node and for the giant axon for $E = 50$ mV. What are the slope and the chord conductance of $i(V)$ for each of these cases?

PROBLEM.   Show that

$$\gamma(V=0) = \frac{c\pi e}{E}(1 - e^{-eE/kT})$$

where γ is the chord conductance of the node at $V = 0$.

Conti *et al.* (1976) measured $i(V)$ from Equations (107.1) and (107.2). This implies a knowledge of m and h for these membranes. From $i(V)$ they calculated π for a single channel. These values were converted into the chord conductance defined above. The average value from 12 nodes depolarized 8–40 mV was

$$\gamma = 7.9 \pm 0.9 \text{ pS}$$

γ was smallest at small depolarizations.

PROBLEM.   Since i is the open-channel current (the instantaneous current) the chord and slope conductance of $i(V)$ do not have the same distinction as they do in Figure 23.1. Discuss the physical meaning of $\gamma(V=0)$ defined above. Write down an experimental protocol for determining the slope conductance.

PROBLEM.   Show that for p small

$$i \simeq \frac{\sigma_I^2}{I}(1 + \cdots)$$

Figure 107.2 is the Na current spectral density at -30 mV (40 mV depolarized from rest). The data points represent the difference between spectra from nodes in TEA–Ringer's and nodes in TEA–TTX–Ringer's. At this potential, channel noise is much larger than $1/f$ noise in the node. The continuous curve in Figure 107.2 is $S_{Na}(f) = S_m(f) + S_h(f)$ previously defined. The two knees in the spectrum correspond to the h process (lower knee) and the m process (upper knee).

PROBLEM.   Both $1/f$ noise and channel noise spectral densities are proportional to I^2. Explain why, then, Na channel $1/f$ noise decreases relative to Na channel Lorentzian noise as the node is depolarized.

PROBLEM.   Conti *et al.* (1975), Section 106, compare their Na current noise spectral densities to a single Lorentzian plus $1/f$ noise. Does this Lorentzian approximate m or h kinetics? What is the justification for using a

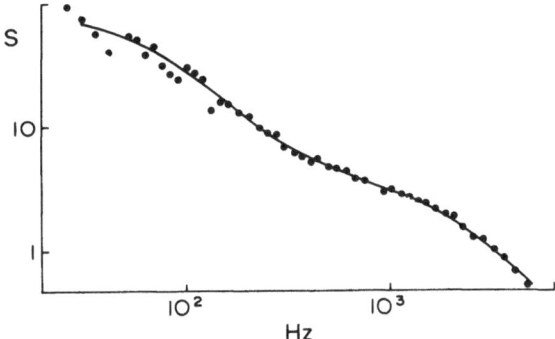

Figure 107.2. Na current noise spectral density $S(f)$ versus frequency ($V = -30$ mV). The data points are taken from Conti *et al.* (1976a) and the solid line is calculated from the Hodgkin-Huxley (1952) model for Na current, as described in the text, with α's and β's appropriate for the frog node of Ranvier. From Clay (1978).

single Lorentzian when m^3h kinetics imply a sum of seven Lorentzians (see Section 77)?

For technical reasons, Conti *et al.* (1976a) state "... not much confidence can be placed in either the data or the fit to the model below 300 Hz." (The 152-msec pulse is too brief for either the high-pass filters used to measure the noise or the nerve membrane gating process to settle completely.) Nevertheless, theirs are the first experiments to separate cleanly the *m* and *h* processes [preliminary reports of *h* noise are given by Anderson and Cahalan (1974), Sjölin and Grampp (1975), and van den Berg *et al.* (1975)]. Recent evidence from this group in which the low-frequency spectra are better resolved indicates that the *m* and *h* process may not be independent as treated above.

Although Conti *et al.* (1976) show potassium noise data they do not write on K channels. Earlier, Begenisich and Stevens (1975), using essentially the same method of pulse voltage clamping but concentrating on the variance of the noise, estimated the single-channel conductance of K channels at 4.0 ± 0.27 pS. Furthermore, γ_K did not depend significantly on membrane voltage, suggesting that K channels have only two conduction states.

PROBLEM. Compare the Begenisich and Stevens value for γ_K for the node with γ_K of Conti *et al.* (1975) for the giant axon. Can this difference be explained by the difference in conductance between Ringer's solution and sea water? If so, what about γ_{Na} determined for the node and for the giant axon?

Yi-der Chen has emphasized the measurement of membrane noise to distinguish between various models of channel kinetics. Chen (1976, 1977a,b) and Chen and Hill (1973) suggest tests, using only the variance, to decide between two-state random switch models and multiconduction state models. The interested reader is referred to these theoretical papers as an aid to designing experiments.

108. Current Noise from the Node of Ranvier II

Na channel noise is difficult to measure because of its fast kinetics and because it inactivates. F. Sigworth has used a method whereby the noise can be studied during the transient wave of Na current produced by a voltage clamp pulse. (Sigworth, 1977, 1978, 1979a,b, 1980a,b).

The preparation is the node of single sciatic nerve fibers from *Rana temporaria* or *R. pipiens*. As in previous work, it is unknown whether the nerves are motor or sensory. No difference was found between species. The nerves were 12–17 μm in diameter. Node resistance ranged between 126 and 35 MΩ; node capacitance ranged between 1.8 and 3.8 pF; the reversal potential for Na ranged between 45 and 75 mV with a mean value of 59 ± 9 mV. All experiments were done from 2 to 5°C.

PROBLEM. The upper frequency limit in Sigworth's experiments was set at 5000 Hz. Calculate the lower frequency limit (see Figure 108.1).

PROBLEM. Assume a typical value of 3 pF for the capacitance of a node. Estimate the effective membrane area from the capacitance. Estimate the specific membrane resistance in Ω cm² for a node with a dc resistance of 80 MΩ. Compare these values with those in Section 107.

Figure 108.1 illustrates Sigworth's method. TEA (20 mmol)–Ringer's is used; a background leakage current has been subtracted from the Na current transients. The top group of traces shows the response to eight identical voltage clamp pulses to −5 mV from a holding potential of −75 mV. Each pulse is 20 msec long. The average Na current transient is calculated from digitized records by the formula

$$I_j(t) = \frac{1}{n} \sum_{k=1}^{n} Y_{jk}(t)$$

where Y_{jk} is the jth point in time of the kth sodium current. The average Na

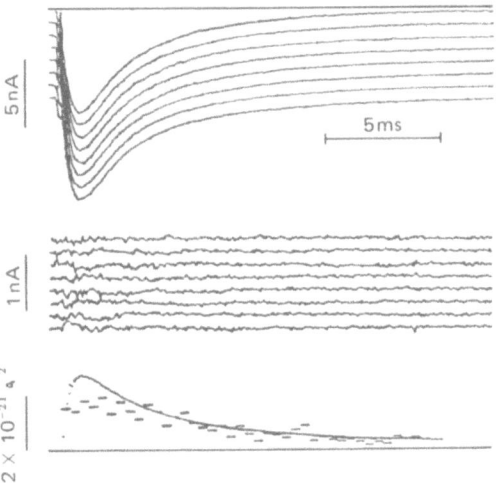

Figure 108.1. An example of the calculation of the mean and the variance of eight successive current responses to a 20-msec depolarization to $V = -5$ mV. Temperature, 2–5°C; 20 mmol TEA–Ringer's solution. Top: Inward Na currents after subtraction of a linear leakage component. Successive records are displaced downward. Middle: The average current from the top eight currents has been subtracted from each record and this difference is displayed at higher gain. The instantaneous residual current (the noise) varies from record to record as expected. The general pattern that emerges is that the noise is largest near the peak of the inward current. Bottom: The bars are the variances for four adjacent points and after subtraction of the thermal noise background variance. The smooth curve is an inverted average current. Divide the variance scale by 0.3 pA to obtain the mean current scale. From 9 records (not those shown above). Sigworth (1979a).

current transient is subtracted from each of the individual transients. This difference is shown at higher gain in the middle group of traces. The noise varies during the time course of the Na current and is largest during the first few milliseconds.

The variance of the noise is calculated from

$$\sigma_j^2(t) = \frac{1}{n-1} \sum_{k=1}^{n} [Y_{jk}(t) - I_j(t)]^2$$

where [ ] is the deviation from the mean at the jth moment. The variance is compared to the Na current (inverted) in the bottom graph of Figure 108.1. This method, which uses an ensemble average rather than a time average, allows one to study the noise during a nonstationary process. Lamb and Simon (1976a,b) used a similar method to study photoreceptor noise.

Although hundreds of pulses were applied in a typical experiment, Sigworth used local means of 4, 6, or 8 transients to do the subtraction. This helped eliminate drift. The largest source of background noise was Johnson noise from the passive access and shunt resistance and Johnson noise from the nodal membrane. Background noise is a function of membrane voltage; it was calculated from a model that includes the time-varying membrane conductance and subtracted from the measured noise at each voltage. This difference was assumed to be due entirely to the open–close kinetics of Na channel noise.

The instantaneous Na current was fitted to the equation

$$I(V) = I_s \frac{1 - \exp[-\beta(V - E)]}{1 + \exp[-\beta_1(V - E_1)]}$$

where E is the Na reversal potential and I_s is the saturation current obtained at large depolarizations. This is also the form of the single open-channel Na current $i(V)$. In that case, i_s is the maximum current passed through one channel, typically 0.15 pA at 120 mV.

PROBLEM. Sigworth found $\beta = 0.029/\text{mV}$, $E = 60$ mV, $\beta_1 = 0.025/\text{mV}$, $E_1 = 10$ mV. A typical value of I_s is 3 μA at 125 mV. Plot $I(V)$ for V between -80 and $+120$ mV. This equation gives a better fit than the constant field equation usually used for the node (see Section 107) but is not intended to have any physical significance.

From the theory of channel noise (Section 73),

$$I = Npi$$

where N is the total number of channels, p is the probability of a channel being open, and i is the current through a single channel. Also (Section 73),

$$\sigma_I^2 = Np(1 - p)i^2$$

Eliminating p between these two equations gives

$$\sigma_I^2 = iI - \frac{1}{N} I^2$$

Sigworth plots σ_I^2 against I (suitably corrected for background noise and leakage current) to obtain estimates of i and N. Typical values for Na are

$$i = -0.55 \text{ pA} \qquad @ -15 \text{ mV}$$
$$N = 20,400$$

(From Sigworth, 1979a; depolarize to -15 mV after 50 msec, -105-mV prepulse; $n = 32$.)

PROBLEM. Calculate the single-channel chord conductance γ at -5 mV for the above data. The formula is

$$\gamma = i(V)/(V - E)$$

Plot $i(V)$ for $i_s = 0.15$ pA and superimpose γ (calculated above) on this graph.

PROBLEM. Plot σ^2 vs. $iI - (1/N)I^2$ using the above values, for I between 0 and 10 nA. Show separately as dotted lines the two terms iI and $-(1/N)I^2$. Discuss the validity of extracting two unknowns (i and N) from one equation. Show that the maximum variance occurs when $I = iN/2$.

The average value of γ at -5 mV is

$$\gamma_{Na} = 6.4 \pm 0.9 \text{ pS} \qquad @ -5 \text{ mV}$$

N ranged between 20,000 and 46,000. Although there is some tendency for the larger N values to be from larger diameter fibers, the results are mixed (Sigworth, 1979a, Table 1).

A major assumption in this work is that the only background noise to be subtracted from the measured variance is Johnson noise. As a control, Sigworth adds 200 nmol TTX and measures the leakage current and the noise during depolarizing pulses. Leakage current is typically less than 1.5 nA; noise variance is typically less than 10^{-20} A^2 and the detectable change in variance during depolarization is less than 2×10^{-23} A^2.

PROBLEM. Compare 2×10^{-23} A^2 with the maximum value of σ^2 calculated from i and N at -15 mV. Why are errors in measured variance expected to be largest near E, the Na reversal potential?

PROBLEM. Estimate the maximum peak-to-peak Na current noise expected during a step to -15 mV for 1 cm^2 of nodal membrane.

Excess $1/f$ noise was negligible at all potentials used in this work. Conti *et al.* (1976) found $1/f$ noise negligible at large depolarizations (> 32 mV) but considerable at small depolarizations (< 32 mV). At 24-mV depolarization from rest, $1/f$ noise is larger than h noise above 50 Hz (see Conti *et al.*, 1976a, Figure 6).

Sigworth concludes there is no change in the average single-channel conductance during activation or during inactivation. As in Conti *et al.* (1976), inactivation is an all or none process. Na channels have only two conduction states, open and closed (Sigworth, 1977). Recent evidence from Sigworth suggests, however, that there may be more than one way to reach the open state.

Lecar and Nossal (1971a,b) gave a theory for the expected fluctuations in the threshold for firing. Their value for the relative spread (see Figure 102-1) of fluctuations was 4% and was due primarily to Na channel noise. This is larger than measured values; however, Lecar and Nossal assumed that $N = 7500$ in the node. Sigworth has calculated the relative spread with N based on his work and found values near 1%. This corresponds to experimental values of Verveen and Pecher which are near 1.2% for 10-μm-diameter fibers. Sigworth concludes that fluctuations in the firing threshold are entirely accounted for by Na channel noise and that Verveen's original hypothesis—that membrane noise underlies the macroscopic phenomenon of random firing for constant stimulus—is correct.

Sigworth also calculated the analogous function to the time average autocorrelation for the nonstationary process, namely

$$C_j(t, t_0) = \frac{1}{n-1} \sum_{k=1}^{n} [Y_{jk}(t) - I_j(t)][Y_{jk}(t_0) - I_j(t_0)]$$

where t_0 is a reference time. The use of this function as a predictor for kinetic models may well be the next chapter in membrane noise analysis, but discussion of its properties is beyond the scope of the present work.

From the relationship $p = I/Ni$, Sigworth shows that the peak open-channel probability at -5 mV is about 0.6 and at $+125$ mV, $p = 0.9$.

PROBLEM. Show that m^3h kinetics imply that

$$p_{\max} = m^3 \left(1 + \frac{r}{3}\right)^{-(3+r)}$$

where

$$r = \theta_m / \theta_h$$

Using values of the rate constants from the literature, Sigworth shows that $p_{\max} = 0.58$ at -5 mV and 0.832 at $+125$ mV.

109. Small-Signal Impedance of Nerve and Heart Cell-Membranes

In Section 82 we derived the impedance of a population of two-state channels embedded in a membrane and in Section 87 we derived the theoretical impedance of the excitable nerve membrane from the linearized HH equations. In the hyperpolarized state the nerve membrane acts like a simple *RC* circuit. Near rest, however, a complicated *RrLC* circuit is necessary to describe the membrane impedance. This is a general feature of all excitable cell membranes, although the frequency and the voltage range in which the more complicated circuit is dominant will be different in different cells. For example, nerve cells at rest have a resonance near 100 Hz; in the node the resonant frequency is near 300 Hz and in the heart cell membrane it is near 1 Hz. (Clapham and DeFelice, 1976; Clapham, 1979; DeFelice and DeHaan, 1977; DeHaan and DeFelice, 1978a,b; Clay *et al.*, 1979). In this section we will compare impedance data from the squid giant axon and from embryonic chick ventricle cells in both the frequency and the time domain.

Figure 109.1 shows data from the Mediterranian squid *Loligo vulgarus*. A small sine wave current superimposed on a steady dc current was injected into the axon. The amplitude of the sine wave current was adjusted so that the voltage response was around 1 mV. The dc current offset the membrane voltage, shown as a parameter in millivolts in the figure. The ratio of the sine wave amplitude of the output voltage to the sine wave amplitude of the input current is plotted against frequency with the dc offset (the membrane potential) as a parameter. In the figure, the ratio is squared and normalized to 1 cm² of membrane area.

PROBLEM. Draw an equivalent circuit of the experimental setup used to collect the data in Figure 109.1. Why is it impossible to do the experiment at depolarizations larger than about −49 mV? Estimate the specific membrane dc resistance at rest and compare with the theoretical model shown in Figure 87.2.

PROBLEM. How is it possible that the membrane impedance at a particular *V* and *f* (e.g., at the resonance point in the −53 mV curve in Figure 109.1) can be higher than the highest dc resistance obtainable at hyperpolarizations (the left hand plateau of the −65 mV curve) where one expects the membrane channels to have essentially closed down? In the theoretical model there is a particular value of *V* and *f* near rest for which

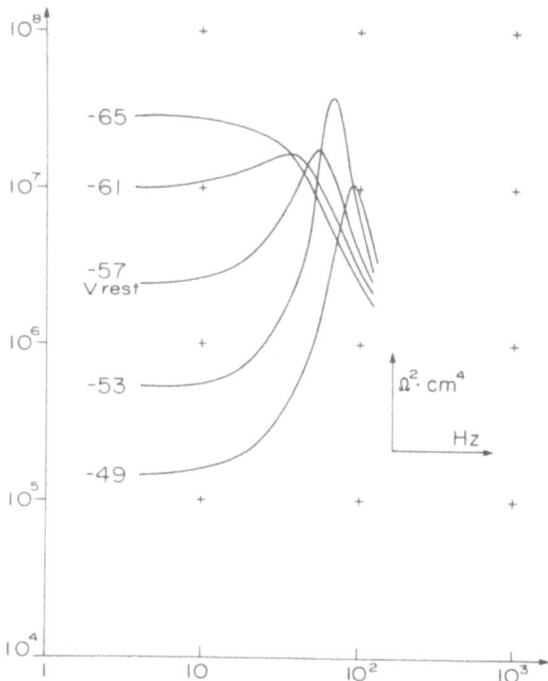

Figure 109.1. Measured impedance (modulus square) of the giant axon membrane of the squid *Loligo vulgarus* at five membrane potentials near rest. $T = 8.5°C$; $A = 0.28$ cm²; NSW. Compare with the theoretical calculations from the HH model shown in Figure 87.2. DeFelice, Wanke, and Conti (unpublished data).

the impedance becomes infinite. (See Figures 87.3 and 87.4. There are two such singularities in the HH axon, represented by the two peaks shown in the figures.)

Figure 109.2 shows data from the Atlantic squid *Loligo pealie*. These experiments, unlike Figure 109.1, were done under voltage clamp. A 1-mV peak-to-peak sine wave is superimposed on the dc potential used to clamp the membrane. The ratio of the sine wave amplitude of the output current to the sine wave amplitude of the input voltage is the admittance of the membrane (the admittance is the inverse of the impedance). In (a) through (c) the membrane was held at its resting potential of −57.5 mV. The admittance at three frequencies is shown: (a) 15 Hz, (b) 73 Hz, and (c) 300 Hz. At very low frequencies (not shown) current and voltage are nearly in phase; at 15-Hz current lags voltage; at 73 Hz (in this particular case) current and voltage are again in phase; at 300 Hz current leads voltage. The in-phase response at 73 Hz is also the frequency for which the current amplitude is a minimum (the impedance is maximum).

PROBLEM. Calculate the membrane impedance for (a) through (c) in Figure 109.2 in Ω cm². Show that in general current lagging voltage implies an inductive reactance, whereas current leading voltage implies a capacitive reactance. The inductive reactance of the squid giant axon membrane was discovered by Cole (see Cole, 1941 and 1968) and was explained quantitatively as a time-variant conductance by Hodgkin and Huxley (1952, p. 538–540). For an excellent discussion of the apparent reactance of time-

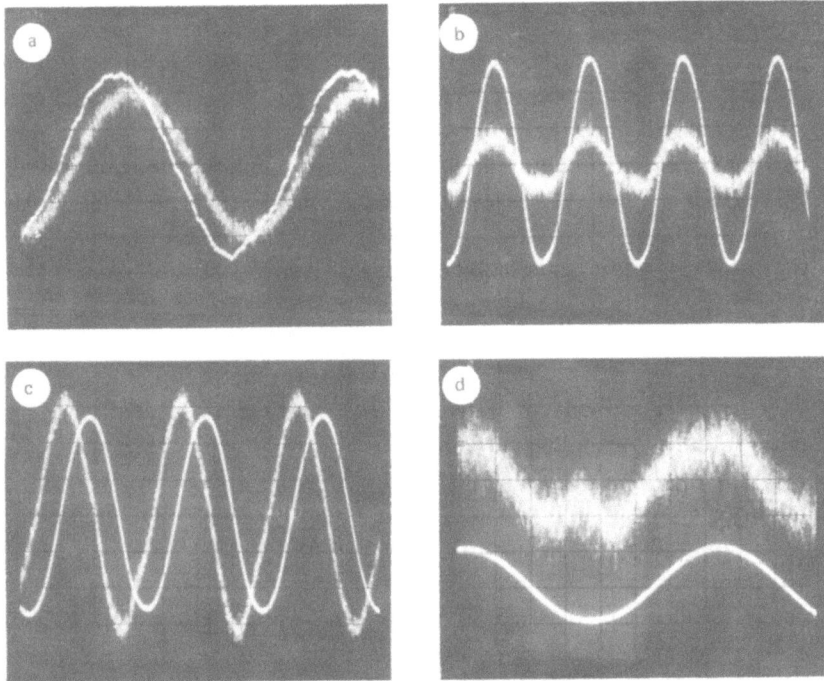

Figure 109.2. The measured admittance of the giant axon membrane of the squid *Loligo pealei*. (a)–(c) show data from the same axon voltage clamped at rest (−57.5 mV) at 9°C. A constant 1 mV peak-to-peak sine wave was added to the clamp potential (smooth traces). The fuzzy traces are the output currents from the membrane. Time scales are different for each frame: (a) 15 Hz, current lags voltage, (b) 73 Hz, current and voltage at zero phase, (c) 300 Hz, current leads voltage. Frame (b) is the frequency at which the minimum current at rest, determined by an rms meter, is obtained, 100 nA/div., 0.24 cm². (d) Different axon at the absolute minimum current at any frequency or voltage. Clamp potential is −55.1 mV temperature, 8°C; 1 mV peak-to-peak, 71.4 Hz, 20 nA/div., 0.24 cm². The speed of the voltage clamp in (d) is lower compared to (a)–(c) to reduce the noise. (DeFelice, Adelman and Mauro, unpublished data.)

variant conductances in excitable membranes see Mauro (1961) and Mauro *et al.* (1970).

Figure 109.2d is from a different axon clamped to the mean voltage and perturbing frequency at which the admittance is lowest. (The experiment is done as follows: At rest, the frequency of the perturbing sine wave is swept and the frequency where the sine wave current amplitude is lowest is found. This is also the point where the current and voltage are in phase. Then the membrane is depolarized by 1 or 2 mV and the frequency is adjusted to find the new minimum. This is repeated until the lowest absolute value of the admittance is obtained.) At the absolute minimum a second frequency component in the output waveform is obvious. Although the input is a pure sinusoid, the output contains frequency components at twice, three times, etc., the driving frequency. The small humps in the valleys of the fundamental wave in Figure 109.2d are primarily due to the second-order harmonic (also seen, less well, in Figure 109.2b). The membrane is highly nonlinear and even a small perturbation like a 1-mV ripple on the command voltage brings out the second- and higher-order terms. These higher-order terms are also present at other frequencies, but are most obvious when the first-order term is a minimum.

PROBLEM. Show by graphical construction that two sine waves, one of frequency f_0 and one of frequency $2f_0$, can add to give a waveform similar to that shown in the current in Figure 109.2d. (Select the phase such that alternate peaks in the $2f_0$ wave coincide with peaks in the f_0 wave.)

The data shown in Figure 109.2 were taken at Woods Hole in May 1976. Since then we have completed a careful study of the second- and higher-order components of the current response to a sinusoidal clamp with D. Clapham. Some of these experiments were guided by unpublished theoretical work of F. Cooley and R. FitzHugh. Cooley calculated the expected response to sine wave clamps from the 1952 HH equations, which predict that the amplitude of the first-order component (the admittance) is zero at two particular voltages and frequencies; at these points only the second- and higher-order components remain. Zero amplitude first-order components were not observed in our experiments; Figure 109.2d is a typical result of an attempt to find this point near rest. FitzHugh computed the current response to a sine wave clamp of the form

$$V(t) = V + V_0 \cos(2\pi f_0 t)$$

The current response is

$$I(t) = I + V_0 Y_1 + V_0^2 Y_2 + \cdots$$

where I is the dc response to V, Y_1 is the admittance $(Z = 1/Y_1)$, Y_2 is the second-order component, and so on. Y_2 has two terms, one at dc and one at $2f_0$. For V_0 less than 1 mV the dc term is negligible. It is the $2f_0$ term of Y_2 that is primarily responsible for the nonsinusoidal shape of the current response in Figure 109.2b and 109.2d.

Figure 109.3 is from Clay *et al.* (1979) and shows the impedance of an excitable heart cell membrane viewed from the time domain. The tissue is reaggregated 7-day chick ventricle in 4.8 mmol external potassium (see DeHaan and DeFelice, 1978a, for a complete description of this preparation). The rough line is the average voltage response to a 4-sec rectangular current pulse injected into the heart cell aggregate. R_i is the input resistance, measured at long times, and C_i is the input capacitance, measured from the leading edge of the response (see Section 83).

PROBLEM. Estimate the absolute values of R_i and C_i from the data given in Figure 109.3. Refer these values to 1 cm² of membrane and compare with giant axon membrane (Section 106) and frog node membrane (Sections 107 and 108) at rest. R_i is also called the steady-state or the slope resistance.

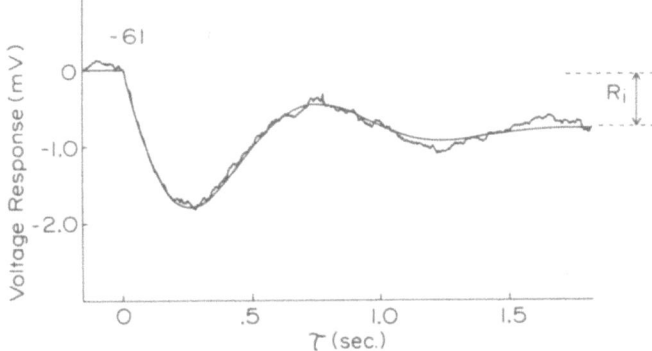

Figure 109.3. The average pulse response from a heart cell aggregate 171 μm in diameter (membrane area about 6.3×10^{-3} cm²). Injected current is a rectangular depolarizing current pulse 0.17 nA high and 4 sec long. The resting potential is −61 mV. Seven-day chick ventricle in 4.8 mmol external K and in the presence of TTX. Temperature: 37°C. Compare with Hodgkin and Huxley (1952), p. 539, Figure 23. Clay *et al.* (1979).

The solid line in Figure 109.3 is from Equation (83.2). Embryonic chick heart ventricle cell membranes have a small-signal impedance that is approximated by Figure 81.1 (with C in parallel). The difference between heart and nerve is in the circuit time constants. The value of L/r used to fit the data in Figure 109.3 is 290 msec.

EXERCISE. Compare Figure 109.3 with Hodgkin and Huxley (1952, Figure 23). Note particularly the time scales and the relative magnitude of the specific resistances.

For additional information on the theory and measurement of membrane impedance see Chandler *et al.* (1962), Sabah and Leibovic (1969), Fishman, Poussart, Moore, and Siebenga (1977), Fishman (1975a,b,c), and Poussart (1976).

110. Photoreceptor Noise. An Early Study

Membrane noise was recognized as an important phenomenon in photoreceptors by Hagins in 1959 (Hagins, 1959). In 1965 Hagins summarized the knowledge in this field in a paper on information flow in photoreceptors. The basic premise was that in order for a "quantal response" (a brief pulse of current that results from an absorbed photon) to be recognized it must be distinguished from the inherent noise in the receptor. The minimum inherent noise was assumed to be thermal noise.

Figure 110.1 is a schematic of the photoreceptor cell used by Hagins to illustrate a quantal response initiated at the outer segment of the cell (the source) being transmitted through the cell body (the channel) to the synapse (the receiver).

If a single photon produces a response in the form of a brief current,

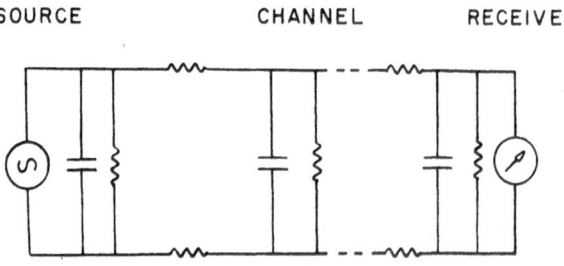

Figure 110.1. Equivalent circuit for a photoreceptor. From Hagins (1965).

large numbers of photons arriving randomly should produce a mean photovoltage and random fluctuations about this mean. This process is described by Campbell's theorem (Chapter 5, Section D). The formula given by Hagins (1965) for the mean photovoltage is

$$\langle V \rangle = n\alpha\langle q \rangle R$$

n is the mean frequency of arrival of the photons, $\langle q \rangle$ is the total charge contained in a single response, and R is the input impedance through which the currents flow. Hagins was making extracellular measurements and therefore R is the effective interstitial impedance. α is the efficiency of excitation of the quantal responses. This expression for $\langle V \rangle$ should be compared with the one derived from a specific assumption about the elementary event by Katz and Miledi in 1972 (Section 90).

PROBLEM. For perfect efficiency of excitation, the mean current is taken by Hagins to be (see Section 78)

$$I = n\langle q \rangle$$

What is the assumption that has been made about the shape of the elementary current event?

The low frequency of the spectral density of the extracellular voltage is given as

$$S(0) = 2n\alpha\langle q^2 \rangle R^2$$

What are the units of $S(0)$? How does the shape of the elementary event influence the expression for $S(0)$?

Two 1-MΩ pipettes, A and B, were placed near the cell bodies of the squid photoreceptor. Two beams of light from the same source were focused on the outer segments opposite each electrode. The response in each electrode, to a flux of 10^5–10^6 photons/sec, was filtered to pass in the band 0.6–3.6 Hz. $\langle V^2 \rangle = \langle (V_A - V_B)^2 \rangle$ was measured. The differential scheme allowed the independent fluctuations due to light absorption to add while the systematic variations in light intensity subtract. In darkness, only amplifier and electrode noise were observed. In light, the noise level increased proportionally to the light intensity as predicted by the shot noise model. The light induced fluctuations had a Gaussian amplitude distribution and the spectral density shown in Figure 110.2.

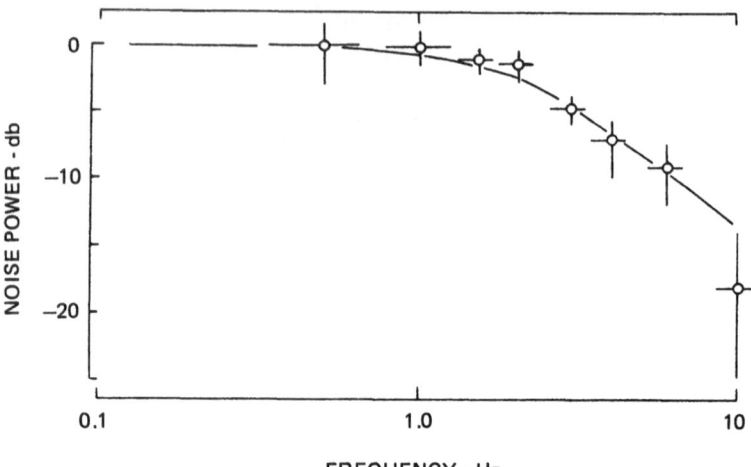

Figure 110.2. From Hagins (1965). Spectral density of 550-nm light-induced noise in the photovoltage, measured extracellularly from squid photoreceptor, and corrected for amplifier and electrode noise. Vertical axis normalized to one (0 dB). The solid curve is a theoretical spectrum for randomly distributed exponentially shaped elementary events.

PROBLEM.   From Figure 110.2, estimate the time constant associated with the elementary events. From such measurements, the lower limit for the value of n can be determined. For a given stimulus intensity, show that

$$\frac{\langle V \rangle^2}{S(0)} = \frac{n\alpha}{2} \frac{\langle q \rangle^2}{\langle q^2 \rangle}$$

Since $\langle q \rangle^2 / \langle q^2 \rangle \leq 1$

$$n\alpha \geq 2 \frac{\langle V \rangle^2}{S(0)}$$

n is measured separately from the known stimulus intensity; using this value α was calculated to be in the range $0.03 < \alpha < 0.30$. The efficiency of production of the quantal responses was compared with values obtained from other experiments and found to be similar.

The method can also be used to calculate a limit for the size of $\langle q \rangle$ if R is known. Thus

$$\langle q \rangle = \frac{1}{n\alpha} \frac{\langle V \rangle}{R} \leq \frac{S(0)}{2\langle V \rangle R}$$

To do this experiment, the squid retina had to be prepared in such a way to make the measurement of R meaningful. This was done by using large areas of retina (0.6 cm^2) and measuring the voltage between its two surfaces and the current flowing in an external circuit. The open-circuit voltage and closed-circuit current are both proportional to light intensity up to 500 photons/sec at 500 nm. The value for R was $140 \ \Omega \cdot \text{cm}^2$ at 10 Hz. These measurements gave $\langle q \rangle \simeq 57,000 \pm 11,000$ electronic charge units per simulating photon. Extreme values from 0.9 to 240,000 were also found. Hagins (1965) summarizes these results as follows:

(a) The photon has a probability of 0.3 of initiating an elementary response.

(b) A response consists of a surge of inward current lasting less than 0.1 sec. The total charge that flows in an average response is 10^5–10^6 electronic charge units.

(c) The current flows in a loop along the outer segment and then out through the membrane to the site of origin in the interstitial spaces of the retina.

(d) The average membrane depolarization produced by a single quantum response is about 20 μV (estimated using the known space constant of the outer segment). This is easily distinguished from background sources, such as thermal noise in the receptor.

Partial replacement of Na in the external bathing solution reduces the receptor current and total replacement of Na reversibly abolishes the response. The model is consistent with the view that the inward elementary current event is carried by Na. The elementary current estimated from the noise measurements is

$$i \simeq 10^6 \times 1.6 \times 10^{-19} \ \text{C}/0.1 \ \text{sec} = 1.6 \times 10^{-12} \ \text{A}$$

The largest value of electronic charge units/photon has been used in this calculation. The driving force for Na was estimated, from known concentration distributions, as the difference between the Nernst potentials for Na and Cl.

PROBLEM. What assumption does the above calculation of driving force make about the driving force on Cl at rest? The driving force for Na was taken as 70 mV; from $i = \gamma(V - E_{\text{Na}})$ we have

$$\gamma = \frac{1.6 \ \text{pA}}{70 \ \text{mV}} = 23 \ \text{pS}$$

Admitting a tenfold uncertainty in this number, Hagins speculated that 23 pS is compatible with channels of molecular dimensions. He emphasizes that this calculation, however simplified, supports the idea that isolated molecular events in a cell membrane can, in principle, account for physiologically significant currents which "count photons."

Hagins compared his results with Fatt and Katz (see Section 90): Mepp's represent the flow of 10^9 electronic charge units and

> ... are known to be caused by quanta of ACh containing thousands of molecules; however, it is possible that single transmitter molecules acting at the postsynaptic end-plate membrane produce conductances of the order of that estimated here for the quantal responses. Thus it may be that the quantal responses exemplify the unit of change in the membrane conductance on the molecular scale.

This study on the invertebrate photoreceptor is an early and extremely illustrative example of the use of the noise method which deduces properties of underlying events that cannot be observed directly.

Suggested Reading. For noise analyses applied to photoreceptors see Dodge *et al.* (1968), Minke *et al.* (1975), Simon *et al.* (1975), Lamb and Simon (1976a,b), Simon and Lamb (1977), Schwartz (1977), Wu and Pak (1978), and Wong (1978). For noise analysis applied to stretch receptors see Firth (1966) and Sjölin and Grampp (1975). For noise analysis applied to mechanoreceptors see Wiederhold (1974, 1976, 1977), DeFelice and Alkon (1977), and Gallin and Wiederhold (1977). For noise analysis applied to active transport and to other types of ionic transport in epithelial tissue see Segal (1972, 1974), Fishman (1973), Lindemann and van Driessche (1977), van Driessche and Lindemann (1976), van Driessche and Borghgraef (1975), Hoshiko (1975, 1977), van Driessche and Gögelein (1978). For noise analysis applied to the fertilization process in sea urchins, see Dale, DeFelice and Taglietti (1978) and DeFelice and Dale (1979).

Appendix: Integral Transforms

CF Equations[a]

TABLE I. *A Table of Fourier Integrals*

$$G(g) = \int_{-\infty}^{\infty} F(f)e^{i2\pi fg}\, df \quad \text{the plus-}i \text{ transform of } F(f)$$

$$F(f) = \int_{-\infty}^{\infty} G(g)e^{-i2\pi fg}\, dg \quad \text{the minus-}i \text{ transform of } G(g)$$

| $F(f)$ | $(p = i2\pi f)$ | $G(g)$ |

Part 2. Elementary combinations and transformations

	$F(f)$	$G(g)$
(201)	$F_1 \pm F_2$	$G_1 \pm G_2$
(202)	$F_1 F_2$	$\int_{-\infty}^{\infty} G_1(x)G_2(g-x)\, dx$ (convolution theorem)
(203)	$\int_{-\infty}^{\infty} F_1(-x)F_2(f+x)\, dx$	$G_1 G_2$
(204)	λF	λG
(205)	$F(af)$	$\dfrac{1}{a} G\left(\dfrac{g}{a}\right)$ (similarity theorem)
(206)	$F(f - f_0)$	$e^{p_0 g}G$
(207)	$e^{-pg_0}F$	$G(g - g_0)$ (shift theorem)
(208)	pF	$\dfrac{dG}{dg}$ (derivative theorem)
(209)	$\dfrac{dF}{dp}$	$-gG$
(210)	$\dfrac{1}{p}F$	$\int_{-\infty}^{g} G\, dg$
(211)	$2\pi i \int_{-\infty}^{f} F\, df$	$-\dfrac{1}{g}G$
(212)	$\dfrac{dF}{d\lambda}$	$\dfrac{dG}{d\lambda}$
(213)	$\int_{\lambda_0}^{\lambda} F\, d\lambda$	$\int_{\lambda_0}^{\lambda} G\, d\lambda$
(214)	$F(-f)$	$G(-g)$
(215)	$F^*(\pm f)$	$G^*(\mp g)$
(219)	$\dfrac{F(f)}{p - p_0}$	$e^{p_0 g} \int_{-\infty}^{g} e^{-p_0 g}G(g)\, dg$
(220)	$\dfrac{pF(f)}{p - p_0}$	$G(g) + p_0 e^{p_0 g} \int_{-\infty}^{g} e^{-p_0 g}G(g)\, dg$

TABLE 1. (continued)

$$G(g) = \int_{-\infty}^{\infty} F(f)e^{i2\pi fg}\,df \quad \text{the plus-}i\text{ transform of }F(f)$$

$$F(f) = \int_{-\infty}^{\infty} G(g)e^{-i2\pi fg}\,dg \quad \text{the minus-}i\text{ transform of }G(g)$$

$F(f)$	$(p = i2\pi f)$	$G(g)$

Part 4. Rational algebraic functions of f

(403.1)	1	$\delta(g)$	(delta function)
(404.1)	p	$\dfrac{d\delta(g)}{dg}$	
(407)	$\lvert p \rvert$	$-\dfrac{1}{\pi g^2}$	
(408.1)	$\dfrac{1}{p^k}$ $\Big\}$ $k(\text{integer}) > 0$	$\pm\dfrac{g^{k-1}}{(k-1)!}$	$0 < \pm g$
(431)	$\dfrac{1}{(p \pm b)^k}$	$\pm\dfrac{1}{(k-1)!}g^{k-1}e^{\mp bg}$	$0 < \pm g$
(438)	$\dfrac{1}{p+b}$	e^{-bg}	$0 < g$
(439)	$\dfrac{1}{p-b}$	$-e^{bg}$	$g < 0$
(442)	$\dfrac{1}{(p \pm b)^2}$	$\pm g e^{\mp bg}$	$0 < \pm g$
(448)	$\dfrac{1}{(p \pm a)(p \pm b)}$	$\dfrac{e^{\mp bg} - e^{\mp ag}}{a - b}$	$0 < \pm g$
(449)	$\dfrac{p}{(p \pm a)(p \pm b)}$	$\pm\dfrac{ae^{\mp ag} - be^{\mp bg}}{a - b}$	$0 < \pm g$
(452)	$\dfrac{1}{(p+a)(p+b)(p+c)}$	$\dfrac{(c-b)e^{-ag} + (a-c)e^{-bg} + (b-a)e^{-cg}}{(a-b)(b-c)(c-a)}$	$0 < g$
(453)	$\dfrac{p}{(p+a)(p+b)(p+c)}$	$\dfrac{a(b-c)e^{-ag} + b(c-a)e^{-bg} + c(a-b)e^{-cg}}{(a-b)(b-c)(c-a)}$	$0 < g$

continued overleaf

TABLE 1. (*continued*)

$$G(g) = \int_{-\infty}^{\infty} F(f)e^{i2\pi fg}\, df \qquad \text{the plus-}i \text{ transform of } F(f)$$

$$F(f) = \int_{-\infty}^{\infty} G(g)e^{-i2\pi fg}\, dg \qquad \text{the minus-}i \text{ transform of } G(g)$$

$F(f)$	$(p = i2\pi f)$	$G(g)$

Part 5. Irrational algebraic functions of f

(521)	$\dfrac{1}{p^a}$	$\text{Re } a \geq 1$	$\dfrac{1}{\Gamma(a)} g^{a-1} \qquad 0 < g$
(522.5)	$\dfrac{1}{|p|^a}$	$\text{Re } a \geq 1$	$\dfrac{|g|^{a-1}}{2\Gamma(a)\cos(\pi a/2)}$
(523)	$\dfrac{1}{|f|^{1/2}}$		$\dfrac{1}{|g|^{1/2}}$

Part 6. Exponential and trigonometric functions of f or $1/f$

(601)	e^{-xp}		$\delta(g - x)$
(604)	$\dfrac{e^{-xp}}{p \pm b}$		$\pm e^{\mp b(g-x)} \qquad \pm x < \pm g$
(606.1)	$\dfrac{e^{-xp}}{p^a}$	$\text{Re } a \geq 1$	$\dfrac{1}{\Gamma(a)}(g - x)^{a-1} \qquad x < g$
(622)	$\dfrac{\sinh Ap}{p}$	$A(\text{real}) > 0$	$\dfrac{1}{2} \qquad |g| < A$
(625)	$\text{sech } \pi f$		$\text{sech } \pi g$
(632)	$e^{-a|p|}$		$\dfrac{a}{\pi(g^2 + a^2)}$

[a] a, b, and c (α, β, and γ in original CF) are real > 0 or complex, real part > 0.

BH Equations[a]

*The following material is reprinted from Nouvelles Tables d'Intégrales Définies
(D. Bierens de Haan, 1939) courtesy of G. E. Stechert & Co.*

$f(x)$	$\int_0^\infty f(x)\,dx$

Table 20

(11) $\dfrac{x^2}{x^6 + px^4 + qx^2 + r}$ $\dfrac{\pi\sqrt{r}}{[a(a^2 - p)\sqrt{r}] - 2r}$

where a is the largest root of the equation

$$(z^2 - p)^2 - 8z\sqrt{r} - 4q = 0$$

Table 150

(1) $x^{p-1}\sin qx$ $\dfrac{1}{q^p}\,\Gamma(p)\sin\dfrac{p\pi}{2}$ $p^2 < 1$

(2) $x^{p-1}\cos qx$ $\dfrac{1}{q^p}\,\Gamma(p)\cos\dfrac{p\pi}{2}$ $p^2 < 1$

Table 151

(1) $\dfrac{\sin px}{x}$ $\begin{cases} \dfrac{\pi}{2} & p > 0 \\[2mm] 0 & p = 0 \\[2mm] -\dfrac{\pi}{2} & p < 0 \end{cases}$

(8) $\dfrac{\sin qx \cos px}{x}$ $\begin{cases} \dfrac{\pi}{2} & q > p \\[2mm] 0 & q < p \\[2mm] \dfrac{\pi}{4} & q = p \end{cases}$

(9) $\dfrac{\sin qx \cos^2 px}{x}$ $\begin{cases} \dfrac{\pi}{2} & q > 2p \\[2mm] \dfrac{3\pi}{8} & q = 2p \\[2mm] \dfrac{\pi}{4} & q < 2p \end{cases}$

continued overleaf

BH Equations (*continued*)

$f(x)$	$\displaystyle\int_0^\infty f(x)\,dx$
(10) $\dfrac{\sin^2 qx \sin px}{x}$	$\dfrac{\pi}{4} \qquad p < 2q$
	$\dfrac{\pi}{8} \qquad p = 2q$
	$0 \qquad p > 2q$
(12) $\dfrac{\sin^2 qx \cos px}{x}$	$\dfrac{1}{8}\ln\dfrac{(p^2 - 4q^2)^2}{p^4}$

Table 157

(1) $\dfrac{\sin qx \sin px}{x^2}$	$\dfrac{p\pi}{2} \qquad p \le q$
	$\dfrac{q\pi}{2} \qquad p \ge q$
(2) $\dfrac{\sin^2 qx \sin px}{x^2}$	$\dfrac{2q + p}{8}\ln(2q + p)^2 - \dfrac{2q - p}{8}\ln(2q - p)^2 - \dfrac{1}{2}p\ln p$
(5) $\dfrac{\sin^2 qx \cos px}{x^2}$	$0 \qquad p \ge 2q$
	$\dfrac{2q - p}{4}\pi \qquad p < 2q$

Table 158

(1) $\dfrac{1 - \cos qx}{x^2}$	$\dfrac{q\pi}{2}$
(2) $\dfrac{\cos qx - \cos px}{x^2}$	$\dfrac{(p - q)\pi}{2}$
(3) $\dfrac{\sin x - x \cos x}{x^2}$	1

BH Equations (*continued*)

	$f(x)$	$\int_0^\infty f(x)\,dx$

Table 160

(1) $\dfrac{\sin px}{q+x}$ $\sin(pq)\mathrm{Ci}(pq) + \cos(pq)\left(\dfrac{\pi}{2} - \mathrm{Si}(pq)\right)$

(2) $\dfrac{\cos px}{q+x}$ $-\cos(pq)\mathrm{Ci}(pq) + \sin(pq)\left(\dfrac{\pi}{2} - \mathrm{Si}(pq)\right)$

(3) $\dfrac{\sin px}{q^2 + x^2}$ $\dfrac{1}{2q}\left[e^{-pq}\mathrm{Ei}(pq) - e^{pq}\mathrm{Ei}(-pq)\right]$

(4) $\dfrac{x \sin px}{q^2 + x^2}$ $\dfrac{\pi}{2}\,e^{-pq}$

(5) $\dfrac{\cos px}{q^2 + x^2}$ $\dfrac{\pi}{2q}\,e^{-pq}$

(6) $\dfrac{x \cos px}{q^2 + x^2}$ $-\dfrac{1}{2}\left[e^{pq}\mathrm{Ei}(-pq) + e^{-pq}\mathrm{Ei}(pq)\right]$

Table 170

(3) $\dfrac{x \sin px}{(q^2 + x^2)^2}$ $\dfrac{\pi}{4q}\,pe^{-pq}$

(4) $\dfrac{x^3 \sin px}{(q^2 + x^2)^2}$ $\dfrac{\pi}{4}\,(2 - pq)e^{-pq}$

(5) $\dfrac{x \sin px}{(q^2 + x^2)^3}$ $\dfrac{\pi}{16q^3}\,(pq + 1)pe^{-pq}$

(7) $\dfrac{\cos px}{(q^2 + x^2)^2}$ $\dfrac{\pi}{4q^3}\,(1 + pq)e^{-pq}$

(8) $\dfrac{x^2 \cos px}{(q^2 + x^2)^2}$ $\dfrac{\pi}{4q}\,(1 - pq)e^{-pq}$

(9) $\dfrac{\cos px}{(q^2 + x^2)^3}$ $\dfrac{\pi}{16q^5}\,(3 + 3pq + p^2q^2)e^{-pq}$

continued overleaf

BH Equations (continued)

$f(x)$	$\int_0^\infty f(x)\,dx$

Table 174

(1) $\dfrac{x \sin px}{(q^2 + x^2)(r^2 + x^2)}$ $\qquad$ $\dfrac{\pi}{2(q^2 - r^2)}(e^{-pr} - e^{-pq})$

(2) $\dfrac{x^3 \sin px}{(q^2 + x^2)(r^2 + x^2)}$ $\qquad$ $\dfrac{\pi}{2(q^2 - r^2)}(q^2 e^{-pq} - r^2 e^{-pr})$

Table 175

(1) $\dfrac{\cos px}{(q^2 + x^2)(r^2 + x^2)}$ $\qquad$ $\dfrac{\pi}{2qr(q^2 - r^2)}(qe^{-pr} - re^{-pq})$

(2) $\dfrac{x^2 \cos px}{(q^2 + x^2)(r^2 + x^2)}$ $\qquad$ $\dfrac{\pi}{2(q^2 - r^2)}(qe^{-pq} - re^{-pr})$

Table 177

(1) $\dfrac{\sin px}{x^{1/2}}$ $\qquad$ $\left(\dfrac{\pi}{2p}\right)^{1/2}$

(2) $\dfrac{\cos px}{x^{1/2}}$ $\qquad$ $\left(\dfrac{\pi}{2p}\right)^{1/2}$

Table 261

(1) $e^{-px} \sin qx$ $\qquad$ $\dfrac{q}{p^2 + q^2}$

(2) $e^{-px} \cos qx$ $\qquad$ $\dfrac{p}{p^2 + q^2}$

Table 264

(1) $\dfrac{\sin px}{e^{qx} + 1}$ $\qquad$ $\dfrac{1}{2p} - \dfrac{1}{q}\dfrac{\pi}{e^{p\pi/q} - e^{-p\pi/q}}$

(2) $\dfrac{\sin px}{e^{qx} - 1}$ $\qquad$ $\dfrac{\pi}{2q}\dfrac{e^{2p\pi/q} + 1}{e^{2p\pi/q} - 1} - \dfrac{1}{2p}$

(6) $\dfrac{\sin px}{e^{qx} - e^{-qx}}$ $\qquad$ $\dfrac{\pi}{4q}\dfrac{e^{p\pi/q} - 1}{e^{p\pi/q} + 1}$

BH Equations (*continued*)

	$f(x)$	$\displaystyle\int_0^\infty f(x)\,dx$
(14)	$\dfrac{\cos px}{e^{qx} + e^{-qx}}$	$\dfrac{\pi}{2q}\,\dfrac{1}{e^{p\pi/2q} + e^{-p\pi/2q}}$
(16)	$\dfrac{e^{qx}\cos px}{(e^{qx} + 1)^2}$	$\dfrac{1}{q^2}\,\dfrac{p\pi}{e^{p\pi/q} - e^{-p\pi/q}}$
(19)	$\dfrac{\sin px\,\sin rx}{e^{qx} + e^{-qx}}$	$\dfrac{\pi}{2q}\,\dfrac{\sinh(p\pi/2q)\,\sinh(r\pi/2q)}{\cosh(p\pi/q) + \cosh(r\pi/q)}$
(20)	$\dfrac{\sin px\,\cos rx}{e^{qx} - e^{-qx}}$	$\dfrac{\pi}{4q}\,\dfrac{\sinh(r\pi/q)}{\cosh(p\pi/q) + \cosh(r\pi/q)}$
(21)	$\dfrac{\cos px\,\cos rx}{e^{qx} + e^{-qx}}$	$\pi\,\dfrac{\cosh(p\pi/2q)}{2q}\,\dfrac{\cosh(r\pi/2q)}{\cosh(p\pi/q) + \cosh(r\pi/q)}$

Table 265

	$f(x)$	$\displaystyle\int_0^\infty f(x)\,dx$	
(1)	$\dfrac{\sinh qx}{\cosh qx}\sin rx$	$\dfrac{\pi}{2q}\,\dfrac{1}{\sinh(r\pi/2q)}$	
(2)	$\dfrac{\sinh px}{\cosh qx}\sin rx$	$\dfrac{\pi}{q}\,\dfrac{\sinh(r\pi/2q)\,\sin(p\pi/2q)}{\cosh(r\pi/q) + \cos(p\pi/q)}$	$p < 2q$
(3)	$\dfrac{\cosh qx}{\sinh qx}\sin rx$	$\dfrac{\pi}{2q}\,\dfrac{e^{r\pi/q} + 1}{e^{r\pi/q} - 1}$	
(4)	$\dfrac{\cosh px}{\sinh qx}\sin rx$	$\dfrac{\pi}{2q}\,\dfrac{\sinh(r\pi/q)}{\cosh(r\pi/q) + \cos(p\pi/q)}$	$p^2 \leq q^2$
(6)	$\dfrac{\cosh px}{\cosh qx}\cos rx$	$\dfrac{\pi}{q}\,\dfrac{\cosh(r\pi/2q)\,\cos(p\pi/2q)}{\cosh(r\pi/q) + \cos(p\pi/q)}$	$p < 2q$
(7)	$\dfrac{\sinh px}{\sinh qx}\cos rx$	$\dfrac{\pi}{2q}\,\dfrac{\sin(p\pi/q)}{\cosh(r\pi/q) + \cos(p\pi/q)}$	$p^2 \leq q^2$

Table 281

	$f(x)$	$\displaystyle\int_0^\infty f(x)\,dx$	
(1)	$e^{-x/k}\sin qx\,\sin rx$	$\dfrac{1}{2}\,\dfrac{k}{1 + (q - r)^2 k^2}$	$\lim k \to \infty$
(2)	$e^{-x/k}\cos qx\,\cos rx$	$\dfrac{1}{2}\,\dfrac{k}{1 + (q - r)^2 k^2}$	$\lim k \to \infty$

continued overleaf

BH Equations (*continued*)

$f(x)$	$\int_0^\infty f(x)\,dx$

Table 361

(1) $xe^{-px}\sin qx$ $\dfrac{2pq}{(p^2+q^2)^2}$

(2) $x^2e^{-px}\sin qx$ $2\,\dfrac{3p^2q-q^3}{(p^2+q^2)^3}$

(5) $xe^{-px}\cos qx$ $\dfrac{p^2-q^2}{(p^2+q^2)^2}$

(6) $x^2e^{-px}\cos qx$ $2\,\dfrac{p^3-3pq^2}{(p^2+q^2)^3}$

Table 364

(1) $\dfrac{x}{2\sinh x}\cos qx$ $\dfrac{1}{2}\,\pi^2\,\dfrac{e^{-q\pi}}{(1+e^{-q\pi})^2}$

(2) $x\,\dfrac{\sinh x}{\cosh x}\cos qx$ $-\dfrac{1}{2}\,\pi^2 e^{-1/2q\pi}\,\dfrac{1+e^{-q\pi}}{(1-e^{-q\pi})^2}$

(3) $x\,\dfrac{\cosh x}{\sinh x}\cos qx$ $-\pi^2\,\dfrac{e^{-q\pi}}{(1-e^{-q\pi})^2}$

(6) $\dfrac{x}{2\cosh \pi x}\sin qx$ $\dfrac{1}{8}\,\dfrac{\sinh(q/2)}{[\cosh(q/2)]^2}$

(7) $\dfrac{x}{2\sinh \pi x}\cos qx$ $\dfrac{1}{2}\,\dfrac{e^q}{(e^q+1)^2}$

Table 365

(1) $\dfrac{e^{-px}}{x}\sin qx$ $\arctan\dfrac{q}{p}$

(2) $\dfrac{e^{-ipx}}{x}\sin qx$ $\dfrac{1}{2}\,i\ln\dfrac{p-q}{p+q}$

(5) $\dfrac{e^{-px}}{x}\sin qx\,\sin rx$ $\dfrac{1}{4}\,\ln\dfrac{p^2+(q+r)^2}{p^2+(q-r)^2}$

(6) $\dfrac{e^{-px}}{x}\sin rx\,\cos qx$ $\dfrac{1}{2}\,\arctan\dfrac{2pr}{p^2+q^2-r^2}$

---

[a] For BH Equations, p, q, and r are real constants.

BMP Equations[a]

The following material is from Tables of Integral Transforms I. Bateman Manuscript Project by Erdélyi. Copyright © 1954, McGraw-Hill Book Co., Inc. Used with the permission of McGraw-Hill Book Co.

Fourier Cosine Transforms

$f(x)$	$g(y) = \int_0^\infty f(x) \cos(xy)\, dx, \ y > 0$

1.1. General

(1) $\quad g(x)$

$$\frac{\pi}{2} f(y)$$

(2) $\quad f(ax)$

$\qquad a > 0$

$$\frac{1}{a} g\left(\frac{y}{a}\right)$$

(3) $\quad f(ax) \cos bx$

$\qquad a, b > 0$

$$\frac{1}{2a}\left[g\left(\frac{y+b}{a}\right) + g\left(\frac{y-b}{a}\right) \right]$$

(4) $\quad f(ax) \sin bx$

$\qquad a, \ b > 0$

$$\frac{1}{2a}\int_0^\infty f(x) \sin\left(\frac{y+b}{a}x\right) dx - \frac{1}{2a}\int_0^\infty f(x) \sin\left(\frac{y-b}{a}x\right) dx$$

(5) $\quad x^{2n} f(x)$

$$(-1)^n \frac{d^{2n}}{dy^{2n}} g(y)$$

(6) $\quad x^{2n+1} f(x)$

$$(-1)^n \frac{d^{2n+1}}{dy^{2n+1}} \int_0^\infty f(x) \sin(xy)\, dx$$

1.2. Algebraic functions

(1) $\quad 1 \quad 0 < x < a$

$\qquad\ \ 0 \quad a < x < \infty$

$$\frac{1}{y} \sin ay$$

(3) $\quad 0 \quad 0 < x < a$

$\qquad \dfrac{1}{x} \quad a < x < \infty$

$$-\text{Ci}(ay)$$

1.3. Powers with arbitrary index

(1) $\quad x^{-\nu} \quad 0 < \text{Re } \nu < 1$

$$\frac{\pi}{2} \frac{\sec (\nu\pi/2)}{\Gamma(\nu)} y^{\nu-1}$$

---

continued overleaf

Fourier Cosine Transforms (*continued*)

$f(x)$	$g(y) = \int_0^\infty f(x)\cos(xy)\,dx,\ y > 0$

1.4. Exponential functions

(2) $\dfrac{1}{x}(e^{-bx} - e^{-ax})$ $\dfrac{1}{2}\ln\dfrac{a^2 + y^2}{b^2 + y^2}$

       $\operatorname{Re} a,\ \operatorname{Re} b > 0$

(11) e^{-ax^2} $\operatorname{Re} a > 0$ $\dfrac{1}{2}(\pi/a)^{1/2}e^{-y^2/4a}$

1.5. Logarithmic functions

(1) $\ln x$ $0 < x < 1$ $-\dfrac{1}{y}\operatorname{Si}(y)$

          0 $1 < x < \infty$

(3) $\dfrac{\ln bx}{x^2 + a^2}\ \operatorname{Re} a > 0$ $\dfrac{\pi}{4a}[2e^{-ay}\ln ab + e^{ay}\operatorname{Ei}(-ay) - e^{-ay}\operatorname{Ei}(ay)]$

(12) $\ln\dfrac{a^2 + x^2}{b^2 + x^2}$ $\operatorname{Re} a > 0$ $\dfrac{\pi}{y}(e^{-by} - e^{-ay})$

                      $\operatorname{Re} b > 0$

1.6. Trigonometric functions

(3) $\dfrac{x\sin ax}{x^2 + b^2}$ $\dfrac{\pi}{2}e^{-ab}\cosh by$ $y < a$

       $a > 0,\ \operatorname{Re} b > 0$ $-\dfrac{\pi}{2}e^{-by}\sinh ab$ $y > a$

(4) $\dfrac{\sin ax}{x(x^2 + b^2)}$ $-\dfrac{\pi}{2b^2}e^{-ab}\cosh by + \dfrac{\pi}{2b^2}$ $y < a$

       $a > 0,\ \operatorname{Re} b > 0$ $\dfrac{\pi}{2b^2}e^{-by}\sinh ab$ $y > a$

(6) $e^{-bx}\sin ax$ $\dfrac{(a + y)/2}{b^2 + (a + y)^2} + \dfrac{(a - y)/2}{b^2 + (a - y)^2}$

       $a > 0,\ \operatorname{Re} b > 0$

(18) $\dfrac{\cos ax}{x^2 + b^2}$ $\dfrac{\pi}{2b}e^{-ab}\cosh by$ $y < a$

       $a > 0,\ \operatorname{Re} b > 0$ $\dfrac{\pi}{2b}e^{-by}\cosh ab$ $y > a$

(19) $e^{-bx}\cos ax$ $\dfrac{b/2}{b^2 + (a - y)^2} + \dfrac{b/2}{b^2 + (a + y)^2}$

       $\operatorname{Re} b > |\operatorname{Im} a|$

Fourier Cosine Transforms (continued)

$f(x)$	$g(y) = \int_0^\infty f(x) \cos(xy)\, dx, \; y > 0$

1.8. Inverse trigonometric functions

(7) $\tan^{-1}(a/x)$ $a > 0$ $\dfrac{1}{2y}[e^{-ay}\,\mathrm{Ei}(ay) - e^{ay}\,\mathrm{Ei}(-ay)]$

(8) $\tan^{-1}(2/x^2)$ $\dfrac{\pi}{y}\,e^{-y}\sin y$

1.9. Hyperbolic functions

(1) $\mathrm{sech}(ax)$
   $\mathrm{Re}\, a > 0$ $\dfrac{\pi}{2a}\,\mathrm{sech}\dfrac{\pi y}{2a}$

(2) $(\mathrm{sech}\, ax)^2$
   $\mathrm{Re}\, a > 0$ $\dfrac{\pi}{2a^2}\,\mathrm{csch}\dfrac{\pi y}{2a}$

(18) $x\,\mathrm{csch}\, ax$
   $\mathrm{Re}\, a > 0$ $\left(\dfrac{\pi}{2a}\right)^2 \mathrm{sech}^2\dfrac{\pi y}{2a}$

(21) $\dfrac{\mathrm{sech}\,\pi x}{1 + x^2}$ $2\cosh(y/2) - [e^y \tan^{-1}(e^{-y/2}) + e^{-y}\tan^{-1}(e^{y/2})]$

(22) $\dfrac{\mathrm{sech}(\pi x/2)}{1 + x^2}$ $(\cosh y)\ln(1 + e^{-2y}) + ye^{-y}$

(29) $\dfrac{x\tanh(\pi x/2)}{1 + x^2}$ $-ye^{-y} - \ln(1 - e^{-2y})\cosh y$

1.11. Gamma function, etc.

(13) $\mathrm{Ci}(ax)$ $a > 0$ $\begin{aligned} &0 \qquad 0 < y < a \\ &-\dfrac{\pi}{2y} \qquad a < y < \infty \end{aligned}$

(15) $\dfrac{\mathrm{Ci}(ax)}{x^2 + b^2}$

   $a, b > 0$ $\dfrac{\pi}{2b}\cosh by\,\overline{\mathrm{Ei}}(-ab) \qquad 0 < y \le a$

   $\dfrac{\pi}{4b}\left\{e^{-by}[\overline{\mathrm{Ei}}(ab) + \mathrm{Ei}(-ab) - \overline{\mathrm{Ei}}(by)] + e^{by}\mathrm{Ei}(-by)\right\}$

   $a \le y < \infty$

(19) $\mathrm{Ei}(-x)$ $-\dfrac{1}{y}\tan^{-1}y$

continued overleaf

Fourier Sine Transforms

$f(x)$	$g(y) = \int_0^\infty f(x) \sin(xy)\, dx, \; y > 0$

2.1. General

(1) Same as 1.1(1)

(2) Same as 1.1(2)

(3) Same as 1.1(3)

(4) $f(ax) \sin bx$

    $a, b > 0$

$$\frac{1}{2a} \int_0^\infty f(x) \cos\left(\frac{y-b}{a} x\right) dx$$

$$-\frac{1}{2a} \int_0^\infty f(x) \cos\left(\frac{y+b}{a} x\right) dx$$

(5) Same as 1.1(5)

(6) $x^{2n+1} f(x)$

$$(-1)^{n+1} \frac{d^{2n+1}}{dy^{2n+1}} \int_0^\infty f(x) \cos(xy)\, dx$$

2.2. Algebraic functions

(1)   1 $0 < x < a$

     0    $a < x < \infty$

$$\frac{1}{y}(1 - \cos ay)$$

(4) $\dfrac{1}{x}$ $0 < x < a$

     0    $a < x < \infty$

$$\mathrm{Si}\,(ay)$$

(5)   0 $0 < x < a$

     $\dfrac{1}{x}$ $a < x < \infty$

$$-\mathrm{Si}(ay) + \frac{\pi}{2}$$

(20) $\dfrac{1}{x(x^2 + a^2)}$ $\mathrm{Re}\, a > 0$

$$\frac{\pi}{2a^2}(1 - e^{-ay})$$

2.3. Powers with arbitrary index

(1) $x^{-\nu}$ $0 < \mathrm{Re}\, \nu < 2$

$$y^{\nu-1} \Gamma(1 - \nu) \cos(\nu\pi/2)$$

Fourier Sine Transforms (*continued*)

$f(x)$	$g(y) = \int_0^\infty f(x) \sin(xy)\,dx, \; y > 0$

2.5. Logarithmic functions[a]

(1) $\ln x$ $0 < x < 1$ $-\dfrac{1}{y}\,[C + \ln y - \mathrm{Ci}(y)]$

          0 $1 < x < \infty$

(2) $\dfrac{\ln x}{x}$ $-\dfrac{\pi}{2}\,(C + \ln y)$

(11) $\ln\left|\dfrac{x+a}{x-a}\right|$ $a > 0$ $\dfrac{\pi}{y}\sin(ay)$

(14) $\dfrac{\ln(1 + a^2 x^2)}{x}$ $a > 0$ $-\pi\,\mathrm{Ei}(-y/a)$

2.6. Trigonometric functions

(1) $\dfrac{1}{x}\sin ax$ $a > 0$ $\dfrac{1}{2}\ln\left|\dfrac{y+a}{y-a}\right|$

(5) $\dfrac{\sin bx}{x^2 + a^2}$ $\mathrm{Re}\,a > 0$ $\dfrac{\pi}{2a}\,e^{-ab}\sinh ay$ $0 < y < b$

                        $b > 0$ $\dfrac{\pi}{2a}\,e^{-ay}\sinh ab$ $b < y < \infty$

2.8. Inverse trigonometric functions

(3) $\tan^{-1}(x/a)$ $a > 0$ $\dfrac{\pi}{2y}\,e^{-ay}$

(5) $\mathrm{ctn}^{-1} ax$ $\mathrm{Re}\,a > 0$ $\dfrac{\pi}{2y}\,(1 - e^{-y/a})$

(8) $\tan^{-1}(2a/x)$ $\mathrm{Re}\,a > 0$ $\dfrac{\pi}{y}\,e^{-ay}\sinh ay$

continued overleaf

[a] C, as used in this section, refers to Euler's constant.

Fourier Sine Transforms (*continued*)

$f(x)$	$g(y) = \displaystyle\int_0^\infty f(x)\sin(xy)\,dx,\ y > 0$

2.9. Hyperbolic functions

(2) $\operatorname{csch} ax$ Re $a > 0$ $\qquad\dfrac{\pi}{2a}\tanh\dfrac{\pi y}{2a}$

(3) $\operatorname{ctnh}(ax/2) - 1$ Re $a > 0$ $\qquad\dfrac{\pi}{a}\operatorname{ctnh}\dfrac{\pi y}{a} - y$

(4) $1 - \tanh(ax/2)$ Re $a > 0$ $\qquad y - \dfrac{\pi}{a}\operatorname{csch}\dfrac{y}{a}$

(13) $\dfrac{1}{x}\operatorname{sech} ax$ Re $a > 0$ $\qquad 2\tan^{-1}(e^{\pi y/2a})$

(15) $\dfrac{\operatorname{csch} \pi x}{1 + x^2}$ $\qquad -\dfrac{1}{2}ye^{-y} + (\sinh y)\ln(1 + e^{-y})$

(16) $\dfrac{\operatorname{csch}(\pi x/2)}{1 + x^2}$ $\qquad e^y\tan^{-1}(e^{-y}) - e^{-y}\tan^{-1}(e^y)$

(20) $\dfrac{\tanh(\pi x/2)}{1 + x^2}$ $\qquad ye^{-y} - (\sinh y)\ln(1 - e^{-2y})$

(29) $e^{-x}\operatorname{csch} x$ $\qquad \dfrac{\pi}{2}\operatorname{ctnh}\dfrac{\pi y}{2} - \dfrac{1}{y}$

2.11. Gamma functions, etc.

(7) $\operatorname{si}(ax)$ $a > 0$ $\qquad 0 \qquad 0 < y < a$

$\operatorname{si}(ax) = \operatorname{Si}(ax) - \dfrac{\pi}{2}$ $\qquad \dfrac{\pi}{2y} \qquad a < y < \infty$

(8) $\dfrac{\operatorname{si}(bx)}{a^2 + x^2}$ $a, b > 0$ $\qquad \dfrac{\pi}{2a}\operatorname{Ei}(-ab)\sinh ay \qquad 0 < y < b$

$\qquad\qquad\qquad\dfrac{\pi}{4a}e^{-ay}[\operatorname{Ei}(-ay) - \overline{\operatorname{Ei}}(ay) - \operatorname{Ei}(-ab) - \operatorname{Ei}(ab)]$

$\qquad\qquad\qquad + \dfrac{\pi}{2a}\operatorname{Ei}(-ay)\sinh ay \qquad b < y < \infty$

(11) $\operatorname{Si}(bx)$ $0 < x < a$ $\qquad (1/2y)\,[\operatorname{Si}(ay + ab) - \operatorname{Si}(ay - ab)$

$\qquad\quad 0 \qquad a < x < \infty$ $\qquad\qquad\qquad - 2\cos ay\,\operatorname{Si}(ab)]$

$\qquad\quad b > 0$

(17) $\operatorname{Ci}(bx)$ $0 < x < a$ $\qquad (1/2y)\,\{-2\cos(ay)\operatorname{Ci}(ab) + \operatorname{Ci}(ay + ab)$

$\qquad\quad 0 \qquad a < x < \infty$ $\qquad\qquad + \operatorname{Ci}(ay - ab) + \log[b^2/(y^2 - b^2)]\}\quad b < y < \infty$

$\qquad\quad b > 0$

(18) $\operatorname{Ei}(-ax)$ $a > 0$ $\qquad (1/2y)\log(1 + y^2/a^2)$

Exponential Fourier Transforms

$f(x)$	$g(y) = \displaystyle\int_{-\infty}^{\infty} f(x)e^{-ixy}\, dx$

3.1. General

(1) $\quad g(x)$ $\qquad\qquad 2\pi f(-y)$

(2) $\quad f^*(x)$ $\qquad\qquad g^*(-y)$

(3) $\quad f(x) = f(-x)$ $\qquad 2\displaystyle\int_0^{\infty} f(x)\cos xy\, dx$

(4) $\quad f(x) = -f(-x)$ $\qquad 2i\displaystyle\int_0^{\infty} f(x)\sin xy\, dx$

(5) $\quad f\left(\dfrac{x}{a} + b\right) \quad a > 0$ $\qquad ae^{iaby}g(ay)$

(6) $\quad f\left(b - \dfrac{x}{a}\right) \quad a > 0$ $\qquad ae^{-iaby}g(-ay)$

(7) $\quad f(ax)e^{ibx} \quad a > 0$ $\qquad \dfrac{1}{a}g\left(\dfrac{y-b}{a}\right)$

(8) $\quad f(ax)\cos bx \quad a > 0$ $\qquad \dfrac{1}{a}\left[g\left(\dfrac{y-b}{a}\right) + g\left(\dfrac{y+b}{a}\right)\right]$

(9) $\quad f(ax)\sin bx \quad a > 0$ $\qquad \dfrac{1}{2ai}\left[g\left(\dfrac{y-b}{a}\right) - g\left(\dfrac{y+b}{a}\right)\right]$

(10) $\quad x^n f(x)$ $\qquad i^n \dfrac{d^n}{dy^n} g(y)$

(11) $\quad f^{(n)}(x) \ (n^{\text{th}} \text{ derivative})$ $\qquad i^n y^n g(y)$

3.2. Elementary functions

(1) $\quad \dfrac{1}{1 + x^2}$ $\qquad\qquad \pi e^{-|y|}$

continued overleaf

Hilbert Transforms

$f(x)$	$g(y) = \dfrac{1}{\pi} \displaystyle\int_{-\infty}^{\infty} \dfrac{f(x)}{x-y}\,dx$

15.1. General formulas

(1)	$f(x)$	$g(y)$
(2)	$g(x)$	$-f(y)$
(3)	$f(a+x) \quad$ (*a* real)	$g(a+y)$
(4)	$f(ax) \quad a > 0$	$g(ay)$
(5)	$f(-ax) \quad a > 0$	$-g(-ay)$
(6)	$xf(x)$	$yg(y) + \dfrac{1}{\pi}\displaystyle\int_{-\infty}^{\infty} f(x)\,dx$
(7)	$(x+a)f(x)$	$(y+a)g(y) + \dfrac{1}{\pi}\displaystyle\int_{-\infty}^{\infty} f(x)\,dx$
(8)	$f'(x)$	$g'(y)$

15.2. Elementary functions

(6)	$\dfrac{1}{x+a} \quad \text{Im } a > 0$	$\dfrac{i}{y+a}$
(7)	$\dfrac{1}{x+a} \quad \text{Im } a < 0$	$\dfrac{-i}{y+a}$
(10)	$\dfrac{1}{x^2 + a^2} \quad \text{Re } a > 0$	$-\dfrac{y}{a(y^2 + a^2)}$
(11)	$\dfrac{x}{x^2 + a^2} \quad \text{Re } a > 0$	$\dfrac{a}{y^2 + a^2}$
(12)	$\dfrac{\lambda x + \mu a}{x^2 + a^2} \quad \text{Re } a > 0$	$\dfrac{\lambda a - \mu y}{y^2 + a^2}$
(37)	$0 \qquad -\infty < x < a$	$-\dfrac{1}{\pi}\,e^{-by}\text{Ei}(by - ab) \qquad -\infty < x < a$
	$e^{-bx} \qquad a < x < \infty$	$-\dfrac{1}{\pi}\,e^{-by}\overline{\text{Ei}}(by - ab) \qquad a < x < \infty$
	$(b > 0)$	

Hilbert Transforms (*continued*)

$f(x)$	$g(y) = \dfrac{1}{\pi} \displaystyle\int_{-\infty}^{\infty} \dfrac{f(x)}{x - y}\,dx$
(38) e^{iax} $a > 0$	ie^{iay}
(43) $\sin ax$ $a > 0$	$\cos ay$
(44) $\dfrac{\sin ax}{x}$ $a > 0$	$\dfrac{\cos ay - 1}{y}$
(47) $\cos ax$ $a > 0$	$-\sin ay$
(48) $\dfrac{1 - \cos ax}{x}$ $a > 0$	$\dfrac{\sin ay}{y}$

15.3. Higher transcendental functions

$e^{-ax}\mathrm{Ei}(ax)$ $-\infty < x < 0$	0 $-\infty < x < 0$
$e^{-ax}\overline{\mathrm{Ei}}(ax)$ $0 < x < \infty$	πe^{-ay} $0 < y < \infty$

GR Equations

3.241

(4)
$$\int_0^\infty \frac{x^{u-1}\,dx}{(p+qx^a)^{n+1}} = \frac{1}{ap^{n+1}} \left(\frac{p}{q}\right)^{u/a} \frac{\Gamma(u/a)\Gamma[1+n-(u/a)]}{\Gamma(1+n)}$$

$$\text{for } 0 < \frac{u}{a} < n+1$$

References

Adrian, R. H. (1969). Rectification in muscle membrane. *Prog. Biophys. Mol. Biol.* **19**, 339–369.

Adrian, R. H. and W. Almers (1973). Measurement of membrane capacity in skeletal muscle. *Nature New Biol.* **242**, 62–64.

Adrian, R. H. and W. Almers (1976). Membrane capacity measurements on frog skeletal muscle in media of low ion content. *J. Physiol.* **237**, 573–605.

Amatniek, E. (1958). Measurement of bioelectric potentials with microelectrodes and neutralized input capacity amplifier. *IRE Trans. Med. Electronics* **10**, 3–14. (The IRE is now the IEEE).

Anderson, C. R. and C. F. Stevens (1973). Voltage clamp analysis of acetylcholine produced end-plate current fluctuations at frog neromuscular junction. *J. Physiol.* **235**, 655–691.

Anderson, C. F. and M. Cahalan (1974). Gating kinetics and conductance of single Na channels in excitable membranes: Studies of current fluctuations in voltage clamped node of Ranvier. *Proc. Inter. Union of Physiol. Sciences* **11** (abstract #8).

Anderson, C. F., S. G. Cull-Candy, and R. Miledi (1976). Glutamate and quisqualate noise in voltage-clamped locust muscle fibres. *Nature* **261**, 151–153.

Anderson, C. R., S. G. Cull-Candy, and R. Miledi (1977). Potential-dependent transition temperature of ionic channels induced by glutamate in locust muscle. *Nature* **268**, 663–665.

Anderson, C. F., S. G. Cull-Candy, and R. Miledi (1978). Glutamate current noise: post-synaptic channel kinetics investigated under voltage clamp. *J. Physiol.* **282**, 219–242.

Arguimbau, L. B. (1948). *Vacuum Tube Circuits*. New York: John Wiley and Sons.

Atkinson, E. (1890). *Elementary Treatise on Physics* or *Ganot's Physics*. New York: William Wood. Translated and edited by E. Atkinson, from *Elements de Physique* by Ganot, 13th edition of the original translation in 1863.

Bacq, Z. M. (1976). *Chemical Transmission of Nerve Impulses. A Historical Sketch*. Oxford: Pergamon Press.

Barker, J. L. and R. N. McBurney (1978). Different properties of single channels activated by GABA and glycine on cultured mouse neurons. *J. Physiol.* **284**, 127–128P.

Bangham, A. D. (1968). Membrane models with phospholipids. *Prog. Biophys. Mol. Biol.* **18**, 31–95.

Barnard College (1974). History of Physics Laboratory: *Volta and Electricity* (motion picture). Edited by Samuel Devons. For additional information write to author.

Barnes, J. A. and S. Jarvis, Jr. (1971). Efficient numerical and analog modeling of flicker noise processes. *Natl. Bur. Stand. (U. S.) Tech. Note* **604** (June 1971).

Bateman Manuscript Project (1954). Ed. A. Erdélyi. *Tables of Integral Transforms, Vols. I and II.* New York: McGraw-Hill.

Bean, R. C., W. C. Shepherd, H. Chan, and J. Eichner (1969). Discrete conductance fluctuations in lipid bilayer protein membranes. *J. Gen. Physiol.* **53**, 741–757.

Begenisich, T. and C. F. Stevens (1975). How many conductance states do potassium currents have? *Biophys. J.* **15**, 843–846.

Bell, D. A. (1960). *Electrical Noise.* New York: D. Van Nostrand.

Bendat, Julius S. (1958). *Principles and Applications of Random Noise Theory.* New York: John Wiley and Sons.

Bendat, J. S. and A. G. Piersol (1966). *Measurement and Analysis of Random Data.* New York: John Wiley and Sons.

Bendat, J. S. and A. G. Piersol (1971). *Random Data: Analysis and Measurement Procedures.* New York: Wiley-Interscience.

Ben-Haim, D., F. Dreyer, and K. Peper (1975). ACh receptor: modification of synaptic gating mechanisms after treatment with a disulfide bond reducing agent. *Pfluegers Arch.* **355**, 19–26.

Benjamin, Park (1898). *A History of Electricity.* New York: Arno Press. A reprint of the original 1898 edition by John Wiley and Sons, New York.

Bennett, William R. (1960). *Electrical Noise.* New York: McGraw-Hill.

Bevan, S. J., B. Katz, and R. Miledi (1975). Membrane potential fluctuations produced by glutamate in nerve cells of the squid. *Proc. R. Soc. London Ser. B.* **191**, 561–565.

Bierens de Haan, D. (1939). *Nouvelles Tables d'Integrales Definies.* New York: G. E. Stechert & Co. With an English translation of the introduction by Prof. J. F. Ritt.

Bird, J. F. (1974a). Neuronal $1/f$ noise and membrane models. *Biophys. J.* **14**, 563–565.

Bird, J. F. (1974b). Noise spectrum analysis of a Markov process vs. random walk computer solutions simulating $1/f$ noise spectra. *J. Appl. Phys.* **45**, 499–500.

Blair, E. A. and J. Erlanger (1932). Responses of axons to brief shocks. *Proc Soc. Exp. Biol. N. Y.* **29**, 926–927.

Blair, E. A. and J. Erlanger (1933). A comparison of the characteristics of axons through their individual electrical responses. *Am. J. Physiol.* **106**, 524–564.

Blaquiére, Austin (1966). *Nonlinear System Analysis.* New York: Academic Press.

Born, Max (1949). *Natural Philosophy of Cause and Chance.* The 1948 Waynflete Lectures delivered in the College of St. Mary Magdalen at Oxford University. New York: Dover. Originally published by Oxford Univ. Press. Republished with new material by Dover Publications, New York in 1964.

Bracewell, Ronald M. (1965). *The Fourier Transform and Its Applications.* New York: McGraw-Hill.

Bridgman, P. W. (1928). Note on the principle of detailed balancing. *Phys. Rev.* **31**, 101-102.

Brock, L. G., J. S. Coombs, and J. C. Eccles (1952). The recording of potentials from motoneurones with an intracellular electrode. *J. Physiol.* **117**, 431–460.

Brookhart, J. M. and V. B. Mountcastle (1977), editors. *Handbook of Physiology*, Section 1, The Nervous System, Excitation and Conduction. New York: Waverly Press.

Burr-Brown Research Corporation (1966). *Handbook of Operational Amplifier Active RC Networks*, 1st edition. Burr-Brown Research Corporation, Tucson, Arizona.

Butler, J. A. V. (1951), editor. *Electrical Phenomena at Interfaces in Chemistry, Physics and Biology*. London: Methuen & Co.

Byerly, William Elwood (1959). *An Elementary Treatise on Fourier's Series*. New York: Dover. Reprinted from the original 1893 edition by Ginn: Boston.

Calvin, W. H. and C. F. Stevens (1967). Synaptic noise as a source of variability in the interval between action potentials. *Science* **155**, 842–844.

Calvin, W. H. and C. F. Stevens (1968). Synaptic noise and other sources of randomness in motoneuron interspike intervals. *J. Neurophysiol.* **31**, 574–587.

Campbell, N. (1909a). The study of discontinuous phenomena. *Proc. Cambridge Philos. Soc.* **15**, 117–136.

Campbell, N. (1909b). Discontinuities in light emission. *Proc. Cambridge Philos. Soc.* **15**, 310–328.

Campbell, N. (1910). Discontinuities in light emission II. *Proc. Cambridge Philos. Soc.* **15**, 513–525.

Campbell, G. A. and M. R. Foster (1948). *Fourier Integrals for Practical Applications*. Princeton, N. J.: Van Nostrand Co.

Carslaw, H. S. (1930). *Introduction to the Theory of Fourier's Series and Integrals*, 3rd edition. New York: Dover. 1st edition published in 1908.

Chandler, W. K., R. FitzHugh, and K. S Cole (1962). Theoretical stability properties of a space-clamped axon. *Biophys. J.* **2**, 105–127.

Chen, Yi-der and T. L. Hill (1973). Fluctuations and noise in kinetic systems: application to K^+ channels in squid axon. *Biophysiol. J.* **13**, 1276–1295.

Chen, Yi-der (1973). Fluctuations and noise in kinetic systems II. Open ensembles at equilibrium and steady state. *J. Chem. Phys.* **59**, 5810–5813.

Chen, Yi-der (1975). Fluctuations and noise in kinetic systems III. *J. Theor. Biol.* **55**, 229–243.

Chen, Yi-der (1976). Differentiation of channel models by noise analysis. *Biophys. J.* **16**, 965–971.

Chen, Yi-der (1977a). Fluctuations and noise in non-linear kinetic systems. *J. Theor. Biol.* **65**, 357–367.

Chen, Yi-der (1977b). Noise analysis of kinetic systems and its applications to membrane channels. In: *Advances in Chemical Physics*. Ed. S. A. Rice.

Churchill, Ruel V. (1972). *Operational Mathematics*. New York: McGraw-Hill.

Clapham D. E. and L. J. DeFelice (1976). The theoretical small signal impedance of the frog node, *Rana pipiens*. *Pfluegers Arch.* **366**, 273–276.

Clapham, D. E. (1979). A whole tissue model of heart cell aggregates: Electrical coupling between cells, membrane impedance, and the extracellular space. Ph. D. thesis, Emory University.

Clay, John R. (1976). A stochastic analysis of the graded excitatory response of nerve membrane. *J. Theor. Biol.* **59**, 141–158.

Clay, John R. (1977). Monte Carlo simulation of membrane noise: an analysis of fluctuations in graded excitation of nerve membrane. *J. Theor. Biol.* **64**, 671–680.

Clay, John R. (1978). Comparison of ion current noise predicted from different models of the Na channel gating mechanism in nerve membrane. *J. Mem. Biol.* **42**, 215–227.

Clay, John R. and M. F. Shlesinger (1976). Theoretical model of the ionic mechanism of $1/f$ noise in nerve membrane. *Biophys. J.* **16**, 121–136.

Clay, John R. and. M. F. Shlesinger (1977a). Unified theory of $1/f$ and conductance noise in nerve membrane. *J. Theor. Biol.* **66**, 763–773.

Clay, J. R. and M. F. Shlesinger (1977b). Random walk analysis of K fluxes associated with nerve impulses. *Proc. Natl. Acad. Sci.* **74**, 5543–5546.

Clay, J. R., L. J. DeFelice, and R. L. DeHaan (1979). Current noise derived from voltage noise and impedance of heart cell aggregate membranes near rest. *Biophys. J.* **28**, 169–184.

Cochran, W. T. and J. W. Cooley (1967). What is a fast Fourier transform? *IEEE Trans. Audio Electroacoust.* **15**, 45–55.

Cole, K. S. (1941). Rectification and inductance in the squid giant axon. *J. Gen. Physiol.* **25**, 29–51.

Cole, K. S. (1968). *Membranes, Ions and Impulses.* Berkeley: University of California Press.

Cole, K. S. (1979). Mostly membranes. *Ann. Rev. Physiol.* **41**, 1–24.

Colquhoun, D. (1975). Analysis of end-plate current fluctuations produced by acetylcholine and acetylmonoethylcholine in rat muscle. *Br. J. Pharmacol.* **58**, 428–429P.

Colquhoun, D., V. E. Dionne, J. H. Steinbach, and C. F. Stevens (1975). Conductance of channels opened by acetylcholine-like drugs in the muscle end-plate. *Nature* **253**, 204–206.

Colquhoun, D., W. Large, and H. P. Rang (1977). An analysis of the action of a false transmitter at the neuromuscular junction. *J. Physiol.* **266**, 361–395.

Colquhoun, D. and A. G. Hawkes (1977). Relaxation and fluctuations of membrane currents that flow through drug operated ion channels. *Proc. R. Soc. London Ser. B* **199**, 231–262.

Conti, F. (1970). Nerve membrane electrical characteristics near rest. *Biophysik* **6**, 257–270.

Conti, F. and E. Wanke (1975). Channel noise in membranes and lipid bilayers. *Q. Rev. Biophys.* **8**, 451–506.

Conti, F., L. J. DeFelice, and E. Wanke (1975). Potassium and sodium ion current noise in the membrane of the squid giant axon. *J. Physiol.* **248**, 45–82.

Conti, F., B. Hille, B. Neumcke, W. Nonner, and R. Stämpfli (1976a). Measurement of the conductance of the sodium channel from current fluctuations at the node of Ranvier. *J. Physiol.* **262**, 699–727.

Conti, F., B. Hille, B. Neumcke, W. Nonner, and R. Stämpfli (1976b). Conductance of the sodium channel in myelinated nerve fibers with modified sodium inactivation. *J. Physiol.* **262**, 729–742.

Crawford, A. C. and R. N. McBurney (1976). On the elementary conductance event produced by L-glutamate and quanta of the natural transmitter at the neuromuscular junctions of *Maia squinado*. *J. Physiol.* **258**, 205–225.

Cull-Candy, S. G., R. Miledi, and A. Trautmann (1978). ACh-induced channels and transmitter release at human endplates. *Nature* **271**, 74-75.

Cull-Candy, S. G., R. Miledi, and A. Trautmann (1979a). ACh receptors in organ-cultured human muscle fibers. *Nature* **277**, 236–238.

Cull-Candy, S. G., R. Miledi, and A. Trautmann (1979b). End-plate currents and ACh noise at normal and myasthenic human end plates. *J. Physiol.* **287**, 247–265.

Dale, B., L. J. DeFelice, and V. Taglietti (1978). Membrane noise and conductance increase during single sperm–egg interactions. *Nature* **275**, 217–219.

Davenport, Wilbur B., Jr. (1970). *Probability and Random Processes*. New York: McGraw-Hill.

Davies, J. T. and E. K. Rideal (1961). *Interfacial Phenomena*. New York: Academic Press.

Davis, Harold T. (1960). *Introduction to Nonlinear Differential and Integral Equations*. New York: Dover. 1962 reprint of the U. S. Atomic Energy Commission report originally published in 1960.

DeFelice, L. J. (1976). $1/f$-resistor noise. *J. Appl. Phys.* **47**, 350–352.

DeFelice, L. J. (1977). Fluctuation analysis in neurobiology. *Int. Rev. Neurobiol.* **20**, 169–208.

DeFelice, L. J. (1978). Oliver Heaviside: A speech by B. van der Pol. *Electron. Power*. **24**, 222–226 (IEE, Great Britain.)

DeFelice, L. J. and J. R. Adair (1973). Electrical noise from single cells. *Biophys. J.* **13**, 72a.

DeFelice, L. J. and D. L. Alkon (1977). Voltage noise from hair cells during mechanical stimulation. *Nature* **269**, 613–615.

DeFelice, L. J. and B. Dale (1979). Voltage response to fertilization and polyspermy in sea urchin eggs. *Develop. Biol.* **72**, 327–341.

DeFelice, L. J. and R. L. DeHaan (1977). Membrane noise and intercellular communication. *Proc. IEEE* **65**, 796–799. (Special Issue on Biological Signals.)

DeFelice, L. J. and D. R. Firth (1971). Spontaneous voltage fluctuations in glass microelectrodes. *IEEE Trans. Biomed. Eng.* BME-**18**, 339–351.

DeFelice, L. J. and. J. P. L. M. Michalides (1972). Electrical noise from synthetic membranes. *J. Membr. Biol.* **9**, 261–290.

DeFelice, L. J. and B. A. Sokol (1976a). A new analysis for membrane noise. The Integral Spectrum. *Biophys. J.* **16**, 827–838.

DeFelice, L. J. and B. A. Sokol (1976b). Correlation analysis of membrane noise. *J. Membr. Biol.* **26**, 405–406.

DeFelice, L. J., E. Wanke, and F. Conti (1975). Potassium and sodium current noise from squid axon membranes. *Fed. Proc. Fed. Am. Soc. Exp. Biol.* **34**, 1338–1342.

DeGoede, J. and A. A. Verveen (1977). Electrical membrane noise: Its origin and interpretation. In: *Electrical phenomena at the biological membrane level*. Ed. E. Roux, pp. 337–348. Amsterdam: Elsevier.

DeGroot, S. R. (1966). *Thermodynamics of Irreversible Processes*. Amsterdam: North-Holland.

DeHaan, R. L. and L. J. DeFelice (1978a). Oscillatory properties and excitability of the heart cell membrane. In: *Theoretical Chemistry: Advances and Perspectives: Periodicities in Chemistry and Biology, vol. 4*. Eds. H. Eyring and D. Henderson, pp. 181–233. New York: Academic Press.

DeHaan, R. L. and L. J. DeFelice (1978b). Electrical noise and rhythmic properties of embryonic heart cell aggregates. *Fed. Proc. Fed. Am. Soc. Exp. Biol.* **37**, 2132–2138.

del Castillo, J. and L. Stark (1952). Local responses in single medullated nerve fibres. *J. Physiol.* **118**, 207–215.

del Castillo, J. and E. E. Suckling (1957). Possible quantal nature of subthreshold responses at single nodes of Ranvier. *Fed. Proc. Fed. Am. Soc. Exp. Biol.* **16**, 29.

Derksen, H. E. (1965). Axon membrane voltage fluctuations. *Acta Physiol. Pharmacol. Neerl.* **13**, 373–466. Ph. D. thesis, University of Leiden.

Derksen, H. E. and A. A. Verveen (1966). Fluctuations of resting membrane potential. *Science* **151**, 1388–1389.

Detwiler, P. B., A. C. Hodgkin, and P. A. McNaughton (1978). A surprising property of electrical spread in the network of rods in the turtle's retina. *Nature* **274**, 562–565.

Dibner, Bern (1952). *Galvani-Volta: A controversy that led to the discovery of useful electricity.* Norwalk, Connecticut: Burndy Library.

Dionne, V. E. and R. L. Ruff (1976). Endplate current fluctuations reveal only one channel type at the frog neuromuscular junction. *Biophys. J.* **16**, 212a.

Dionne, V. E. and R. L. Ruff (1977). Endplate current fluctuations reveal only one channel type at frog neuromuscular junction. *Nature* **266**, 263–265.

Dodge, F. A., B. W. Knight and J. Toyoda (1968). Voltage noise in *Limulus* visual cells. *Science* **160**, 88–90.

Dorset, D. L. and H. M. Fishman (1975). Excess electrical noise during current flow through porous membranes separating ionic solutions. *J. Membr. Biol.* **21**, 291–309.

Draper, M. H. and S. Weidmann (1951). Cardiac resting and action potentials recorded with an intracellular electrode. *J. Physiol.* **115**, 74–94.

Dreyer, F. and K. Peper (1975). The density, cooperativity, conductance and open time of acetylcholine receptors in the frog neuromuscular junction. *Pfluegers Arch.* **355**, R79.

Dreyer, F., Chr. Walther, and K. Peper (1975). Receptors in the postsynaptic membrane of normal and denervated frog skeletal muscle fibers: their reactions to ACh and other cholinergic agonists. *Pfluegers Arch.* **359**, R71.

Dreyer, F., K.-D. Müller, K. Peper, and R. Sterz (1976a). Temperature dependance of ionic channel properties of ACh-receptors at frog and mouse neuromuscular junctions. *Pfluegers Arch.* **365**, R36.

Dreyer, F., Chr. Walther, and K. Peper (1976b). Junctional and extrajunctional acetylcholine receptors in normal and denervated frog muscle fibers. (Noise analysis experiments with different agonists.) *Pfluegers Arch.* **366**, 1–9.

Dreyer, F. K.-D. Müller, K. Peper, and R. Sterz (1976). The M. omohyoideus of the mouse as a convenient mammalian muscle preparation. (A study of junctional and extrajunctional acetylcholine receptors by noise analysis and cooperativity.) *Pfluegers Arch.* **367**, 115–122.

Ehrenstein, G. M. (1971). Excitability in lipid bilayer membranes. In: *Biophysics and Physiology of Excitable Membranes*, pp. 463–476. Ed. W. J. Adelman, Jr. New York: Van Nostrand Reinhold.

Ehrenstein, G. (1976). Ion channels in nerve membrane. *Physics Today* **29**, 33–39.

Ehrenstein, G. and H. Lecar (1972). The mechanism of signal transmission in nerve axons. In: *Annual Review of Biophysics and Bioengineering I*, 347–368. Ed. M. F. Morales. Palo Alto, California: Annual Reviews, Inc.

Ehrenstein, G. and H. Lecar (1977). Electrically gated ionic channels in lipid bilayers. *Quart. Rev. Biophys.* **10**, 1–34.

Ehrenstein, G., H. Lecar, and R. Nossal (1970). The nature of the negative resistance in bimolecular lipid membranes containing excitability—inducing material. *J. Gen. Physiol.* **55**, 119–133.

Ehrenstein, G., R. Blumenthal, R. Latorre, and H. Lecar (1974). Kinetics of the opening and closing of individual EIM channels in a lipid bilayer. *J. Gen. Physiol.* **63**, 707–721.

Eisenberg, M., J. E. Hall, and C. A. Mead (1973). The nature of the voltage-dependent conductance induced by alamethicin in black lipid membranes. *J. Membr. Biol.* **14**, 143–146.

Eisenberg, R. S. and E. A. Johnson (1970). Three dimensional electrical field problems in Physiology. *Prog. Biophys. Mol. Biol.* **20**, 1–65.

Eisenberg, R. S., R. T. Mathias, and J. L. Rae (1977). Measurement, modeling and analysis of the linear electrical properties of cells. *Ann. N. Y. Acad. Sci.* **303**, 342–354.

Eisenberg, R. S., V. Barcilon, and R. T. Mathias (1979). Electrical properties of a spherical syncytium. *Biophys. J.* **25**, 151–180.

Erdélyi, A. (1954). *Tables of Integral Transforms I and II. Bateman Manuscript Project.* New York: McGraw-Hill.

Ermishkin, L. N., Kh. M. Kasumov, and V. M. Potzeluyev (1976). Single ionic channels induced in lipid bilayers by polyene antibiotics: amphotericin-B and nystatine. *Nature* **262**, 698–699.

Falk, G. and P. Fatt (1964). Linear electrical properties of striated muscle fibers observed with intracellular electrodes. *Proc. R. Soc. London Ser. B.* **160**, 69–123.

Fatt, P. (1961). Intracellular microelectrodes. In: *Methods in Medical Research*, Vol. 9, pp. 381–404. Ed. J. H. Quastel. Chicago: Year Book Medical.

Fatt, P. (1964). An analysis of the transverse electrical impedance of striated muscle. *Proc. R. Soc. London Ser. B. Biol. Sci.* **159**, 606–651.

Fatt, P. and B. Katz (1950). Some observations on biological noise. *Nature* **166**, 597–598.

Fatt, P. and B. Katz (1951). An analysis of the end-plate potential recorded with an intracellular electrode. *J. Physiol.* **115**, 320–369.

Fatt, P. and B. Katz (1952). Spontaneous subthreshold activity at motor nerve endings. *J. Physiol.* **117**, 109–128.

Feher, G. (1978). Emerging techniques: Fluctuation spectroscopy. In: *Trends in Biochemical Sciences* 3, pp. 111–113. Amsterdam: Elsevier/North-Holland.

Feher, G. and M. Weissman (1973). Fluctuation spectroscopy: determination of chemical reaction kinetics from the frequency spectrum of fluctuations. *Proc. Natl. Acad. Sci.* **70**, 870–875.

Feller, W. (1960). *An Introduction to Probability Theory and Its Applications*, Vols. 1 and 2. New York: Wiley & Sons.

Ferrier, J. M., C. Morvan, W. J. Lucas and J. Dainty (1979). Plasmalemma voltage noise in *Chara corallina. Plant Physiol.* **63**, 709–714.

Ferris, C. D. (1974). *Introduction to Bioelectrodes.* New York: Plenum Press.

Filkelstein, A. and A. Mauro (1963). Equivalent circuits as related to ionic systems. *Biophys. J.* **3**, 215–237.

Firth, D. R. (1966). Interspike interval fluctuations in the crayfish stretch receptor. *Biophys. J.* **6**, 201–215.

Firth, D. R. and L. J. DeFelice (1971). Electrical resistance and volume flow in glass microelectrodes. *Can. J. Physiol. Pharmacol.* **49**, 436–447.

Fishman, H. M. (1973a). Relaxation spectra of potassium channel noise from squid axon membranes. *Proc. Natl. Acad. Sci. USA*, **70**, 876–879.

Fishman, H. M. (1973b). Low impedance capillary electrode for wide-band recording of membrane potential in large axons. *IEEE Trans. Bio-Med. Electron.* **20**. 380–382.

Fishman, H. M. (1975a). Patch voltage clamp of squid axon membrane. *J. Membr. Biol.* **24**, 265–277.

Fishman, H. M. (1975b). Axon membrane noise measurements. *Fed. Proc. Fed. Am. Soc. Exp. Biol.* **34**, 1330–1337.

Fishman, H. M. (1975c). Complex impedance measurements of squid axon membrane via input–output cross correlation function. In: *Proceedings of the 1st Symposium on Testing and Identification of Nonlinear Systems*, pp. 257–274. Eds. G. D. McCann and P. Z. Marmarelis. Pasadena: California Institute of Technology.

Fishman, H. M. and D. L. Dorset (1973). Comments on electrical fluctuations associated with active transport. *Biophys. J.* **13**, 1339–1342.

Fishman, H. M., L. E. Moore, and D. J. M. Poussart (1975). Potassium-ion conduct ion noise in squid axon membrane. *J. Membr. Biol.* **24**, 305–328.

Fishman, H. M., L. E. Moore, and D. J. M. Poussart (1976). Identification of Na conduction noise in squid axon. *Biol. Bull.* **151**, 408–409.

Fishman, H. M., L. E. Moore, and D. Poussart (1977). Ion movements and kinetics in squid axon II Spontaneous electrical fluctuations. *Ann. N. Y. Acad. Sci.* **303**, 399–423.

Fishman, H. M., D. J. M. Poussart, and L. E. Moore (1975). Noise measurements in squid axon membrane. *J. Membr. Biol.* **24**, 281–304.

Fishman, H. M., D. J. M. Poussart, L. E. Moore, and E. Siebenga (1977). Potassium conduction description from the low frequency impedance and admittance of squid axon. *J. Membr. Biol.* **32**, 255–290.

FitzHugh, R. (1965). A kinetic model of the conductance changes in nerve membrane. *J. Cell Comp. Physiol.* **66**, 111–118.

FitzHugh, R. (1969). Mathematical models of excitation and propagation in nerve. In: *Biological Engineering*. Ed. H. P. Schwann. New York: McGraw-Hill.

Flasterstein, A. H. (1966a). Voltage fluctuations of metal–electrolyte interfaces in electrophysiology. *Med. Biol. Eng.* **4**, 583–588.

Flasterstein, A. H. (1966b). A general analysis of voltage fluctuations of metal–electrolyte interfaces. *Med. Biol. Eng.* **4**, 589–594.

Frankenhaeuser, B. (1960). Na permeability in toad nerve and in squid nerve. *J. Physiol.* **152**, 159–166.

Franklin, Benjamin (1941). *Experiments and Observations on Electricity*. Cambridge, Massachusetts: Harvard University Press. Edited, with a critical and historical introduction by I. Bernard Cohen, from the 1774 5th English edition.

Franklin, P. (1944). *Methods of Advanced Calculus*. New York: McGraw-Hill.

Frehland, E. (1976). $1/f$ noise generated by constrained diffusion through a flux-controlling boundary. *Z. Naturforsch.* **31a**, 942–948.

Frehland, E. (1977). Diffusion as a source of $1/f$ noise. *J. Membr. Biol.* **32**, 195–196.

Frehland, E. and P. Läuger (1974). Ion transport through pores: transient phenomena. *J. Theor. Biol.* **47**, 189–207.

Fukuyo, H. (1977). Chairman of the Proceedings of the Symposium on $1/f$ fluctuations, Tokyo, 11–13 July 1977. For further information write the author.

Fürth, R. (1956), editor. *Investigations on the Theory of the Brownian Movement* by Albert Einstein. New York: Dover Publications.

Gabor, D. (1950). Communication theory and physics. *Philos. Mag.* **41**, 1161–1187.

Gallin, E. K. and M. L. Wiederhold (1977). Responses of Aplysia statocyst receptor cells to physiologic stimulation. *J. Physiol.* **266**, 123–127.

Galvani, L. (1791). *Effects of Electricity on Muscular Motion*. Translated by Margaret Glover Foley. With notes and critical introduction by I. Bernard Cohen.

Ganot (1890). *Elementary, Treatise on Physics*. New York: William Wood. Translated and edited by E. Atkinson, from *Elements de Physique*, 13th edition of the original translation in 1863.

Geddes, L. A. (1972). *Electrodes and the Measurement of Bioelectric Events*. New York: Wiley-Interscience.

de Groot, S. R. (1966). *Thermodynamics of Irreversible Processes*. Amsterdam: North-Holland Publishing Co.

Geller, M. and E. W. Ng (1969). Table of integrals of the exponential integral. *J. Res. Natl. Bur. Stand.* **73B**, 191–210.

Girault, M. (1966). *Stochastic Processes*. Heidelberg: Springer-Verlag.

Goldman, D. E. (1943). Potential, impedance and rectification in cell membrane. *J. Gen. Physiol.* **27**, 37–60.

Goldman, Stanford (1948). *Frequency Analysis, Modulation and Noise*. New York: Dover Publications. 1967 reprint of the original 1948 edition by McGraw-Hill, New York.

Goldman, Stanford (1949). *Laplace Transform Theory and Electrical Transients*. New York: Dover. 1966 reprinting of the 5th edition published by General Publishing Co., Ontario, Canada.

Goldman, Stanford (1953). *Information Theory*. New York: Dover Publications 1968. Don Mills, Ontario: General Publishing Co.

Gordon, L. G. M. and D. A. Haydon (1972). The unit conductance channel of alamethicin. *Biochim. Biophys. Acta*, **255** 1014–1018.

Gradshteyn, I. S. and I. M. Ryzhik (1965). *Table of Integrals, Series and Products*. New York: Academic Press.

Graham, J. and R. W. Gerard (1946). Membrane potentials and excitation of impaled single muscle fibers. *J. Cell Comp. Physiol.* **28**, 99–117.

Green, M. E. (1974). Noise spectra of ion transport across an onion membrane. *J. Phys. Chem.* **78**, 761–762.

Green, M. E. (1976). Diffusion and $1/f$-noise. *J. Membr. Biol.* **28**, 181–186.

Green, M. E. (1977). Diffusion and $1/f$-noise, part II. *J. Membr. Biol.* **32**, 197–199.

Green, M. E. and M. Yafuso (1969). A study of the noise generated during ion transport across membranes. *J. Phys. Chem.* **72**, 4072–4078.

Guggenheim, E. A. (1967). *Boltzmann's Distribution Law*. Amsterdam: North-Holland Publishing Co., 4th printing.

Hagins, W. A. (1959). The minimum membrane current required to excite a photoreceptor. Abstracts of the 3rd Annual Meeting, Biophysical Society, Paper 01.

Hagins, W. A. (1965). Electrical signs of information flow in photoreceptors. Cold Spring Harbor Symposia on Quantitative Biology **30**, 403–418.

Handel, P. H. (1971). Turbulence theory for the current carriers in solids and a theory of $1/f$-noise. *Phys. Rev. A*3, 2066–2073.

Handel, P. H. (1975). $1/f$-noise, an infrared phenomenon. *Phys. Rev. Lett.* **34**, 1492–1495.

Halford, D. (1968). A general mechanical model for $1F1$-αspectral density random noise with special reference to flicker noise $1/1F1$. *Proc. IEEE* **56**, 251–258.

Haus, H. A., *et al.* (1959). Representation of noise in linear two ports. *Proc. IRE* **47**, 69–74.

Heiden, C. (1969). Power spectrum of stochastic pulse sequences with correlation between the pulse parameters. *Phys. Rev.* **188**, 319–326.

Hill, T. L. (1965). *Statistical Mechanics: Principles and Selected Applications*. New York: McGraw-Hill.

Hill, T. L. (1968). *Thermodynamics for Chemists and Biologists*. New York: Addison-Wesley.

Hill, T. L. and Y.-D. Chen (1972). On the theory of ion transport across nerve membrane. IV. Noise from the open-close kinetics of K^+ channels. *Biophys. J.* **12**, 948–959.

Hladky, S. B. and D. A. Haydon (1970). Discreteness of conductance change in bimolecular lipid membranes in the presence of certain antibiotics. *Nature* **225**, 451–453.

Hladky, S. B. and D. A. Haydon (1972). Ion transfer across lipid membranes in the presence of gramicidin A. I. Studies of the unit conductance channel. *Biochim. Biophys. Acta* **274**, 294–312.

Hochstadt, Harry (1973). *Integral Equations*. New York: John Wiley and Sons.

Hodgkin, A. L. (1951). The ionic basis of electrical activity in nerve and muscle. *Biol. Rev.* **26**, 339–409.

Hodgkin, A. L. (1954). A note on conduction velocity. *J. Physiol.* **125**, 221–224.

Hodgkin, A. L. and B. Katz (1949). The effect of sodium ions on the electrical activity of the giant axon of the squid. *J. Physiol.* **108**, 37–77.

Hodgkin, A. L. and A. F. Huxley (1952). A quantitative description of membrane current and its application to conduction and excitation in nerve. *J. Physiol.* **117**, 500–544.

Hodgkin, A. L. and S. Nakajima (1972). The effect of diameter on the electrical constants of frog skeletal muscle fibers. *J. Physiol.* **221**, 105.

Hoff, H. E. (1936). Galvani and the pre-Galvanian electrophysiologists. *Ann. Sci.* **1**, 157–172.

Holden, A. V. (1976a). Flicker noise and structural changes in nerve membrane. *J. Theor. Biol.* **57**, 243–246.

Holden, A. V. (1976b). Lecture notes in biomathematics Vol. 12: *Models of the Stochastic Activity of Neurons*. Heidelberg: Springer-Verlag, N. Y.

Holden, A. V. and J. E. Rubio (1976). A model for flicker noise in nerve membrane. *Biol. Cyber.* **24**, 227–236.

Hooge, F. N. (1969a). $1/f$ noise is no surface effect. *Phys. Lett.* **26A**, 139–140.

Hooge, F. N. (1969b). Contact noise. *Phys. Lett.* **29A**, 642–643.

Hooge, F. N. (1970). $1/f$ noise in the conductance of ions in aqueous solutions. *Phys. Lett.* **33A**, 169–170.

Hooge, F. N. (1972). Discussions of recent experiments on $1/f$ noise. *Physica* **60**, 130–144.

Hooge, F. N. (1976). $1/f$ noise. *Physica* **83B**, 14–23.

Hooge, F. N. and J. L. M. Gaal (1971). Fluctuations with a $1/f$-spectrum in the conductance of ionic solutions and in the voltage of concentration cells. *Philips Res. Rep.* **26**, 77–90.

Hoshiko, T. (1975), Power density spectrum of frog skin. *J. Gen. Phys.* **66**, 16a.

Hoshiko, T. (1977), Equivalent inductance in frog skin patch clamp current noise. *Biophys. J.* **17**, 22a.

Jack, J. J. B., D. Noble, and R. W. Tsien (1975). *Electric Current Flow in Excitable Cells*. London: Oxford University Press.

Jackson, J. D. (1975). *Classical Electrodynamics*, 2nd edition. New York: John Wiley and Sons.

Jahnke, Eugene and Fritz Emde (1945). *Tables of Functions with Formulae and Curves*. New York: Dover. 4th edition.

Jeans, Sir James (1925). *The Dynamic Theory of Gases*. Reprinted in 1954 by Dover, New York. 4th edition.

Johnson, J. B. (1925). The Schottky effect in low frequency circuits. *Phys. Rev.* **26**, 71–85.

Johnson, J. B. (1927). Thermal agitation of electricity in conductors. *Phys. Rev.* **29**, 367–368.

Johnson, J. B. (1928). Thermal agitation of electricity in conductors. *Phys. Rev.* **32**, 97–109.

Kasai, M. and J. P. Changeaux (1971). *In vitro* excitation of purified membrane fragments by cholinergic agonists. III. Comparison of the doser–response curves to decamethonium with the corresponding binding curves of decamethonium to the cholinergic receptor. *J. Membr. Biol* **6**, 58–60.

Katchalsky, A. and P. F. Curran (1965). *Nonequilibrium Thermodynamics in Biophysics.* Cambridge, Massachusetts: Harvard University Press.

Katz, B. (1966). *Nerve, Muscle and Synapse.* New York: McGraw-Hill.

Katz, B. and R. Miledi (1970). Membrane noise produced by acetylcholine. *Nature* **226**, 962–963.

Katz, B. and R. Miledi (1971). Further observations on acetylcholine noise. *Nature New Biol.* **232**, 124–126.

Katz, B. and R. Miledi (1972). The statistical nature of the acetylcholine potential and its molecular components. *J. Physiol.* **224**, 665–699.

Katz, B. and R. Miledi (1973). The characteristics of "end-plate" noise produced by different depolarizing drugs. *J. Physiol.* **230**, 707–717.

Katz, B. and R. Miledi (1975). The effect of procaine on the action of acetylcholine at the neuromuscular junction. *J. Physiol.* **249**, 269–284.

King, R. (1966). *Electrical Noise.* London: Chapman and Hall.

Kolb, H.-A. and G. Boheim (1978). Analysis of the multi-pore system of alamethicin in a lipid membrane. II. Autocorrelation analysis and power spectral density. *J. Membr. Biol.* **38**, 151–191.

Kolb, H.-A. and E. Bamberg (1977). Influence of membrane thickness and ion concentration of the properties of the gramicidin A channel: autocorrelation, spectral power density, relaxation and single-channel studies. *Biochim. Biophys. Acta* **464**, 127–141.

Kolb, H.-A. and E. Frehland (1979). Noise-current generated by carrier-mediated ion transport at non-equilibrium. *J. Membr. Biol.* (submitted).

Kolb, H.-A. and P. Läuger (1977). Electrical noise from lipid bilayer membranes in the presence of hydrophobic ions. *J. Membr. Biol.* **37**, 321–345.

Kolb, H.-A. and P. Läuger (1978). Spectral analysis of current noise generated by carrier-mediated ion transport. *J. Membr. Biol* **41**, 167–187.

Kolb, H.-A., P. Läuger and E. Bamberg (1975). Correlation analysis of electrical noise in lipid bilayer membranes: kinetics of gramicidin channels. *J. Membr. Biol.* **20**, 133–154.

Kolb, H.-A., P. Läuger and E. Bamberg (1976). Comments on "Correlation analysis of membrane noise." *J. Membr. Biol.* **26**, 407.

Krawczyk, S. (1978). Ionic channel formation in a living cell membrane. *Nature* **273**, 56–57.

Lamb, T. D. and E. J. Simon (1976a). Power spectral measurements of noise in the turtle retina. *J. Physiol.* **263**, 103P–105P.

Lamb, T. D. and E. J. Simon (1976b). The relation between intercellular coupling and electrical noise in turtle photoreceptors. *J. Physiol.* **263**, 257–286.

Landau, E. M. and D. Ben Haim (1974). Acetylcholine noise: analysis after chemical modification. *Science* **185**, 944–946.

Landau, L. D. and E. M. Lifshitz (1959). *Course of Theoretical Physics, Vol. 5: Statistical Physics.* Oxford: Pergamon Press. Reading, Massachusetts: Addison Wesley.

Lassen, U. V. and Sten-Knudsen, O. (1968). Direct measurements of membrane potential and membrane resistance of human red cells. *J. Physiol.* **195**, 681–696.

Latorre, R., G. Ehrenstein, and H. Lecar (1972). Ion transport through excitability-inducing material (EIM) channels in lipid bilayer membranes. *J. Gen. Physiol.* **60**, 72–85.

Läuger, P. (1975). Shot noise in ion channels. *Biochim. Biophys. Acta* **413**, 1–10.

Lecar, H. and R. Nossal (1971a). Theory of threshold fluctuations in nerves: I. Relationships between electrical noise and fluctuations in axon firing. *Biophys. J.* **11**, 1048–1067.

Lecar, H. and R. Nossal (1971b). Theory of threshold fluctuations in nerve: II. Analysis of various sources of membrane noise. *Biophys. J.* **11**, 1068–1084.

Lecar, H., G. Ehrenstein, and R. Latorre (1975). Mechanism for channel gating in excitable bilayers. *Ann. N. Y. Acad. Sci.* **264**, 304–313.

Lee, Y. W. (1960). *Statistical Theory of Communication*. New York: John Wiley and Sons.

Lieberstein, H. M. (1973). *Mathematical Physiology Part II. Toward a Mathematical Description of the Electrical Behavior of Electrically Active Cells*. New York: American Elsevier.

Leibovic, K. N. (1972). *Nervous System Theory*. New York: Academic Press.

Lighthill, M. J. (1959). *Introduction to Fourier Analysis and Generalized Functions*. London: Cambridge University Press.

Lindemann, B. and W. Van Driessche (1977). Sodium specific membrane channels of frog skin are pores: current fluctuations reveal high turnover. *Science* **195**, 292–294.

Ling, G. and R. W. Gerard (1949). The normal membrane potential of frog sartorius fibers. *J. Cell. Comp. Physiol.* **34**, 383–396.

Loewenstein, W. R. (1978). Chairman of "Membrane Channels," a symposium published in *Fed. Proc. Fed. Am. Soc. Exp. Biol.* **37**, 2626–2657.

Lorentz, H. A. (1919). *Stralingstheorie*. Leiden: E. J. Brill.

Lowand, A. (1940). *Tables of sine, cosine and exponential integrals*, Vol. I. Washington, D. C.: National Bureau of Standards.

Lündström, I. and D. McQueen (1974). A proposed $1/f$ noise mechanism in nerve cell membrane. *J. Theor. Biol.* **45**, 405–409.

MacDonald, D. K. C. (1962). *Noise and Fluctuations: An Introduction*. New York: John Wiley and Sons.

MacInnes, D. A. (1961). *The Principles of Electrochemistry*. New York: Dover. Corrected version of the 1947 printing of the work originally published by Reinhold in 1939.

Mathias, R. T., J. L. Rae, and R. S. Eisenberg (1979). Electrical properties of structural components of the crystalline lens. *Biophys. J.* **25**, 181–201.

Mauro, A. (1961). Anomalous impedance, a phenomenological property of time-variant resistance. An analytic review. *Biophys. J.* **1**, 353–372.

Mauro, A. (1969). The role of the Voltaic pile in the Galvani–Volta controversy concerning animal *vs* metallic electricity. *J. Hist. Med. Allied Sciences* **24**, 140–150.

Mauro, A. and M. Rossetto (1976). In: *Electrobiology of Nerve, Synapse and Muscle*, pp. 37–44. Eds. J. P. Reuben, D. P. Purpura, M. V. L. Bennet, and E. R. Kandel. New York: Raven Press, N. Y.

Mauro, A., F. Conti, F. Dodge, and R. Schor (1970). Subthreshold behavior and phenomelogical impedance of the squid giant axon. *J. Gen. Physiol.* **55**, 497–523.

Mauro, A., R. P. Nanavati, and E. Heyer (1972a). Time-variant conductance of bilayer membranes treated with monazomycin and alamethicin. *Proc. Natl. Acad. Sci.* **69**, 3742–3744.

Mauro, A. F., A. R. Freeman, J. W. Cooley, and A. Cass (1972b). Propagated oscillatory response and classical electronic response of squid giant axon. *Biophysik* **8**, 118–132.

McBurney, R. N. and Jo L. Baker (1978). GABA-induced conductance fluctuations in cultured spinal motoneurons. *Nature* **274**, 596–597.

Michalides, J. P. L. M., R. A. M. Wallaart, and L. J. DeFelice (1973). Electrical noise from PVC-membranes. *Pfluegers Arch.* **341**, 97–104.

Minke, B., C.-F. Wu, and W. L. Pak (1975). Induction of photoreceptor voltage noise in the dark in *Drosophila* mutant. *Nature* **258**, 84–87.

Moelwyn-Hughes, E. A. (1961). *Physical Chemistry.* New York: Macmillan Co.

Molenaar, W., J. de Goede and A. A. Verveen (1976). Membrane noise in Paramecium. *Nature* **260**, 344–346.

Monnier, A. M. and H. H. Jasper (1932). Recherche de la relation entre les potentiels d'action élémentaires et la chronaxie de subordination. *C. R. Soc. Biol. Paris* **122**, 87–91.

Moore, L. E. and E. Neher (1976). Fluctuation and relaxation analysis of monazomycin induced conductance in black lipid membranes. *J. Membr. Biol.* **27**, 347.

Motchenbacher, C. D. and F. C. Fitchen (1973). *Low Noise Electronic Design.* New York: John Wiley and Sons.

Moullin, Eric Balliol (1938). *Spontaneous Fluctuations of Voltage due to Brownian Motions of Electricity, Shot Effect and Kindred Phenomena.* Oxford Engineering Science Series. London: Oxford University Press.

Mueller, P. and D. O. Rudin (1963). Induced excitability in reconstituted cell membrane structure. *J. Theor. Biol.* **4**, 268–280.

Mueller, P., D. O. Rudin, H. T. Tien, and W. W. Westcott (1962). Reconstitution of all membrane structure *in vitro* and its transformation into an excitable system. *Nature* **194**, 979–981.

Nastuk, W. L. (1950). Electrical activity of single muscle fibers at the neuromuscular junction. *Fed. Proc. Fed. Am. Soc. Exp. Biol.* **9**, 67.

Nastuk, W. L. and A. L. Hodgkin (1950). The electrical activity of single muscle fibres. *J. Cell. Comp. Physiol.* **35**, 39–74.

Neher, C. and B. Sakmann (1975). Voltage dependence of drug-induced conductance in frog neuromuscular junction. *Proc. Natl. Acad. Sci. USA* **72**, 2140–2144.

Neher, E. and B. Sakmann (1976a). Single-channel currents recorded from membrane of denervated frog muscle fibres. *Nature* **260**, 799–802.

Neher, E. and B. Sakmann (1976b). Noise analysis of drug induced voltage clamp currents in denervated frog muscle fibres. *J. Physiol.* **258**, 705–729.

Neher, E., B. Sakmann, and J. H. Steinbach (1978). The extracellular patch clamp: A method for resolving currents, through individual open channels in biological membranes. *Pfluegers Arch.* **375**, 219–228.

Neher, E. and C. F. Stevens (1977). Conductance fluctuation and ionic pores in membranes. *Ann. Rev. Biophys. Bioeng.* **6**, 345–381.

Neher, E. and J. H. Steinbach (1978). Local anesthetics transiently block currents through single ACh-receptor channels. *J. Physiol.* **277**, 153–176.

Neher, E. and H. P. Zingsheim (1974). The properties of ionic channels measured by noise analysis in thin lipid membranes. *Pflueg. Arch.* **351**, 61–67.

Neumcke, B. (1975). $1/f$ membrane noise generated by diffusion processes in unstirred solution layers. *Biophys. Struct. Mechan.* **1**, 295–309.

Neumcke, B. (1978). $1/f$ noise in membranes. *Biophys. Struct. Mechanism* **4**, 179–199.

Newton, Sir Isaac (1730). *Opticks.* New York: Dover Publications, Inc., 1952, Based on the fourth London edition.

Noma, A. and W. Trautwein (1978). Relaxation and noise analysis of the ACh-induced K current in the sino-atrial node. *J. Physiol.* **284**, 97–98P.

Nonner, W., F. Conti, B. Hille, B. Neumcke, and R. Stämpfli (1976). Current noise and the conductance of single Na channels. *Pfluegers Arch.* **362**, R27.

Nyquist, H. (1927). Thermal agitation in conductors. *Phys. Rev.* **29**, 614.

Nyquist, H. (1928). Thermal agitation of electrical charge in conductors. *Phys. Rev.* **32**, 110–113.

Offner, F. F. (1971a). $1/f$ fluctuation in membrane potential as related to membrane theory. *Biophys. J.* **11**, 123–124.

Offner, F. F. (1971b). Quantitative measure of $1/f$ noise and membrane theory. *Biophys. J.* **11**, 969–971.

Offner, F. F. (1972). Comments on "Modified random-walk model of $1/f$ noise." *J. Appl. Phys.* **43**, 1277–1278.

Papoulis, A. (1965). *Probability, Random Variables and Stochastic Processes.* New York: McGraw-Hill.

Patlak, C. S. (1960). Derivation of an equation for the diffusion potential. *Nature* **188**, 944–955.

Patlak, C. S., K. A. F. Gration, and P. N. R. Usherwood (1979). Single glutamate-activated channels in locust muscle. *Nature* **278**, 643–645.

Pecher, C. (1936). Etude statistique des variations spontanées de l'excitabilité d'une fibre nerveuse. *C. R. Soc. Biol. Paris* **122**, 87–91.

Pecher, Ch. (1937). Fluctuations indépendantes de l'excitabilité de deux fibres d'un même nerf. *C. R. Soc. Biol. Paris* **124**, 839–842.

Pecher, Ch. (1939). La fluctuation d'excitabilité de la fibre nerveuse. *Arch. Internal. Physiol.* **49**, 129–152.

Peper, K., F. Dreyer, and K.-D. Müller (1975). Analysis of cooperativity of drug-receptor interaction by quantitative iontophoresis at frog motor end plates. *Cold Spring Harbor Symp. Quant. Biol.* **40**, 187–192.

Pfeiffer, Paul E. (1965). *Concepts of Probability Theory.* New York: McGraw-Hill.

Pierce, J. R. (1956). Physical Sources of Noise *IRE Proc.* **44**, 601–608. The IRE is now the IEEE.

Plonsey, Robert (1969). *Bioelectric Phenomena.* New York: McGraw-Hill. Introduction and 1st chapter by D. G. Fleming.

Poussart, D. (1969). Nerve membrane current noise: Direct measurements under voltage clamp. *Proc. Natl. Acad. Sci.* **64**, 95–99.

Poussart, D. (1970). Current Noise in Nerve Membrane. Ph. D. thesis. MIT, Cambridge, Massachusetts.

Poussart, D. (1971). Membrane current noise in lobster axon under voltage clamp. *Biophys. J.* **11**, 211–234.

Poussart, D. J. M. (1973). Low-level average power measurements: noise figure improvement through parallel or series connection of noisy amplifiers. *Rev. Sci. Instrum.* **44**, 1049–1052.

Poussart, D. (1977). Comments on the analysis of membrane noise by the integral spectrum procedure. *Biophys. J.* **18**, 134–137.

Poussart, D. J. M., L. E. Moore, and H. M. Fishman (1976). Maximally-fast measurement of complex admittance of squid axon through digital Fourier transform. *Biol. Bull.* **151**, 425–426.

Poussart, D., L. E. Moore, and H. M. Fishman (1977). Ion movements and kinetics in squid axon. I. Complex admittance. *Ann. N. Y. Acad. Sci.* **303**, 355–379.

Prigogine, I. (1955). *Introduction to Thermodynamics of Irreversible Processes.* Springfield, Illinois: Charles C. Thomas.

Rice, S. O. (1944). Mathematical analysis of random noise. *Bell Syst. Tech. J.* **23**, 282–304. Reprinted in: *Selected Papers on Noise and Stochastic Processes.* Ed. N. Wax. New York: Dover (1954).

Richardson, J. M. (1950). The linear theory of fluctuations arriving from diffusion mechanisms—an attempt at a theory of contact noise. *Bell. Syst. Tech. J.* **29**, 117–141.

Robinson, R. A. and R. H. Stokes (1955). *Electrolyte Solutions: The Measurement and Interpretation of Conductance, Chemical Potential and Diffusion in Solutions of Simple Electrolytes.* London: Butterworths. New York: Academic Press.

Romine, W. O., G. R. Sherette, G. R. Brown, and R. J. Bradley (1977). Evidence that nystatine may not form channels in thin lipid membranes. *Biophys. J.* **17**, 269–274.

Ruff, R. (1976). Local anesthetic alteration of miniature endplate currents and endplate current fluctuations. *Biophys. J.* **16**, 433–439.

Ruff, R. (1977). A quantitative analysis of local anaesthetic alteration of miniature endplate currents and end-plate current fluctuations. *J. Physiol.* **264**, 89–124.

Sabah, N. H. and K. N. Leibovic (1969), *Biophys. J.* **9**, 1206–1222.

Sachs, F. and H. Lecar (1973). Acetylcholine noise in tissue culture muscle cells. *Nature New Biol.* **1246**, 214–216.

Sachs, F. and H. Lecar (1977). Acetylcholine-induced current fluctuations in tissue cultured muscle cells under voltage clamp. *Biophys. J.* **17**, 129–143.

Sauvé, R. and E. Bamberg (1978). $1/f$ noise in black lipid membranes induced by ionic channels formed by chemically dimerized Gramicidin A. *J. Membr. Biol.* **43**, 317–333.

Schick, K. L. (1974). Power spectra of pulse sequences and implications for membrane fluctuations. *Acta Biother.* (*Leiden*) **23**, 1–17.

Schick, K. L. and A. A. Verveen (1974). $1/f$ noise with a low frequency white noise limit. *Nature* **251**, 599–601.

Schneider, M. (1970). Linear electrical properties of the transverse tubules and surface membrane of skeletal muscle fibers. *J. Gen. Physiol.* **56**, 640–671.

Schoenberg, M., G. Dominquez, and H. A. Fozzard (1975). Effect of diameter on membrane capacitance and conductance of sleep cardiac purkinje fibers. *J. Gen. Physiol.* **65**, 441–458.

Schönfeld, H. (1955). Beitrag zum $1/f$-gesetz beim Rauschen von Halbleitern. *Z. Naturforsch* **10a**, 291–300.

Schottky, W. (1918). Über spontane Stromschwankungen in verschiedenen Elektrizitätsleitern. *Ann. Phys.* **57**, 541–567.

Schottky, W. (1926). Small-shot effect and flicker effect. *Phys. Rev.* **28**, 74–103.

Schuetze, S. M., E. F. Frank, and G. D. Fischbach (1978). Channel open time and metabolic stability of synaptic and extrasynaptic ACh receptors on cultured chick myotubules. *Proc. Natl. Acad. Sci. USA* **75**, 520–523.

Schwartz, E. A. (1977). Voltage noise observed in rods of the turtle retina. *J. Physiol.* **272**, 217–246.

Segal, J. R. (1972). Electrical fluctuations associated with active transport. *Biophys. J.* **12**, 1371–1390.

Segal, J. R. (1974). Reply to "Comments of Electrical fluctuations associated with active transport." *Biophys. J.* **14**, 513.

Siebenga, E. and A. A. Verveen (1970). Noise voltage of axonal membrane. *Pfluegers Arch.* **318**, 267.

Siebenga, E. and A. A. Verveen (1971). The dependence of the $1/f$ noise intensity of the node of Ranvier on membrane potential. Proceeding of the First European Biophysics Congress, pp. 219–223. For information write to the author.

Siebenga, E. and A. A. Verveen (1972). Membrane noise and ion transport in the node of Ranvier. In: *Biomembranes* 3, pp. 473–482. Eds. F. Kreuzer and J. F. G. Slegers. New York: Plenum.

Siebenga, E. and A. W. A. Meyer (1973). Membrane noise voltage analyses. *Pfluegers Arch.* **343**, 165–171.

Siebenga, E., A. W. A. Meyer, and A. A. Verveen (1973). Membrane shot-noise in electrically depolarized nodes of Ranvier. *Pfluegers Arch.* **341**, 87–96.

Siebenga, E., J. de Goede, and A. A. Verveen (1974). The influence of TTX, DNP and TEA on membrane flicker noise and shot effect noise of the frog node of Ranvier. *Pfluegers Arch.* **351**, 25–34.

Siebenga, E. (1974). Some investigations on the voltage fluctuations in the node of Ranvier of *Rana temporaria*. Thesis, Rijksuniversiteit te Leiden, 1974.

Sigworth, F. J. (1977). Na channels in nerve apparently have two conductance states. *Nature* **270**, 265–267.

Sigworth, F. J. (1978). Fraction of Na channels open at peak conductance. *Biophys. J.* **21**, 41a.

Sigworth, F. J. (1979a). Analysis of nonstationary Na current fluctuations in frog myelinated nerve. Thesis, Yale University. May 1979.

Sigworth, F. J. (1979b). The Cole-Moore delay: Cooperativity among K channels? *Biophys. J.* **25**, 196a.

Sigworth, F. J. (1980a). The variance of Na current fluctuations at the node of Ranvier. *J. Physiol.* (in press).

Sigworth, F. J. (1980b). The conductance of Na channels under conditions of reduced current at the node of Ranvier. *J. Physiol.* (in press).

Simon. E. J., T. D. Lamb, and A. L. Hodgkin (1975). Spontaneous voltage fluctuations in retinal cones and bipolar cells. *Nature* **256**, 661–662.

Simon, E. J. and T. D. Lamb (1977). Electrical noise in turtle cones. In: *Vertebrate Photoreception*. Eds. H. B. Barlow and P. Fatt. London: Academic Press.

Sjölin, L. and W. Grampp (1975). Membrane noise in slowly adapting stretch receptor neurone of lobster. *Nature* **257**, 696–697.

Smith, R. A. (1961). *Wave Mechanics of Crystalline Solids*. Chapman and Hall: London.

Stämpfli, R. and B. Hille (1976). Electrophysiology of the Peripheral Myelinated Nerve. In: *Frog Neurobiology*. Eds. R. Llinás and W. Precht. Heidelberg: Springer-Verlag.

Stevens, C. F. (1972). Inferences about membrane properties from electrical noise measurements. *Biophys. J.* **12**, 1028–1047.

Stevens, C. F. (1975a). Principles and applications of fluctuation analysis: a non-mathematical introduction. *Fed. Proc. Fed. Am. Soc. Exp. Biol.* **34**. 1364–1370.

Stevens, C. F. (1975b). *Cold Spring Harbor Symp. Quant. Biol.* **40**, 169–173.

Stevens, C. F. (1977). Study of membrane permeability changes by fluctuation analysis. *Nature* **270**, 391–396.

Strandberg, N. W. P. (1977). Use of octave filters in the measurement of random noise spectral density. *Biophys. J.* **18**, 132–134.

Teorell, T. (1953). Transport processes and electrical phenomena in ionic membranes. *Prog. Biophys. Biophys. Chem.* **3**, 305–369.

Tolman, Richard C. (1925). The principle of microscopic reversibility. *Proc. Nat'l. Acad. Sci. USA* **11**, 436–439.

Urry, D. W. (1971). The gramicidin A transmembrane channel: a proposed $\pi_{(L_1D)}$ helix. *Proc. Natl. Acad. Sci. USA* **68**, 672–676.

Urry, D. W., M. C. Goodall, J. D. Glickson, and D. F. Meyers (1971). The gramicidin A transmembrane channel: characteristics of head-to-head dimerized $\pi_{(L_1D)}$ helix. *Proc. Natl. Acad. Sci. USA* **68**, 1907–1911.

Valdiosera, R., C. Clausen, and R. S. Eisenberg (1974a). Measurement of the impedance of frog skeletal muscle fibers. *Biophys. J.* **14**, 295–315.

Valdiosera, R., C. Clausen, and R. S. Eisenberg (1974b). Impedance of frog skeletal muscle fibers in various solutions. *J. Gen. Physiol.* **63**, 460–491.

van den Berg, R. J. (1978). Electrical Fluctuations in Myelinated Nerve Membrane Thesis, University of Leyden. For additional information write to the author.

van den Berg, R. J., J. de Goede, and A. A. Verveen (1975). Conductance fluctuations in Ranvier nodes. *Pfluegers Arch.* **360**, 17–23.

van den Berg, R. J., E. Siebenga, and G. de Bruin (1977). Potassium ion noise currents and inactivation in voltage clamped node of Ranvier. *Nature* **256**, 177–179.

van den Berg, R. J. and R. E. Beekman (1977). How many ions do ionic pores of the Ranvier node contain? 18th Dutch Fed. Meeting, p. 127. For information write to author.

van der Ziel, A. (1954). *Noise.* Englewood Cliffs, New Jersey: Prentice-Hall.

van der Ziel, A. (1959). *Fluctuation Phenomena in Semiconductors.* London: Butterworths Scientific Publications. New York: Academic Press.

van der Ziel, Aldert (1970). Noise: Sources, Characterization, Measurement. Englewood Cliffs, New Jersey: Prentice-Hall.

van der Ziel, A. (1976). *Noise in Measurements.* New York: Wiley-Interscience.

van der Ziel, A. (1979). Second International Symposium on $1/f$ Fluctuations 17–19 March 1979, Gainesville, Florida. For information write to the author.

van Driessche, W. and R. Borghgraef (1975). Noise generated during ion transport across frog skin. *Arch. Int. Physiol. Biochim.* **83**, 140–142.

van Driessche, W. and H. Gögelein (1975). K channels in the apical membrane of the toad gallbladder. *Nature* **275**, 665–667.

van Driessche, W. and B. Lindemann (1976). Fluctuations of Na-current through the Na-selective membrane of frog skin. *Pfluegers Arch.* **362**, R28.

van Vliet, K. M. and J. Farrett (1965). *Fluctuation Phenomena in Solids.* New York: Academic Press.

Verveen, A. A. (1961). Fluctuation in Excitability. Research Report on Signal Transmission in Nerve Fibers. Ph. D. thesis, Netherlands Central Institute for Brain Research, Amsterdam.

Verveen, A. A. (1962). Axon diameter and fluctuation in excitability. *Acta Morphol. Neerl.-Scand.* **5**, 79–85.

Verveen, A. A. (1977). Noise as a phenomenon in neurophysiology. *Pfluegers Arch.* **327**, 227.

Verveen, A. A. and H. E. Derksen (1965). Fluctuations in membrane potential of axons and the problem of coding. *Kybernetik* **2**, 152–160.

Verveen, A. A., H. E. Derksen, and K. Schick (1967). Voltage fluctuations of neural membranes. *Nature* **216**, 588–589.

Verveen, A. A. and H. E. Derksen (1968). Fluctuation phenomena in nerve membrane. *Proc. IEEE* **56**, 906–916.

Verveen, A. A. and H. E. Derksen (1969). Amplitude distribution of axon membrane voltage noise. *Acta. Physiol. Pharmacol. Neerl.* **15**, 353–379.

Verveen, A. A. and L. J. DeFelice (1974). Membrane Noise. *Prog. Biophys. Mol. Biol.* **28**, 189–265.

Voss, R. F. and J. Clark (1976). $1/f$ noise from systems in thermal equilibrium. *Phys. Rev. Lett.* **36**, 42–45.

Wanke, E. (1975). Monazomycin and nystatin channel noise. Abstracts of the 5th International Biophysics Congress, P-368, Copenhagen.

Wanke, E. and G. Prestipino (1976). Monozomycin channel noise. *Biochim. Biophys. Acta* **436**, 721–726.

Wanke, E., L. J. DeFelice, and F. Conti (1974). Voltage noise, current noise and impedance in space clamped squid giant axon. *Pfluegers Arch.* **347**, 63–74.

Wax, Nelson (1954), editor. *Selected Papers on Noise and Stochastic Processes.* New York: Dover.

Weidmann, S. (1952). The electrical constants of purkinje fibers. *J. Physiol.* **118**, 348–360.

Weissman, M. B. (1976). Models for $1/f$ noise in nerve membrane. *Biophys. J.* **16**, 1105–1108.

Weissman, M. B. (1977). A model for $1/f$ noise from singularities in the fluctuation weighting function. *J. Appl. Phys.* **48**, 1705–1707.

Wiederhold, M. L. (1974). *Aplysia* statocyst receptor cells: intracellular responses to physiological stimuli. *Brain Res.* **78**, 490–494.

Wiederhold, M. L. (1976). Mechanosensory transduction in "sensory" and "motile" cilia. *Ann. Rev. Biophys. Bioeng.* **5**, 39–62.

Wiederhold, M. L. (1977). Rectification in *Aplysia* Statocyst recepter cells. *J. Physiol.* **266**, 139–156.

Wolf, D. (1978), editor. Proceedings of the 5th International Conference on Noise, Bad Nauheim, Germany, March 13–16 1978. *Noise in Physical Systems.* Heidelberg: Springer-Verlag. For information on previous Symposia in this series write to the author.

Woodbury, J. W. and W. Crill (1951). On the problem of impulse conduction in the heart. In: *Nervous Inhibition.* New York. Pergamon Press.

Wong, F. (1978). Nature of light-induced conductance changes in ventral photoreceptors of *Limulus. Nature* **275**, 76–79.

Wu, Chun-Fong and W. Pak (1978). Light-induced voltage noise in the photoreceptor of *Drosophila melanogaster, J. Gen. Physiol.* **71**, 249–268.

Yafuso, M. and M. E. Green (1971). Noise spectra associated with hydrochloric acid transport through some cation-exchange membranes. *J. Phys. Chem.* **75**, 654–662.

Zingsheim, H. P. and E. Neher (1974). The equivalence of fluctuation analysis and chemical relaxation measurements: a kinetic study of ion pore information in thin lipid membranes. *Biophys. Chem.* **2**, 197–207.

Index